A — BRITISH GEOLOGICAL SURVEY
Natural Environment Research Council

GEOTHERMAL ENERGY —THE POTENTIAL IN THE UNITED KINGDOM

Editors: R. A. Downing and D. A. Gray

The programme was supported by the Department of Energy
and the Commission of the European Communities
EUR — 9505 — EN

1835 Geological Survey of Great Britain
150 Years of Service to the Nation
1985 British Geological Survey

A volume commemorating
the 150th anniversary of
the British Geological Survey

London: Her Majesty's Stationery Office 1986

First published 1986

ISBN 0 11 884366 4

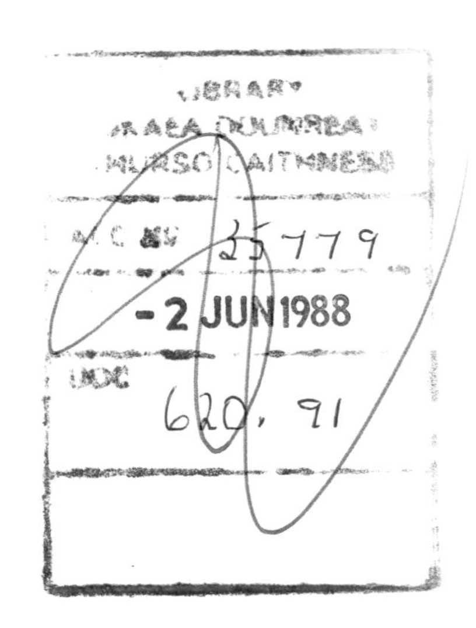

Preface

The marked increase in energy prices in the mid-1970s focused attention on the fact that conventional energy sources are finite. This realisation led to an acceleration of investigations into the feasibility of using sources of energy other than fossil fuels and nuclear energy. Of these so-called renewable forms of energy only geothermal resources are currently making a significant contribution to the world's energy needs.

The United Kingdom is not a region that is immediately associated with geothermal energy. But the geothermal gradient, which is similar to the world average, gives temperatures of 60° to 75°C at depths of 2 km in many areas, sufficient for a wide range of energy demands. Furthermore, geothermal resources are being developed in the Paris and Aquitaine basins in France, as well as in the Pannonian Basin in Hungary, under somewhat similar geological conditions to those that exist in the United Kingdom.

In view of this, the Department of Energy, in collaboration with the Commission of the European Communities, invited the British Geological Survey to investigate the geothermal potential of the UK. The main financial support was from the Department with substantial contributions from the Commission under contracts EG-A1-041-80-UK(H) and EG-AY-112-UK(H), and this assistance is gratefully acknowledged.

The programme began in 1977 with the principal objective of estimating the geothermal potential of the United Kingdom by the mid-1980s. That objective had been achieved by the end of March 1984, when the second four-year programme was completed, and this book reports the work undertaken and the conclusions reached up to that date, with the addition of the results of a well drilled at Cleethorpes in 1984.

The thermal energy available was investigated for each of two components; that which can be derived from naturally occurring hot brines in water-bearing rocks (the low enthalpy 'Aquifer' component) and that which may be accessible from hot rocks which do not transmit significant volumes of water in the natural state (the 'Hot Dry Rock' component).

The programme has confirmed that the total thermal energy resources which exist in the sub-surface are massive. However, the United Kingdom, unlike most territories in Western Europe, is currently self-sufficient in conventional energy resources and the present, temporary global glut of hydrocarbons has produced an international energy pricing policy which militates against the introduction of the renewables. Accordingly, the aquifer resources remain a potentially useful strategic energy supplement with only local development likely in the short to medium term. The potential of the Hot Dry Rock system is more widespread, but its realisation must await the proving of the present experimental technology and its conversion to an economically viable system.

The work programme reviewed in the book was carried out under the general direction of Mr D. A. Gray and supervised by Dr R. A. Downing. Dr D. W. Holliday was responsible for coordinating the studies on sedimentary basins, Mr I. F. Smith for low enthalpy geophysical studies and Mr M. K. Lee for the Hot Dry Rock programme. Many individuals were involved in the programme but in particular the following made significant contributions: Mr D. J. Allen, Mr M. J. Arthur, Dr J. A. Barker, Mr J. R. P. Bennett, Mr M. A. E. Browne, Mr W. G. Burgess, Dr A. J. Burley, Mr R. M. Carruthers, Miss M. A. Cradock-Hartopp, Mr A. C. Cripps, Dr W. M. Edmunds, Dr C. J. Evans, Mr R. B. Evans, Mr I. N. Gale, Miss R. L. Hargeaves, Dr S. Holloway, Mr M. T. Houghton, Dr G. A. Kirby, Dr I. E. Penn, Mr M. Price, Mr K. E. Rollin, Mr K. Smith and Dr L. P. Thomas.

Their involvement is acknowledged by reference to the reports they have written in the series 'Investigation of the Geothermal Potential of the UK', a list of which is given in Appendix 4. The various chapters in the book have been compiled from these reports.

Parts of the programme were sub-contracted to universities. Professor G. C. Brown and Dr P. C. Webb of the Open University were responsible for the measurement of heat production and the geochemistry of granites in northern Britain. Mr J. Wheildon and his colleagues at the Imperial College of Science and Technology in the University of London, including Dr A. Thomas-Betts, Mr J. S. Gebski, Mr G. King and Mr C. N. Crook, were responsible for studies of heat flow. Dr J. N. Andrews of the University of Bath, assisted by Dr R. L. F. Kay and Mr D. J. Lees, investigated the radioelement and inert gas compositions of geothermal brines and the rocks containing them, thereby contributing to knowledge of the geochemistry of these fluids, their sub-surface flow paths and ages.

Prior to and during the earlier stages of the programme, extensive heat flow observations were made by Professor E. R. Oxburgh and his colleagues, then at Oxford University, and by Mr J. Wheildon and his group at Imperial College; the collaboration of both groups in providing access to their results is gratefully acknowledged.

During the drilling and testing of deep exploration wells at Marchwood, Larne and at Western Esplanade in Southampton, Mr D. McIntyre of Drilcon Ltd provided invaluable assistance to the staff of the British Geological

Survey and the Department of Energy responsible for the work.

Considerable assistance in the heat flow measurement programme was received from the National Coal Board. Many hydrocarbon exploration companies permitted measurements for geothermal purposes, additional to their own programmes, to be undertaken in their exploration boreholes and, in particular, the valuable assistance of the following is gratefully acknowledged: Amoco (UK) Explorations Co., Bearcat Explorations (UK) Ltd; British Gas Corporation; British Petroleum Company plc; Carless Exploration Ltd; Conoco North Sea Inc.; Ultramar Exploration Ltd.

The support of staff of the Department of Energy, of the Energy Technology Support Unit of the Atomic Energy Research Establishment, Harwell, and of the Commission of the European Communities has been greatly appreciated, especially that from Dr D. Fairmaner, Mr W. Macpherson and Mr J. R. V. Brooks (Department of Energy), Dr P. Ungemach and Dr K. Louwrier (Commission of the European Communities) and Dr J. D. Garnish (Energy Technology Support Unit) who commented most helpfully on the text. Finally, Miss M. B. Simmons and Dr T. J. Dhonau gave much assistance with the preparation of the text for typesetting and printing and with checking the proofs; their help is gratefully acknowledged.

D. A. Gray
Programmes Director C
British Geological Survey
Keyworth
Nottingham NG12 5GG
United Kingdom

January 1985

Contents

	page
Notation	xii
Chapter 1 **Introduction**	1
R.A. Downing and D.A. Gray	
The nature of geothermal energy	1
Geothermal energy systems	1
Basic geological structure of the UK in relation to geothermal energy prospects	3
Geothermal energy in the UK	4
Investigation of the geothermal potential of the UK	4
Chapter 2 **Heat flow**	8
J. Wheildon and K.E. Rollin	
Introduction	8
Basic concepts	8
Sub-surface temperatures	9
Measurement techniques	11
Heat flow calculation	14
The heat flow field in the UK	16
Chapter 3 **Hot Dry Rock**	21
M.K. Lee	
The Hot Dry Rock concept	21
Background to HDR prospects in the UK	21
Identification of granitic targets	22
Cornubian batholith	25
Northern England	27
Eastern Highlands of Scotland	32
Other Caledonian granites	36
The relationship between heat flow and heat production in British granites	37
Overview of HDR potential in Britain	40
Chapter 4 **Mesozoic basins**	42
I.F. Smith	
Introduction	42
Geological history of the post-Carboniferous sedimentary basins	42
Aquifers with geothermal potential	44
East Yorkshire and Lincolnshire Basin	44
Wessex Basin	52
Worcester Basin	62
Cheshire and West Lancashire basins	68
Carlisle Basin	77
Northern Ireland	80
Chapter 5 **Devonian and Carboniferous basins**	84
D.W. Holliday	
Introduction	84
Palaeogeography and geological history	85
Estimation of sub-surface temperatures	87
General geology and hydrogeology	88
Coal Measures	88
Millstone Grit	96
Carboniferous Limestone	99
Dinantian Clastic Rocks of Northern Britain	104
Old Red Sandstone	105
Variscan Fold Belt	109
Northern Ireland	109
Regional prospects in Upper Palaeozoic rocks	109
Chapter 6 **Geochemistry of geothermal waters in the UK**	111
W.M. Edmunds	
The role of geochemistry in geothermal evaluation	111
Geothermal fluids in the sedimentary basins	113
Origins of formation waters in the sedimentary basins	119
Geothermal waters in granite	120
Geochemical considerations in the utilisation of low enthalpy geothermal brines	122
Chapter 7 **Modelling of low enthalpy geothermal schemes**	124
J.A. Barker	
Introduction	124
Basic principles	124
Single-well schemes	125
Doublet schemes	127
Multiple-well schemes	131
Summary	131
Chapter 8 **Assessment of the geothermal resources**	132
I.N. Gale and K.E. Rollin	
Definitions and terminology	132
Accessible Resource Base	133
Hot Dry Rock Accessible Resource Base	136
Low enthalpy resources of deep aquifers	138
Geothermal resources of the Permo-Triassic sandstones	140
Geothermal resources of the Lower Cretaceous sandstones	147

Chapter 9 **Engineering and economic aspects of low enthalpy development** 149
R.A. Downing

Chapter 10 **Review of the geothermal potential of the UK** 152
R.A. Downing and D.A. Gray

Hot Dry Rock resource 153
Low enthalpy aquifers 153
Summary of low enthalpy resources 158
The risk factor 159
Conclusions 160

Geothermal map of the United Kingdom
(in back pocket)

References 162

Appendix 1 **Conversion factors for selected energy units and equivalent values of some principal sources of energy** 171

Appendix 2 **Observed heat flow values in the UK** 172

Appendix 3 **Glossary of technical terms** 176

Appendix 4 **Reports issued by the British Geological Survey on the geothermal potential of the United Kingdom.** 178

Index 180

List of figures

	page
1.1 The basic geological structure of the United Kingdom.	3
2.1 Effects of thermal conductivity on temperature gradient and sub-surface temperature assuming a constant heat flow of 60 mW/m² and negligible heat production within sediments.	10
2.2 The effects of crustal heat production on surface heat flow and sub-surface temperature in granite and sedimentary basement assuming the same sub-crustal heat flow and crustal thermal conductivity.	10
2.3 Thermal conductivity of Permo-Triassic mudstones plotted against depth of sample.	15
2.4 Equilibrium temperature log, thermal gradients, thermal conductivities and heat flow interpretation for the heat flow borehole at Thornton Cleveleys in west Lancashire.	17
2.5 Graph of temperature against thermal resistance in the Thornton Cleveleys Borehole.	17
2.6 Histogram of heat flow data for the United Kingdom (a) contrasted with the coverage for continental Europe (b).	17
2.7 Plot of observed against estimated heat flow for 60 boreholes.	18
2.8 Heat flow map of the UK.	19
2.9 The relationship of heat flow to the form of the Carnmenellis Granite.	20
3.1 Conceptual diagram of a two borehole hot dry rock system.	21
3.2 Bouguer gravity anomaly map of the UK.	23
3.3 Summary of heat production data for selected British granites.	24
3.4 Three-dimensional model of the granite batholith of south-west England showing the distribution and values of heat flow measurements.	26
3.5 Predicted temperature-depth profile for the Carnmenellis Granite calculated from a one-dimensional heat transfer model.	27
3.6 Three-dimensional model of the Lake District granite batholith showing location of heat flow boreholes.	27
3.7 Summary of uranium and thorium contents and their contribution to heat production of Lake District granites.	29
3.8 Predicted temperature-depth profiles for granites and basement rocks in northern England.	30
3.9 Location of granite batholiths and heat flow values in northern England.	30
3.10 Location of granites and heat flow values in the Eastern Highlands of Scotland.	32
3.11 Predicted temperature-depth profiles for the Ballater Granite and the Moine/Dalradian basement.	34
3.12 Relationship between heat flow and heat production for British granites and basement rocks.	37
3.13 Comparison of predicted sub-surface temperature profiles for British granites and basement rocks.	40
4.1 Major Mesozoic basins in the UK.	42
4.2 Generalised correlation of stratigraphical nomenclature for the major Permo-Triassic basins.	43
4.3 Geological map of the East Yorkshire and Lincolnshire Basin.	45
4.4 Isopachytes of the Sherwood Sandstone Group in the East Yorkshire and Lincolnshire Basin.	47
4.5 Structure contours on the top of the Sherwood Sandstone Group in the East Yorkshire and Lincolnshire Basin.	48
4.6 Estimated temperature at the mid-point of the Sherwood Sandstone Group in the East Yorkshire and Lincolnshire Basin.	49
4.7 Structure contours on the base of the Permian in the East Yorkshire and Lincolnshire Basin.	50
4.8 Isopachytes and lithofacies of the Basal Permian Sands and Breccia in the East Yorkshire and Lincolnshire Basin.	51
4.9 Estimated temperature at the base of the Permian in the East Yorkshire and Lincolnshire Basin.	52
4.10 General structure of the Wessex Basin.	54
4.11 Structure contours on the top of the Sherwood Sandstone Group in the Wessex Basin.	55

4.12 Isopachytes of the Sherwood Sandstone Group in the Wessex Basin. 55

4.13 Facies variations in the Sherwood Sandstone Group in the Wessex Basin. 56

4.14 Estimated transmissivity of the Sherwood Sandstone Group in the Wessex Basin. 59

4.15 Salinity of groundwater in the Sherwood Sandstone Group in the Wessex Basin. 59

4.16 Estimated temperature at the centre of the Sherwood Sandstone Group in the Wessex Basin. 60

4.17 Depth to the top and thickness of the Lower Greensand between Bournemouth and Worthing. 61

4.18 Geological map of the Worcester Basin. 62

4.19 Correlation of Permo-Triassic rocks in the Worcester Basin. 63

4.20 Bouguer gravity anomaly map of the Worcester Basin. 63

4.21 Structure contours and temperature at the base of the Mercia Mudstone Group in the Worcester Basin. 64

4.22 Structure contours and temperature at the base of the Permo-Triassic in the Worcester Basin. 64

4.23 Isopachytes of the Permo-Triassic sandstones in the Worcester Basin. 65

4.24 Geology of the Cheshire and West Lancashire basins. 69

4.25 General correlation of the Permo-Triassic between the Cheshire and West Lancashire basins. 69

4.26 Structure contours on and estimated temperatures at the top of the Tarporley Siltstone in the Cheshire Basin. 70

4.27 Gravity interpretation of the structure of the Cheshire Basin. 70

4.28 Structure contours on and estimated temperatures at the base of the Manchester Marl in the Cheshire Basin. 71

4.29 Structure contours on and estimated temperatures at the base of the Permo-Triassic in the Cheshire Basin. 72

4.30 Porosity-permeability relationships for the Triassic sandstones in the Northern Irish Sea Basin. 74

4.31 Probability distributions of permeability in the Triassic sandstones in Cheshire and Shropshire. 75

4.32 Geological map of the Carlisle Basin and adjacent areas. 76

4.33 Geological log of the Silloth No. 1 Borehole. 78

4.34 Bouguer gravity anomaly map of the Carlisle Basin and adjacent areas. 78

4.35 Permo-Triassic basins in Northern Ireland. 80

4.36 Estimated temperatures at the top of the Sherwood Sandstone in Northern Ireland. 81

5.1 Distribution of Devonian and Carboniferous strata in the UK. 84

5.2 Generalised successions of the Devonian and Carboniferous rocks in Britain. 85

5.3 East Midlands: 90
(a) Structure contours and estimated temperatures at the base of the Permo-Triassic.
(b) Structure contours and estimated temperatures at the base of the Coal Measures.
(c) Isopachytes of the Coal Measures.

5.4 Chloride concentration of groundwaters from the Carboniferous of the East Midlands. 91

5.5 North-west Midlands: 92
(a) Structure contours and estimated temperatures at the base of the Permo-Triassic.
(b) Structure contours and estimated temperatures at the base of the Coal Measures.
(c) Isopachytes of the Coal Measures.

5.6 Midland Valley of Scotland: 93
(a) Structure contours at the base of the Coal Measures (top of Passage Group)
(b) Estimated temperatures at the base of the Coal Measures (top of Passage Group).
(c) Isopachytes of the Passage Group.

5.7 South Wales: 94
(a) Structure contours and estimated temperatures at the base of the Upper Coal Measures.
(b) Structure contours and estimated temperatures at the base of the Coal Measures.
(c) Isopachytes of the Coal Measures.

5.8a Palaeogeographical map of the floor beneath Mesozoic formations and structure contours on the base of the Coal Measures in Berkshire and south Oxfordshire. 95

5.8b Structure contours on the base of the Coal Measures in the Kent Coalfield. 95

5.9 East Midlands: 97
(a) Structure contours and estimated temperatures at the top of the Carboniferous Limestone.
(b) Isopachytes of the Millstone Grit.
(c) Calculated fresh-water head of groundwater in the Millstone Grit.

5.10 North-west Midlands: 99
(a) Isopachytes of the Millstone Grit.
(b) Structure contours on the top of the Carboniferous Limestone.
(c) Estimated temperatures at the top of the Carboniferous Limestone.

5.11 South Wales: 99
(a) Isopachytes of the Millstone Grit.
(b) Structure contours and estimated temperatures at the base of the Millstone Grit.
(c) Structure contours and estimated temperatures at the base of the Carboniferous Limestone.

5.12 Distribution of Palaeozoic rocks, major structural features and position of thermal springs in the Bath-Bristol area. 103

5.13 Facies and thickness of the Fell Sandstone Group with estimated structure contours and temperatures at the top of the Group. 105

5.14 Midland Valley of Scotland: 107
(a) Structure contours and estimated temperatures at the top of the Dinantian.
(b) Structure contours and estimated temperatures at the base of the Dinantian.
(c) Isopachytes of the Upper Old Red Sandstone.

6.1 Composition of fluids produced during four drill-stem tests on the Sherwood Sandstone in the Marchwood Well. 113

6.2 Variation relative to chloride of Na, Ca, Mg, K, Sr, Li, Br, I, B and Rb for formation waters in sedimentary basins in the UK. 114

6.3 Isotope compositions ($\delta^{18}O$ and δD) for formation waters from the Wessex and other sedimentary basins in the UK. 116

6.4 Conceptual model of the flow path of thermal water in the Bath-Bristol area. 117

6.5 Conceptual model of groundwater circulation in the Carnmenellis Granite, Cornwall. 121

6.6 Na/Cl and Li/Cl plots of thermal and non-thermal mine-waters and shallow groundwaters in the Cornubian granite. 121

7.1 Various paths across the viable operating region of a geothermal scheme. 124

7.2 Drawdown variation with time in a well of radius r_w in the vicinity of an impermeable linear barrier at a distance $B/2$ or, equivalently, an identical well at distance B. 125

7.3 Drawdown for a well at the centre of a circular reservoir of radius R. 126

7.4 Drawdown in a well with a fully-penetrating vertical fracture of length $2b$. 126

7.5 (a) Section through a simple geothermal doublet showing the steady-state potentiometric surface. 127
(b) Flow net for the same doublet.

7.6 Variation of abstraction temperature with time for a simple doublet. 128

7.7 Parallel fracture arrangements studied by Andrews and others (1981); (a) two-dimensional flow between rectangular fractures which fully penetrate a confined aquifer, (b) three-dimensional flow between disc-shaped fractures. 129

7.8 Minimum fracture lengths required to limit the drawdown to s_{max} for a breakthrough time t_B. 130

7.9 Minimum fracture radii required to limit the drawdown to s_{max} for various breakthrough times. 130

8.1 Model of the crust in the UK used for the calculation of the Accessible Resource Base. 134

8.2 Estimated temperature at a depth of 7 km. 135

8.3 Accessible Resource Base. 136

8.4 Depth to the 100°C isotherm in the UK. 138

8.5 Hot Dry Rock Accessible Resource Base. 139

8.6 Unit power of the Sherwood Sandstone and the Basal Permian Sands in the East Yorkshire and Lincolnshire Basin. 141

8.7 Potential geothermal fields in the East Yorkshire and Lincolnshire Basin. 141

8.8 Geothermal Resources of the Sherwood Sandstone Group in the Wessex Basin. 143

8.9 Unit power of the Sherwood Sandstone Group in the Wessex Basin. 143

8.10 Potential geothermal fields in the Wessex Basin. 143

8.11 Generalised north-south section through the Worcester Basin. 144

10.1a Depths at which 200°C is attained in the UK. 152

10.1b Depths at which 150°C is attained in the UK. 152

10.2 Distribution of the total heat store in the Carboniferous rocks of the East Midlands. 157

10.3 Potential low enthalpy geothermal fields in the UK. 159

List of tables

1.1 Geothermal energy programme of the British Geological Survey, 1977–1984. 5

2.1 Sub-surface temperatures for Dartmoor and west Wales assuming a variety of heat production distributions in the crust. 12

2.2 Thermal conductivities of some lithostratigraphical formations in the UK. 13

2.3 Thermal conductivities of Hercynian and Caledonian granites in the UK. 14

3.1 Summary of thermal data for granites in south-west England. 25

3.2 Heat flow results from the Shap and Skiddaw boreholes. 28

3.3 Average radioelement content and heat production of core from the Shap and Skiddaw boreholes. 28

3.4 Summary of properties relating to the radiothermal potential of Lake District granites. 28

3.5 Summary of thermal data for granites in northern England. 31

3.6 Heat flow results from the boreholes in the Eastern Highlands. 32

3.7 Average radioelement content and heat production of core from the boreholes in the Eastern Highlands. 33

3.8 Summary of radiothermal properties of granites in the Eastern Highlands. 33

3.9 Heat production and radioelement data from surface sampling of Caledonian granites. 35

3.10 Heat production and radioelement data from boreholes in Caledonian granites. 36

3.11 Heat flow and heat production data for granites and basement rocks in the UK. 38

4.1 Geological succession in the East Yorkshire and Lincolnshire Basin. 46

4.2 Geological succession in the Wessex Basin. 53

4.3 Permo-Triassic sandstones: hydraulic properties at outcrop in south-west England. 58

4.4 Values of thermal conductivity adopted for major lithological groups in the Wessex Basin. 61

4.5 Temperature measurements and calculated heat flow values for boreholes in the Worcester Basin. 66

4.6 Hydrogeological units of the Permo-Triassic in the Worcester Basin. 66

4.7 Analyses of representative groundwaters from the Triassic and Permian sandstones in the Kempsey Borehole. 68

4.8 Depth to the base and thickness of Permo-Triassic strata as recorded in boreholes in the Cheshire and West Lancashire basins. 71

5.1 Aquifer properties of a typical sandstone in the Middle Coal Measures of South Wales. 95

5.2 Aquifer properties of typical sandstones in the Upper Coal Measures of the Oxford Coalfield. 95

5.3 Aquifer properties of typical sandstones in the Old Red Sandstone of the Midland Valley of Scotland. 108

6.1 Application of geochemistry in the evaluation of low enthalpy geothermal resources. 111

6.2 Chemical parameters and their relevance to geothermal investigations as illustrated by reference to the Bath-Bristol area. 112

6.3 Representative analyses of thermal waters and brines. 116

6.4 Saturation indices for minerals in various formation waters calculated using the WATQF program. 122

8.1 Regional distribution of the Accessible Resource Base. 137

8.2 Regional distribution of the Hot Dry Rock Accessible Resource Base. 137

8.3a Geothermal Resources of the East Yorkshire and Lincolnshire Basin. 140

8.3b Identified Resources of the East Yorkshire and Lincolnshire Basin. 140

8.4 Estimated thermal yields of aquifers at population centres in east Yorkshire and Humberside. 142

8.5a Geothermal Resources of the Sherwood Sandstone Group in the Wessex Basin. 142

8.5b Identified Resources of the Sherwood Sandstone Group in the Wessex Basin. 142

8.6 Estimated thermal yields at selected sites in the Wessex Basin when a drawdown of 100 m is imposed. 143

8.7a Geothermal Resources of the Worcester Basin. 144

8.7b Identified Resources of the Worcester Basin. 144

8.8 Unit power at selected sites in the Worcester Basin. 144

8.9a Geothermal Resources of the Cheshire and West Lancashire basins. 145

8.9b Identified Resources of the Cheshire and West Lancashire basins. 145

8.10a Geothermal Resources of the Carlisle Basin. 146

8.10b Identified Resources of the Carlisle Basin. 146

8.11a Geothermal Resources of the Sherwood Sandstone Group in the Northern Ireland Basins. 147

8.11b Identified Resources of the Sherwood Sandstone Group in the Northern Ireland Basins. 147

10.1 Summary of the Geothermal Resources of the Permo-Triassic sandstones. 154

10.2 Summary of the Identified Resources of the Permo-Triassic sandstones. 154

10.3 Summary of the Identified Resources at temperatures greater than 60°C. 155

10.4 Summary of the Identified Resources at temperatures between 40 and 60°C. 155

10.5 Identified resources of the Permo-Triassic sandstones expressed in units equivalent to million tonnes of coal. 158

10.6 Results of investigations of the Permo-Triassic sandstones in hydrocarbon exploration boreholes. 159

Notation

a	fracture separation in a doublet scheme
A	heat production, generally radioactive heat production
A_0	surface heat production
A_R	area of reservoir or aquifer
A_z	radioactive heat production at depth z
b	half length or radius of a vertical fracture
B	separation of identical abstraction wells, or twice the distance to a linear, impermeable boundary or flow barrier
c	specific heat
c_a	specific heat of an aquifer i.e. of the rock matrix and the water it contains
c_f	specific heat of moving material (magma, liquid, gas)
c_0	specific heat at 0°C
c_m	specific heat of rock matrix
c_r	specific heat of rock adjacent to an aquifer
c_w	specific heat of water
c_z	specific heat at depth z
C_A	shape factor for a bounded reservoir or aquifer
d	half the well separation for a simple doublet scheme
D	slope of the line defining the linear relationship between q_0 and A_0 (D is a function of the thickness of the upper crustal radioactive layer)
f	function appearing in Equation 2.5, derived from A_z
F	factor incorporated in the maximum recovery factor used to estimate the Identified Resources. Values of 0.25 and 0.1 have been assumed for doublet and single well systems respectively in the UK
h	aquifer or reservoir thickness
H_0	geothermal resource
k	intrinsic permeability of a rock formation
K	hydraulic conductivity
N	viscosity ratio
P	power output
q	vertical heat flow
q_0	surface heat flow
q_x	see Equation 7.14
q^*	heat flow from the lower crust and mantle
Q	volumetric pumping rate from a well
r_w	well radius
R	reservoir or aquifer radius
R_e	radius of an equipotential contour for a simple doublet
s	drawdown of water level in an aquifer or reservoir
s_0	steady-state drawdown of the water level in the abstraction well of a doublet
s_w	drawdown of the water level in an abstraction well
s_{max}	maximum drawdown
S	storage coefficient of an aquifer or reservoir
t	time or time since the start of pumping
t_B	breakthrough time for a doublet scheme
t_{max}	time at which the drawdown reaches s_{max}
T	transmissivity of an aquifer or reservoir
v	vertical velocity of moving material (magma, liquid, gas); positive upwards
V	volume of an aquifer or reservoir
W	Theis well-function
z	depth
x_e	$=(R_e^2 + d^2)^{1/2}$
α	see Equation 7.21
β	heat capacity ratio, see Equation 7.19
θ	temperature
θ_a	abstraction temperature
θ_g	mean annual ground temperature
θ_i	injection temperature for a doublet
θ_0	initial reservoir temperature
θ_m	mean reservoir temperature
θ_{min}	minimum allowable abstraction temperature
θ_{rej}	reject temperature of heat extraction system
θ_z	temperature at depth z
λ	thermal conductivity
λ_0	thermal conductivity at surface temperature
λ_r	thermal conductivity of rock surrounding an aquifer
λ_z	thermal conductivity at depth z

λ_θ	thermal conductivity at temperature θ
Λ	see Equation 7.20
μ	dynamic viscosity
ϱ	density
ϱ_a	density of an aquifer, i.e. of the rock matrix and the water it contains
ϱ_f	density of moving material (magma, fluid, gas)
ϱ_m	density of the rock matrix
ϱ_r	density of rock adjacent to an aquifer
ϱ_w	density of water
τ_f	$= 4Tt/Sb^2$
ϕ	porosity

a′, b′, u and w are constants; $a' = b' + \theta_g$

In this volume the unit megawatt (MW) generally refers to thermal power. In the few places where it refers instead to electrical power, this is clearly indicated. Geothermal resources are generally expressed in joules but, as the joule represents a small amount of energy, exajoules (10^{18} joules) are used. A recognisable scale is given to the figures if it is borne in mind that the annual use of electricity in the UK is about 10^{18} joules and the total annual coal consumption is about 3×10^{18} joules.

Chapter 1 Introduction

R. A. Downing and D. A. Gray

THE NATURE OF GEOTHERMAL ENERGY

The source of geothermal energy used by mankind is the earth's natural heat which is derived primarily from the decay of the long-lived radioactive isotopes of uranium, thorium and potassium but probably with a contribution from a slight cooling of the earth. The earth acts as a heat engine and at its surface there is a continuous heat flux comprising the heat flow from the mantle and lower crust supplemented by heat production from the radioactive isotopes which are largely concentrated in the upper crust. As a consequence a study of heat flow is a principal exploratory technique for the identification and location of geothermal resources as it permits the prediction of temperatures to depths below those reached by shallow drilling.

Heat is transferred in the crust by conduction through the rocks and, locally, by convection in moving fluids, including groundwater and molten rock (or magma). Over most of the surface of the continents the main transfer mechanism is conduction, but the conductive heat flow is very diffuse, averaging no more than 60 mW/m^2. Although global maps of heat flow have been produced (Chapman and Pollock in Bott, 1982) the detail with which heat flow distribution is known varies considerably and is based on some 5400 world-wide measurements. In general, however, it can be said of continental heat flow that it is highest in the regions of active tectonism and volcanism and lowest over the Precambrian shields. Some anomalies in the pattern of heat flow in the intermediate areas between the two extremes can be related to the occurrence of extensive sedimentary basins. The strata in these include both insulating clays, which reduce the rate of heat loss, and permeable rocks in which groundwater acts as a heat transfer medium.

In the active volcanic zones of the earth, surface heat flow is much higher than the average and can attain values of more than 2 W/m^2 (Rybach and Muffler, 1981). Such high values cannot be explained by conduction alone and the additional energy is provided by convective hydrothermal systems. In these, meteoric water penetrates to depths varying from a few hundred metres to several kilometres, where it is heated by a body of magma before rising to the surface again. The principal driving force is the density difference between the cold recharge and the rising hot water. Away from these active volcanic zones, local variations in the regional patterns of conductive heat flow can be caused by differences in heat production reflecting lateral variations in the occurrence of the radioactive isotopes ^{238}U, ^{235}U, ^{232}Th and ^{40}K, as well as by convective transfer resulting from the movement of groundwater.

GEOTHERMAL ENERGY SYSTEMS

Systems permitting the exploitation of geothermal energy are generally classified as either high enthalpy, with temperatures greater than 150 to 200°C, or low enthalpy, with temperatures of less than 150°C and commonly less than 100°C. In both cases an aqueous solution extracts heat from the hot rocks, which act as a heat exchanger, and the fluid transports it to the surface. Most of the heat is stored in the rocks themselves rather than the fluids they contain and the low thermal conductivity of the rocks emphasises the need for a heat transfer medium.

High enthalpy resources have been developed successfully since 1904 when electricity was generated from a small dynamo driven by natural steam at Lardarello in Italy. The subsequent development of the Lardarello Field and similar schemes elsewhere, use naturally occurring steam or hot water, in convective hydrothermal systems, as the heat transfer medium.

In the 1970s a new concept was proposed whereby a high enthalpy system would be developed in hot but dry rocks which contain insufficient natural fluid to transfer the energy to the surface; this is known as the 'Hot Dry Rock' (HDR) system. Experimental work began at Fenton Hill near Los Alamos, New Mexico and has been followed by other research programmes, notably in Cornwall, England.

By contrast, low enthalpy systems use natural groundwater in permeable formations (or aquifers) which are heated by conductive heat flow. To produce useful energy, aquifers need to be tapped at depths of 2 to 3 kilometres to produce water at temperatures greater than 60°C, or at 1 to 1.5 kilometres for water at more than 40°C.

Irrespective of which system is developed, heat is invariably abstracted from commercial geothermal reservoirs at greater rates than that at which it is being replenished, so that the energy in such reservoirs represents a finite resource with a life-span measured in decades. In high enthalpy systems this is because the hot water flow, occurring naturally prior to the development of the geothermal resource, is generally less than the abstraction rate from the exploitation wells, so that the artificial rate of energy withdrawal is higher than the natural flux. In low enthalpy reservoirs the same result occurs because the low thermal conductivity of rocks limits the rate at which heat is transferred to the reservoir.

High enthalpy systems

The most commonly recognised form of geothermal energy is associated with high enthalpy systems that are found in the active tectonic and volcanic zones associated with plate boundaries as, for example, along the Pacific coastlines and the Alpine zone of Europe and Asia. Developed geothermal fields exist in such areas, including Italy, eastern Africa, western USA, central America, Japan, the Philippines and New Zealand. In these regions heat flow through the crust is very high in areas of relatively limited extent. Commonly these anomalies are caused by the intrusion of granitic magmas into the upper crust at depths of several kilometres. These heat the meteoric water circulating in fractures and faults through the overlying rocks in the hydrothermal circulation systems previously referred to.

High enthalpy resources may exist as either water or vapour-dominated systems depending upon whether water or steam is the principal transport medium. Where temperatures exceed 200°C these resources are generally used for the generation of electricity. In the world as a whole some 3200 megawatts (MW) of electrical power were being generated in 1983 from such resources (Di Pippo, 1984). The higher the temperature of the resource, the more efficient is the conversion to electric power, using conventional turbines, and the lower the cost of development. Current experiments on equipment using the organic Rankine Cycle are leading to power generation at lower temperatures, and this trend is likely to lead to more widespread use of geothermal resources for power generation which could overlap into the low enthalpy area.

Low enthalpy systems

Low enthalpy resources are found where heat flow is more diffuse with values around the world mean of 50 to 60 mW/m^2. These resources are found typically in permeable sedimentary rocks in more stable geological areas, such as north-west Europe, and hence suitable conditions are of more widespread occurrence than in the high enthalpy zones. To date low enthalpy resources have only been developed in limited areas, of which the Paris Basin in France and the Pannonian Basin in Hungary are well-known examples, with others in North America and the USSR. The temperature of water in such sedimentary reservoirs is governed by steady-state conductive heat flow through the upper crust, but this heat flux is commonly redistributed by groundwater flow which can cause anomalies on a regional scale.

Low enthalpy resources are used for space heating as well as for industrial and horticultural processes; applications often referred to as the 'direct use' of geothermal energy. As the groundwaters that provide the means of using the heat are commonly saline, their thermal energy is extracted by heat exchangers and the heat-depleted brine is discharged to waste or reinjected into the ground after use. For direct use of its thermal energy, the water must generally be hotter than 60°C. The reject temperature is typically 30°C but more energy can be extracted by using heat pumps, when reject temperatures may be much lower, of the order of 10°C. The design of heat pumps is steadily improving as a consequence of current research and development so that waters with temperatures as low as 20°C may in the near future be more widely used as energy sources. In the UK this would mean that aquifers lying at depths greater than 500 m would contain water with geothermal potential.

As different processes require different temperatures, the energy can be used initially for purposes that require higher temperatures and then successively passed down or 'cascaded' through other applications that require lower temperatures. For example, energy could be extracted for space heating and the rejected water used subsequently for horticulture or fish-farming, or for therapeutic purposes.

In and around Paris, in France, over 30 low enthalpy schemes are in commercial production for space heating of large blocks of flats. The water is derived from a fractured Jurassic limestone at depths of about 2 km. The aquifer is very permeable and average yields from individual wells, which commonly discharge at the surface under pressure, are 200 m^3/hour at temperatures of up to 85°C which provide between 5 and 8 MW of thermal energy per well. The schemes comprise two wells, referred to as a 'doublet'. One well is for production and the other is to reinject the heat-depleted brine into the aquifer. The French are investigating the feasibility of exploiting the deeper Permo-Triassic sandstones in the Paris Basin; these sandstones are the principal potential geothermal reservoir in the UK.

Shallow groundwaters from aquifers no more than a few tens of metres below the ground surface can also be used as sources of energy to heat domestic and horticultural buildings. The temperatures may be no more than 10°C and the reject temperature after passing through a heat pump can be as low as 3°C. However, the use of groundwaters at temperatures of less than 20°C as energy sources is outside the scope of this book and such applications are not considered.

Hot Dry Rock systems

Thermal anomalies that give rise to high enthalpy geothermal fields are of limited extent and occurrence and low enthalpy resources, although more widespread, are restricted to rock formations at depths of less than 3 to 4 kilometres. Most rocks at depths greater than this are essentially impermeable and deep circulation flow systems are the exception rather than the rule. The feasibility of using the large amounts of heat stored in deep impermeable rocks, by means of the Hot Dry Rock concept, is being investigated by large-scale field experiments in New Mexico, in the USA, by the Los Alamos National Laboratory, and in Cornwall, by the Camborne School of Mines. The programmes involve drilling two boreholes to depths where appropriate temperatures occur, and creating an artificial fracture system between them by injecting water under high pressure (referred to as hydraulic fracturing). Water is then circulated down one borehole to extract heat from the fractured zone, which acts as a natural heat exchanger, before returning to the surface as hot water or steam. At some sites successful development will need more than two boreholes; at the site in Cornwall, for example, a third borehole has been drilled. For controlled hydraulic fracturing between two or more boreholes a knowledge of the orientation of the in situ stress-field at the depth of the reservoir is essential. The original objective of the HDR concept was to produce water at temperatures of more than 200°C for the generation of electricity, but it may be possible to develop the technology at lower temperatures, say 100°C, if the concept can be pursued to economic reality.

BASIC GEOLOGICAL STRUCTURE OF THE UK IN RELATION TO GEOTHERMAL ENERGY PROSPECTS

The geological framework of the UK has a significant bearing on heat flow through the upper crust and, because of variations in thermal conductivity of different rock types, on the geothermal gradient and hence the occurrence and recognition of regions favourable for geothermal exploration. Physiographic or tectonic features quite commonly delineate different heat-flow provinces.

The Precambrian rocks of the UK formed the basement of two separate plates during the Lower Palaeozoic. In north-west Scotland, a fragment of the North Atlantic Shield is represented (Figure 1.1). This was originally part of the Canadian and Greenland Precambrian shields or North American Craton. A few outcrops of Precambrian rocks in Central England which have relatively young ages indicate islands caught up in the closure of the Lower Palaeozoic oceans between the three continents of Canada-Greenland, Fennoscandia and Gondwanaland. They occur in a triangular area which has been called the Midlands Microcraton. These Precambrian rocks probably form the deep basement in southern Britain below a cover of Palaeozoic sediments and are different in character from the rocks north of the Highland Boundary Fault.

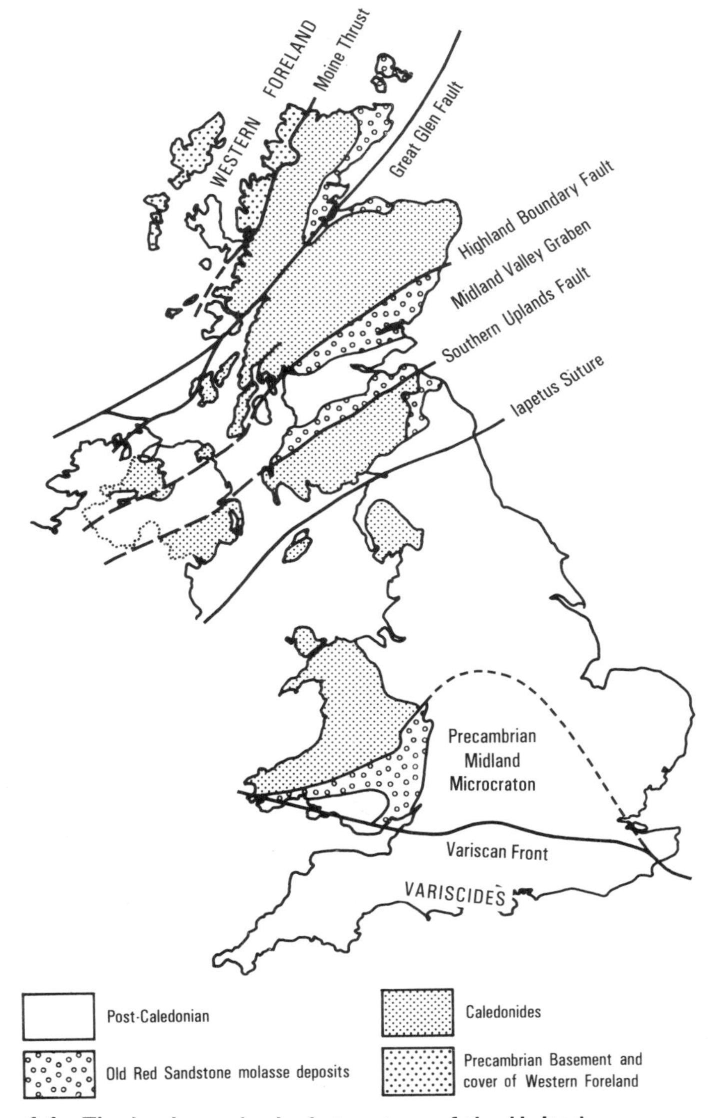

1.1 The basic geological structure of the United Kingdom

The oldest rocks known beneath a large part of the British Isles form part of the complex Caledonian orogenic belt which includes sediments and volcanic rocks ranging in age from Precambrian to Silurian. These rocks were deposited in a Caledonian ocean and were folded at the end of the early Palaeozoic. Two branches of the Caledonian orogen meet along the line of the Pennines; the branch trending NE to SW was deformed during the closure of the Iapetus Ocean and the NW to SE trending belt results from closure of a sea or ocean in the area which is now East Anglia and the southern North Sea. The line of closure of the Iapetus Ocean, known as the Iapetus Suture, is believed to cross England from the Solway Firth to the east coast near the Scottish Border (Figure 1.1). The geothermal significance of this event is that regions north and south of the suture may lie in different heat-flow provinces.

Two further events of geothermal significance took place in late Silurian to early Devonian times: the initiation of the graben that now forms the Midland Valley of Scotland, an area with some geothermal potential, and the emplacement of granite plutons both north and south of the Iapetus Suture (see the Geothermal Map). Some of these plutons have both a high heat production and a large volume, and their distribution has a major bearing on the present day heat-flow pattern and Hot Dry Rock geothermal potential.

The erosion of the continent formed by the Caledonian orogeny gave rise to the thick detrital deposits of the Old Red Sandstone which formed in internal molasse basins in northern Britain and on coastal alluvial plains in the south (Figure 1.1). The intermontane basins that formed in Scotland and northern England included the Orcadian Basin, which extended from north-east Scotland to Orkney and Shetland, as well as the basins in the Midland Valley of Scotland and the Borders region of south-east Scotland and north-east England (Anderton and others, 1979). In southern Britain thick sandstones and siltstones crop out in South Wales and the Welsh Borderlands and occur beneath cover under southern England. Although from their mode of formation these sediments are potential aquifers, extensive cementation has generally destroyed their porosities and they have low permeabilities. However, in the Midland Valley basin they are more permeable and even where now buried beneath younger sediments may have water-bearing potential.

At the end of early Palaeozoic times, the Caledonian orogenic system stabilised. To the south of the Caledonides and the Midlands Microcraton a new orogenic cycle, the Variscan or Hercynian, developed. The Variscan Front, which marks the northern limit of intense structural deformation, extends ESE to WNW from South Wales passing to the south of London (Figure 1.1). On the foreland to the north of the deformed zone, Lower Carboniferous rocks were deposited in fault-controlled basins. These Lower Carboniferous sediments were succeeded by a widespread blanket of fluvio-deltaic Silesian sediments. The Carboniferous sequences contain many sandstones and limestones, which are potential geothermal aquifers where they now occur in the structural basins that subsequently developed as a consequence of earth movements in the late Carboniferous. The Carboniferous Limestone contains the hydrothermal circulation systems that give rise to the warm springs at Bath and Bristol and in South Wales and Derbyshire; no doubt others remain undetected beneath younger sediments and the high heat flow associated with the Eakring structure in the east Midlands could be but one example. At the close of the Carboniferous, granites with high heat production were intruded in south-west England (Figures 3.3 and 3.4) and their presence is reflected by a zone of high heat flow.

After the Hercynian orogenesis was completed, in late Carboniferous and early Permian times, Britain lay within the new continental mass of Pangaea. Tensional stresses developed and fault-controlled depositional basins formed in Permian and Triassic times. Sediments eroded from the surrounding uplands were deposited in, for example, the East Yorkshire and Lincolnshire, the Wessex, Worcester, Cheshire and West Lancashire basins (Figure 4.1). These tended to be marginal depositional areas of the major offshore sedimentary basins in the North Sea, Irish Sea and Western Approaches. The Permo-Triassic sandstones deposited in these basins are the major low enthalpy geothermal reservoirs in the UK.

During the remainder of the Mesozoic and early part of the Tertiary, detrital sediments and limestones accumulated in fault-controlled basins over much of Britain. These sediments include many aquifers but the Lower Greensand is the only one of significance that occurs at depths great enough for temperatures to exceed 20° to 30°C and then only in a small region along the south coast. However, these Mesozoic and Tertiary sediments, where they are preserved, as in east Lincolnshire, Yorkshire, Wessex and Northern Ireland, provide a thick insulating cover of low thermal conductivity over the Permo-Triassic sandstones and thereby increase the geothermal gradient to the top of these reservoirs.

GEOTHERMAL ENERGY IN THE UK

The UK is part of the stable foreland of Europe and is remote from active plate boundaries, so that development of conventional high enthalpy systems for electricity generation is not possible. The last major period of volcanic activity was in the Tertiary and the igneous intrusions associated with this period have long since cooled to equilibrium temperatures.

Visible manifestations of geothermal activity are restricted to a few thermal springs of which the best known are at Bath, Bristol, Buxton and Matlock. These waters are meteoric in origin but have been heated by circulation to appreciable depths below the surface.

The average heat flow in the UK is about 54 mW/m^2 and is similar to the world mean of 60 mW/m^2. The average geothermal gradient, determined from bottom hole temperatures measured in boreholes, is about 26°C/km. However, this value for the geothermal gradient is above the true average because a disproportionate number of boreholes have been drilled into low conductivity sedimentary rocks or on positive heat flow anomalies. A more likely average value in basement rocks is less than 20°C/km.

The heat flow in a number of regions, principally in south-west England and localised areas in northern England, is above average and maxima of more than 100 mW/m^2 are attained. The cause is the presence in the upper crust of granites which have a high heat production because of the decay processes of the long-lived radioactive isotopes of uranium, thorium and potassium. In these areas the geothermal gradient exceeds 30°C/km and similar above-average values are also attained in areas of near-normal heat flow but where thick insulating sediments of low thermal conductivity occur, as in eastern England. Assuming a mean ground surface temperature of 10°C, a gradient of 30°C/km gives a temperature of 70°C at a depth of 2 km and even 25°C/km gives 60°C at the same depth, temperatures that are high enough for direct use of geothermal energy.

In the UK there are two possible sources of geothermal energy. The first is hot groundwater in permeable rocks in deep sedimentary basins and the second, Hot Dry Rocks at depths of 3 km or more. A primary exploration objective is to detect and delineate aquifers with adequate permeability and porosity at depths where temperatures are suitable for the economic development of the hot-water resources they contain. Favourable aquifers are likely to occur in deep sedimentary basins, particularly the Mesozoic basins, where they coincide with regions of above-average heat flow. The heat in these systems is provided by steady-state conduction from an underlying heat source and in stable areas such as the UK, this heat source is simply the background heat flow from the earth's interior supplemented in some regions by heat produced by radioactive decay of elements principally in granitic rocks. Areas of high heat flow can indicate rising groundwater which acts as a heat transfer medium. If the formidable well-engineering, rock mechanics, hydraulic and geochemical problems associated with the HDR concept can be successfully overcome, the potential resources are far larger and more widespread than the natural hot water in aquifers. But, at the present time, hot groundwaters provide the only possible source of geothermal energy in the UK that can be developed with currently available technology. An important factor to bear in mind when considering geothermal heat as an energy source is that heat cannot be transported economically over long distances and a heat load or a power generating facility is required that is within no more than a few kilometres of the well-site.

INVESTIGATION OF THE GEOTHERMAL POTENTIAL OF THE UK

The Arab oil embargo in the mid-seventies, together with the consequent large increase in oil and other energy prices, led to a much wider appreciation that conventional sources of energy are finite and with time will become more scarce and costly. Attention was directed to the so-called alternative sources of energy, perhaps more accurately referred to as the 'renewables', although large-scale development of geothermal energy involves the mining of heat and is not renewable in acceptable time-spans. Against this background the British Geological Survey compiled a report on the available data and prepared outline proposals for an investigation of the geothermal potential of the United Kingdom (Dunham, 1974a). These were submitted to the Department of Energy and led to the publication of an overview of the prospects for this form of energy in the UK (Garnish, 1976). Subsequently, in 1977, the Department invited the Survey to investigate the geothermal potential of the UK; the programme also received financial support from the Commission of the European Communities (C.E.C.). Following discussion with both the Department and the CEC and also with the Energy Technology Support Unit, acting for the Department, the programme described below was agreed and implemented in two phases, from 1977 to 1979 and from 1979 to 1984. The various projects in the programme are given in Table 1.1.

Table 1.1 Geothermal energy programme of the British Geological Survey, 1977–1984

1. Low enthalpy potential of deep sedimentary basins
   - (a) Mesozoic basins
   - (b) Upper Palaeozoic basins
   - (c) Investigations in commercial boreholes
   - (d) Drilling boreholes to measure heat flow
   - (e) Drilling deep exploration boreholes.
2. Potential for HDR development, with particular reference to Caledonian granites
   - (a) Lake District
   - (b) Eastern Highlands of Scotland
   - (c) Identification of buried granites with high radio-element contents.
3. Assessment of the geothermal resources.

In the stable geological environment of Britain, exploration for geothermal resources was directed towards identifying permeable rocks in deep sedimentary basins where the geothermal gradient ensures that water is at a high enough temperature to make development of economic interest. Thus the central theme of the programme was a detailed investigation of the major Mesozoic and Upper Palaeozoic sedimentary basins to identify and delineate those major aquifers having geothermal potential, by assessing the aquifer properties and the quality and temperature of the water they contain. The various basins studied are described in Chapters 4 and 5. The programme involved the collation and interpretation of all available geological data and the interpretation of geological structure from geophysical studies, primarily gravity and seismic surveys. Aquifer properties were assessed by laboratory analysis of rock cores, interpretation of geophysical well logs and, where possible, drill-stem tests.

The groundwaters in sedimentary basins that may have potential for geothermal development are invariably brines. The chemical nature of the brines introduces problems because of their corrosive character and the tendency for salts to precipitate, both in the well and in pipes at the surface, as temperatures and pressures are reduced. Consideration has to be given to the means of overcoming such problems when designing a geothermal development project. Similarly disposal of the geothermal brine after use must not have deleterious effects on the environment. Therefore, in areas away from the coast it is necessary to reinject it into the ground through injection wells. All these problems have been overcome successfully overseas where geothermal energy is being developed commercially. In the UK to date, however, only geothermal exploration wells have been drilled and these have been located on the coast where disposal directly to the sea has been feasible.

A basis of any geothermal exploration programme is an understanding of the areal distribution of heat flow through the upper crust. Research into this topic has been undertaken over a long period of time by many workers. Recent studies have been made by Richardson and Oxburgh (1979) and Wheildon and others (1980) and these existing data were supplemented between 1980 and 1984 by heat flow studies in 17 boreholes, 200 to 300 m deep, specifically drilled for this purpose, as well as in 6 boreholes drilled for other purposes but completed in a form suitable for the measurements of equilibrium rock temperatures, and in 12 selected, deep oil or geothermal exploration boreholes which provided details of rock strata penetrated but only a non-equilibrium temperature at the bottom of the boreholes when drilling had been completed. The geological studies and the heat flow measurements provided the basic data from which the most favourable areas for exploration could be identified.

As already mentioned, firm information about potential geothermal aquifers is limited. The mineral exploration boreholes that provided the basic data had been drilled to meet specific objectives that rarely coincided with the objectives set for geothermal studies. After preliminary studies during the early part of the programme, two deep exploration wells were drilled to provide more reliable data about the nature and properties of potential aquifers in the Mesozoic Basins. The first of these was at Marchwood, near Southampton, and the second was at Larne in Northern Ireland. Both were wildcats in the sense that they were sited in areas where the deep geology was uncertain with information based primarily on gravity surveys and geological interpretation. A third exploration well was drilled recently at Cleethorpes on the south Humberside coast. Following the successful testing of the Marchwood Well, the Department of Energy funded a development well in the centre of Southampton which may be used by Southampton City Council as a field-trial to heat civic offices and the City Baths. The Larne Well proved that the temperatures required could be achieved, as the bottom hole temperature at a depth of 2876 m was 91°C, but the permeability of the formations was too low to yield an adequate supply of water. The well at Cleethorpes proved that the Sherwood Sandstone is a very permeable aquifer in Humberside and east Lincolnshire.

The information and experience provided by these wells was supplemented by tests of possible geothermal reservoirs intersected in oil and coal exploration boreholes. Some 12 boreholes were examined in this manner. Geophysical logs were run, cores were cut and favourable horizons were investigated with drill-stem tests. Through the cooperation of the hydrocarbon industry and the National Coal Board these studies proved to be very cost-effective. Moreover, they demonstrated that where adequate thicknesses occur the geothermal potential of the Permo-Triassic sandstones is significant.

The main investigation in Europe of the feasibility of applying the Hot Dry Rock concept is being carried out in Cornwall by the Camborne School of Mines on behalf of the Department of Energy and the CEC. These investigations are concentrated on the Hercynian granites, where high values of heat production boost the heat flow to approximately twice the average for the UK. The Geological Survey's programme has been concerned with investigating the large number of granites of Caledonian age which may provide similar favourable conditions.

Studies in the UK and elsewhere suggest that two of the most important criteria for high heat flow over crystalline intrusions are high radioactive heat production, which tends to occur in the most evolved granites, and large intrusive volume as indicated by a strong negative Bouguer gravity anomaly, which suggests that the granitic intrusion extends to depths of many kilometres. Data from a large number of Caledonian intrusions in Britain show that intrusions with a high heat production (greater than 3.5 $\mu W/m^3$) accompanied by large negative gravity anomalies are almost entirely confined to the Eastern Highlands of Scotland and northern England. The heat flow and heat production of the Weardale Granite had already been measured in the Rookhope Borehole so attention was concentrated on the Lake District and Eastern Highlands. Boreholes were drilled to depths of 300 m to measure heat flow in the Shap and Skiddaw granites in the Lake District and in the Cairngorm, Mount Battock, Bennachie and Ballater granites in the Eastern Highlands. Detailed geochemical, geophysical and thermal modelling studies were made.

The information derived from these various studies has been used as a basis for making a first assessment of the geothermal resources of the UK. The Accessible Resource Base, which is a base-line estimate of the total heat stored in the upper 7 km of the crust, has been calculated. More practical estimates of the potential low enthalpy geothermal resources have also been made and the most favourable exploration areas identified.

The data base used in the study is extensive. The basic geological information was derived from the geological memoirs and maps published by the British Geological Survey. Geophysical information is based on BGS Bouguer Gravity Anomaly maps and aeromagnetic maps of the UK. Structural interpretation has relied heavily on the extensive seismic surveys undertaken mainly in the course of oil exploration, especially in the East Yorkshire and Lincolnshire, the Cheshire and the Wessex basins. Some seismic surveys have also been made in the West Lancashire, Carlisle and Worcester basins but only very limited surveys have been made in the Midland Valley of Scotland and Northern Ireland. A seismic survey was commissioned near Glasgow to study the deep structure of the Carboniferous and Old Red Sandstone.

Many deep boreholes (that is, drilled to depths of more than 1000 m) exist in the East Yorkshire and Lincolnshire Basin but the numbers are much more limited in the remaining Mesozoic basins. Because of this the information about the hydrogeology and aquifer properties of potential geothermal reservoirs is limited in the deeper parts of the Mesozoic basins, although the extensive offshore exploration programmes in the North Sea and Irish Sea have provided valuable indications of the nature of the rocks at depth.

The structure of the Mesozoic basins has been derived primarily from information provided by deep boreholes and generalised interpretation of reflection seismic data where available. However, it is only in the East Yorkshire and Lincolnshire Basin, because of the extensive sub-surface data, that the structure is relatively well known. In the Wessex Basin there are few deep boreholes (apart from those in a small area around the Dorset oilfield); although there is good seismic coverage it is of little use in determining the structure beneath the Jurassic, as seismic reflections below this horizon are weak and discontinous. Accordingly, the postulated structure of the Permo-Triassic is based on indirect geological and gravity evidence of the positions of major faults which probably controlled deposition in the area. There is not a similar problem in the Cheshire basin, where sub-surface data provide good evidence of the post-Carboniferous structure, which is supported by an interpretation of the gravity data. The deep structure of the Midland Valley of Scotland is not well understood and knowledge of it is based largely on surface geological data together with gravity data and the results of a north-south refraction seismic profile. The shapes of the Mesozoic basins beneath the cover of Tertiary basalt in Northern Ireland have been derived mainly from gravity data.

Interpretation of the structure of Upper Palaeozoic basins is comprehensive in coal-mining areas, but known in much less detail at depths too great for mining. A considerable amount of data is available from oil exploration programmes in the East Midlands and to a limited extent in other regions. In view of the present interest in onshore oil exploration in the UK, it is likely that further deep boreholes will be drilled in major Mesozoic and Upper Palaeozoic basins within the next few years, and this will substantially improve our understanding of their structure.

The available temperature and heat flow data have been summarised in a 'Catalogue of geothermal data for the land area of the UK'. Heat flows have been observed at some 188 sites. Much of the information about the deep geology of Britain, including thermal data, has become available as a result of oil and coal exploration programmes, and the results have been interpreted in the regional context.

In the UK, geothermal studies are at the stage of assessing the economic feasibility of developing the resource that undoubtedly exists. The following chapters summarise the current knowledge about the geothermal potential, which can be used as a basis for decisions. They record the results of a programme that was concerned with interpreting geological and geophysical data to elucidate the deep structure of selected areas of Britain. This has been related to hydraulic properties of the most favoured aquifers and to the chemical composition of their pore fluids with the overall objective of indicating the potential for geothermal development.

A generalised synthesis of information relating to geothermal energy in the UK is presented as a Geothermal Map which forms the Plate in this volume. This map shows heat flow contours, the position of the principal radiothermal granites, the form of the Mesozoic basins and the areas considered to be favourable for low enthalpy development of the Permo-Triassic sandstones. Supplementary diagrams include the estimated temperature at a depth of 2 km, the estimated depth to the 100°C isotherm and the relationship between temperature and depth from corrected borehole measurements.

Chapter 2 **Heat flow**

J. Wheildon and K. E. Rollin

INTRODUCTION

The temperature in the earth's crust increases with depth and the rate of increase is referred to as the geothermal gradient. The inset on the Geothermal Map (see Plate) shows a plot of all observed, corrected borehole and mine temperatures in the UK plotted against depth and this indicates that the average geothermal gradient for the onshore area to depths of about 4 km is close to 26°C/km. However, the data contained in this inset were derived mainly from hydrocarbon boreholes situated in the onshore sedimentary basins of lowland Britain so that the sample is not strictly representative of the whole land area. In these basins the thermal conductivity is relatively low and the gradient is above the average for the UK as a whole, which is probably less than 20°C/km.

The data include many observations of a single temperature from one borehole, so that the scatter of the points away from the mean gradient is a general indication of the range of geothermal gradients which can be observed in the upper few kilometres of the UK crust; the range is about 15 to 40°C/km. The variation is largely determined by differences in the vertical flux of heat from the earth's crust and the thermal conductivity of the crust. It is only with a knowledge of the variation of these two parameters, both in depth and laterally, that any prediction of sub-surface temperatures is possible. This emphasises the importance of heat flow for any programme of geothermal exploration or resource calculation.

For both low enthalpy and HDR geothermal targets a high surface heat flow is both an exploration criterion and an aid to exploitation. In both systems the causes of the high heat flow may be a combination of factors but often a single significant factor can be identified or inferred. For instance, a high surface heat flow above a thick sedimentary basin with no obvious contribution from convective effects must suggest a heat source below the sediments. However, the recognition of high heat production at the surface cannot be assumed to indicate a high surface heat flow, as will be seen in Chapter 3. The recent geothermal studies in the newer Caledonian granites have shown that the vertical distribution of heat production varies substantially so that a measurement of surface heat production in a granite is not in itself a reliable predictor of heat flow, and hence geothermal gradients.

For low enthalpy targets high heat flow is not an essential pre-requisite for exploitation. The important factor is the occurrence of suitable aquifers with adequate permeability, thickness and temperatures, but, nevertheless, elevated heat flow is still a desirable factor. Low enthalpy resources occur in deep sedimentary basins where permeable rocks of sufficient thickness contain water at temperatures high enough for economic development. In the UK, Mesozoic basins containing Permo-Triassic and younger rocks exist as onshore extensions of major offshore basins, and offer the best prospects for this type of resource. In these environments the surface heat flow is not especially high, normally in the range 50 to 70 mW/m^2. However, the thermal conductivity of sediments in these basins is commonly low enough to increase the geothermal gradient to levels at which temperatures suitable for development are attained within 2 to 3 kilometres of the surface.

The requirements for HDR development contrast sharply. Hot dry rocks are universally available within the upper crust at depths of more than a few kilometres. However, temperatures suitable for development at reasonable depths are only encountered where the heat flow is high enough, or the insulating effect of sedimentary rocks great enough, to overcome the disadvantage of the high thermal conductivities that are typical of the intrusive and metasedimentary basement rocks currently considered to be HDR target formations. This severely restricts the potential targets for HDR exploitation in the U.K. to a few proven areas of high heat flow and possibly impermeable formations beneath Mesozoic basins.

BASIC CONCEPTS

Although the temperature increases with depth, the geothermal gradient in any area is not a simple function of depth and is determined by several important terms. For the one dimensional (vertical) case the general equations which govern the variation of temperature with depth are shown below:

$$q = \lambda \frac{\partial \theta}{\partial z} , \qquad 2.1$$

$$\frac{\partial q}{\partial z} = \frac{\partial}{\partial z}\left[\lambda \frac{\partial \theta}{\partial z}\right] = - A - \varrho_f c_f v \frac{\partial \theta}{\partial z} + \varrho c \frac{\partial \theta}{\partial t} , \qquad 2.2$$

where q = vertical heat flow
λ = thermal conductivity
θ = temperature

z = depth
A = heat production
ϱ_f = density of moving material (magma, liquids, gases)
c_f = specific heat of moving material
v = vertical velocity of moving material (positive upwards)
ϱ = density of static material (rocks)
c = specific heat of static material
t = time.

The terms on the right-hand side of Equation 2.2 represent the main factors affecting the temperature gradient, namely:
— the heat production of crustal rocks
— the movement of heat by hot magmas, gases and liquids
— the change of heat store with time.

Generally the temporal changes are ignored as are the convective effects of hot magmas and fluids, so that for steady state heat conduction, where heat production and thermal conductivity depend only on depth, Poisson's equation is used to represent the terrestrial heat flow:

$$\frac{d}{dz}\left[\lambda \frac{d\theta}{dz}\right] = -A \qquad 2.3$$

A note of caution is appropriate here: in Equation 2.2 the second term on the right allows for the effect of groundwater flow and the possibility of such disturbance must not be overlooked. In substantial areas of the UK the conductive heat flow may be modified by a convective component which can easily go unnoticed in heat flow determination. This effect will be reviewed more fully later.

Surface heat flow can be considered to comprise two main components: a background heat flow from the lower crust and mantle, and a contribution from uranium, thorium and radioactive potassium present in minute amounts in crustal rocks. Assuming that the background heat flow is reasonably constant over a large area, most of the variability in surface heat flow is caused by variations in the heat production of the upper crust due to radioactive elements. These occur in disseminated form in many sediments but are mainly concentrated in silica-rich granites which have heat production values locally up to 7 $\mu W/m^3$. The actual heat production of the upper crust cannot be sampled except at the surface and a heat flow measurement is the only method of examining the integrated effect of the vertical distribution of heat production.

SUB-SURFACE TEMPERATURES

The prediction of sub-surface temperatures requires the integration of Equation 2.3 with various assumptions about the vertical distribution of thermal conductivity and heat production. The variation of thermal conductivity with depth in a uniform layer can be equated to the variation of conductivity with temperature since the latter generally increases with depth. Over the range of temperatures of geothermal interest, thermal conductivity shows an inverse dependence on temperature (Birch and Clark, 1940; Kappelmeyer and Haenel, 1974). This can be expressed in the form:

$$\lambda_z = \lambda_0 \, a'/(b' + \theta_z) \ , \qquad 2.4$$

where λ_z = thermal conductivity at depth z
λ_0 = surface thermal conductivity
θ_z = temperature at depth z
$a' = b' + \theta_g$
and b' is an empirical constant.

The values of constants a' and b' depend on the temperature at which surface conductivity is measured and on the temperature scale used. Assuming conductivities are measured at mean surface temperature (taken as 10°C) and θ_z is measured in degrees Celsius, then suitable values for a' and b' are 833.33 and 823.33 respectively.

Various models of the distribution of heat production have been used to solve Equation 2.3 and the main types are discussed briefly below.

Constant vertical distribution of heat production

A thick sedimentary formation might be expected to have a constant heat production with depth. Substituting Equation 2.4 in 2.3 and using the function $A_z = A_0$, Equation 2.3 can be integrated twice to give a solution for the subsurface temperature:

$$\theta_z = a' \exp[(q_0 z - f(z))/(a' \lambda_0)] - b' \ , \qquad 2.5$$

where $f(z)$ is the solution of $f''(z) = A_z$
with boundary conditions $f(0) = f'(0) = 0$.
For a constant vertical distribution of heat production

$$f(z) = A_0 z^2/2 \ . \qquad 2.6$$

This can only be used to depths of z such that $z < q_0/A_0$. For a layered sedimentary sequence, with changes of conductivity and heat production, a summed calculation of the temperature increments for each separate layer is required. Frequently the heat production in sedimentary rocks is assumed to be negligible, and at shallow depths the temperature dependence of conductivity is ignored so that Equation 2.3 is integrated to give:

$$\theta_z = \theta_g + q \int_0^z \frac{dz}{\lambda_z} \ . \qquad 2.7a$$

By assuming a horizontal layered crust with n layers to depth z of thickness t_i and thermal conductivity λ_i, this equation is often simplified to:

$$\theta_z = \theta_g + q \sum_{i=1}^{n} \frac{t_i}{\lambda_i} \ . \qquad 2.7b$$

This simple equation is commonly used to predict underground temperatures in areas of negligible heat production. It is also the equation used to calculate heat flow from temperature and thermal conductivity measurements. It indicates that high underground temperatures are caused by low rock thermal conductivities and high heat flow.

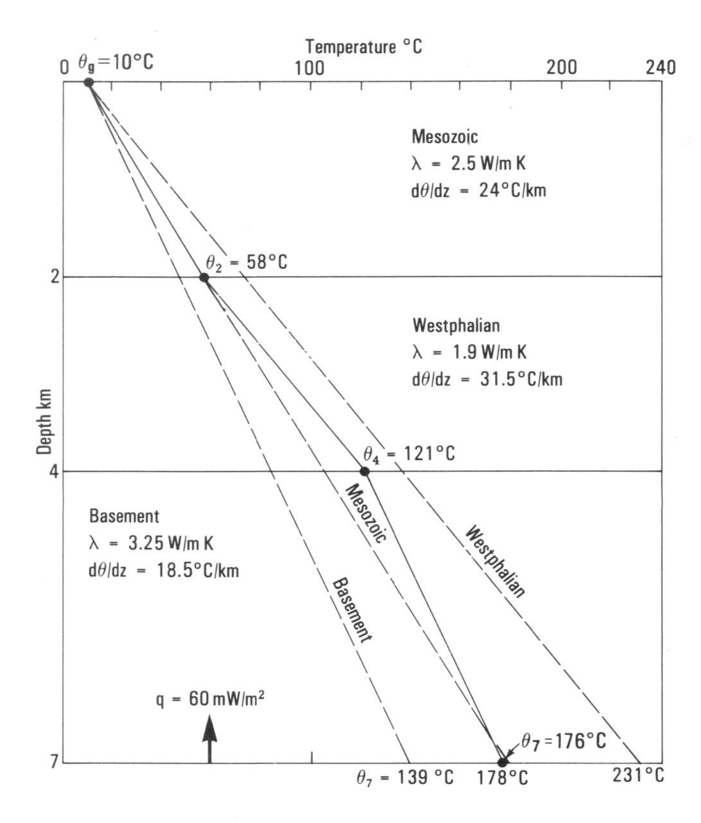

2.1 Effects of thermal conductivity on temperature gradient and sub-surface temperature assuming a constant heat flow of 60 mW/m² and negligible heat production within sediments. The solid line indicates the temperature profile to a depth of 7 km through 2 km of Mesozoic strata, 2 km of Westphalian and 3 km of basement with the mean conductivities shown. The dashed lines would be the temperature profiles and temperatures at 7 km if any of the three units extended through the entire 7 km.

To illustrate the influence of these two properties, Figure 2.1 shows the effect of conductivity variations on the geothermal gradient and sub-surface temperatures for a constant heat flow of 60 mW/m², while Figure 2.2 shows the effect of a radioactive source in the upper crust on the surface heat flow and on subsurface temperatures, assuming similar lower crustal heat flows and crustal conductivities.

Exponential distribution of heat production

An exponential model for the decrease of heat production with depth has been a consequence of the surprising linear relationship between A_0 and q_0 which was first observed over plutonic rocks in the eastern United States by Birch and others (1968) and Roy and others (1968a):

$$q_0 = q^* + A_0 D \quad , \qquad 2.8$$

where q^* = a background or 'reduced' heat flow from the lower crust and mantle and
D = a dimension of length representing a depth of radioactive enrichment.

Both D and q^* are assumed constant over large geographical areas termed geothermal provinces. Typical valves of D from geothermal provinces throughout the world are in the range 4 to 16 km. Lachenbruch (1970) examined possible distributions of A which would satisfy Equation 2.8 and favoured an exponential decrease such that:

$$A_z = A_0 \exp(-z/D) \qquad 2.9$$

In this distribution D is, strictly, the depth at which the heat production is 1/e of the surface value and the exponential decrease must continue to a depth of at least $3D$. This distribution apparently satisfies the condition of differential erosion and is consistent with thermodynamic theory (Rybach, 1976). Since its inception, Equation 2.8 has been the source of much debate. Lachenbruch and Bunker (1971) indicated that in the western United States observed heat production data to depths of 3 km fitted the exponential model but did not preclude other distributions.

Further support for an exponential decrease of heat production in North America was given by Swanberg (1972) for the Idaho batholith; while for the basement of the Eastern Alps Hawkesworth (1974) indicated that measured heat productions cannot distinguish between linear and exponential decreases but that any depth factor D was significantly smaller than the value of D inferred from the relationship between q_0 and A_0. Similarly Bunker and others (1975) in the west Australian shield showed that observed heat productions in the upper 2 km of the crust suggest a near constant A_z rather than a decremental condition and certainly not an exponential decrease with D equal to 4.5 km as inferred from the relationship between q_0 and A_0. Whatever the distribution of heat production, several provinces throughout the world appear to have a linear relationship between q_0 and A_0 and generally the older the continental crust the smaller the value of D. This suggested to Heier (1978) that redistribution of heat producing elements in the crust is an important factor.

For data over granites and Palaeozoic basement in the UK, Richardson and Oxburgh (1978) proposed values of

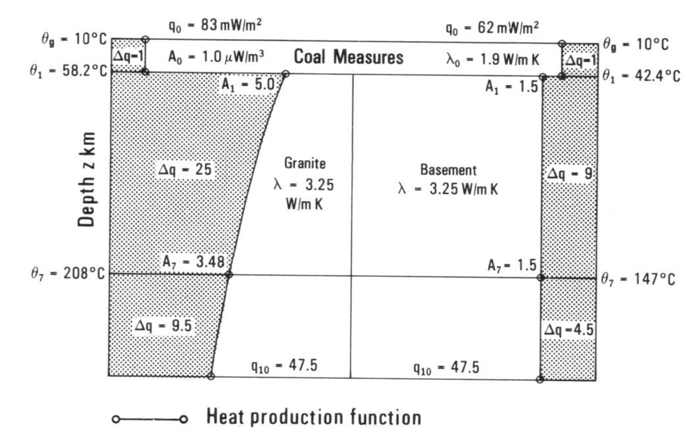

2.2 The effects of crustal heat production on surface heat flow and sub-surface temperature in granite and sedimentary basement assuming the same sub-crustal heat flow and crustal thermal conductivity. The shaded areas indicate the contributions to the surface heat flow (Δq) for the depth ranges 0 to 1 km, 1 to 7 km and 7 to 10 km. For the granite an exponential decrease of heat production has been assumed with D at 16.6 km.

16.6 km for D and 27 mW/m^2 for q^*. England and others (1980) suggested that heat conduction and refraction effects for isolated plutons tended to lead to an overestimate of the true value of q^* and an underestimate of the true depth scale D when derived from the relationship between q_0 and A_0. This means that the true depth scales are even more dissimilar to depth scale derived from other geophysical methods, such as gravity interpretations of radioactive granites.

Jaupart (1983) has further undermined the validity of the relationship between q_0 and A_0. He suggested that the true depth scale of radioactive enrichment is much smaller than that derived from the relationship, mainly because of differential alteration and leaching of the radioactive elements. He also suggested that linearity of the plot of q_0 against A_0 is enhanced by the effects of horizontal conduction between adjacent crustal blocks of contrasting heat production. More recently, reviews of local heat flow anomalies and the plot of q_0 and A_0 suggest that in some areas there is no clear linear relationship. For example, Lucazeau and others (1984) showed that in the Massif Central the heat flow anomaly has a much longer wavelength than the gravity anomaly and the q_0-A_0 data show no correlation.

The original q_0-A_0 dataset for the UK, from which the parameters D of 16.6 km and q^* of 27 mW/m^2 were defined, has now been substantially enlarged. Richardson and Oxburgh (1978) originally suggested that the high value of D indicated that the granites in southern Britain are much less fractionated than petrographically similar granites elsewhere. New data from Scotland suggest that in northern Britain the linear relationship indicates a much smaller value for D. This is consistent with values of D reported for other provinces within the internal zones of orogenic belts (Pollack and Chapman, 1977).

The plot of q_0 against A_0 for the UK dataset is discussed in more detail in Chapter 3 and is shown in Figure 3.12. Whatever the physical validity of the empirical relationship between q_0 and A_0, and the local value of D, equations 2.8 and 2.9 have frequently been used to solve Equation 2.3 and provide an equation for θ_z in the form of Equation 2.5 with:

$$f(z) = A_0 D[z - D(1 - \exp(-z/D))] \quad . \qquad 2.10$$

The effects of an erroneous value for D on the value of θ_z will be discussed later.

Linear distribution of heat production

The scatter of observations of heat production within the top few kilometres of the crust is generally such that neither the exponential nor linear models are definitely proven. In its simplest form the linear model is:

$$A_z = A_0 - u z \qquad 2.11$$

where u is a constant which may vary across provinces or between different intrusions, and might have values close to 0.3. For a granite with a surface heat production A_0 of 6 μW/m^3 this would imply zero heat production at a depth of 20 km. Solving Equation 2.3 using 2.11 gives Equation 2.5 with:

$$f(z) = \frac{A_0 z^2}{2} - \frac{u z^3}{6} \quad . \qquad 2.12$$

Inverse distribution of heat production

As an alternative to the exponential distribution of heat production an inverse relationship in which A_z is proportional to $(1/z)$ can be considered:

$$A_z = \frac{A_0}{(wz + 1)} \quad , \qquad 2.13$$

where w is a constant describing the rate of decrease of heat production with depth and which might have a value close to 0.1.

Using 2.13 to solve Equation 2.3 gives Equation 2.5 with:

$$f(z) = \frac{A_0}{w^2}\left\{1 - (wz + 1)\,[1 - \ln(wz + 1)]\right\} \quad . \qquad 2.14$$

Sub-surface temperatures from different models

Table 2.1 shows sub-surface temperatures which might be expected for two geothermal terrains in the UK. It shows the calculated temperatures for each of the above four models of heat production for a site on the Dartmoor granite, where heat flow is greater than 100 mW/m^2 and heat production is about 5μW/m^3, and for a site in west Wales with a heat flow close to 60 mW/m^2 and at which the metasedimentary basement has a low heat production. The constant (step) model and the linear model have limitations for z so that the heat production function does not become negative. The first constant of integration for the exponential model is taken as the background heat flow q^* so that the use of this substitution is only valid in areas where there is a proven linear relationship between A_0 and q_0. The inverse relationship requires no assumptions of the validity of the A_0 and q_0 relationship but, as with the linear model, does require some estimate of the rate of decrease of heat production with depth. This must be either inferred or taken from observational data, although such data are not generally available for many areas. Nevertheless, the inverse function does allow for a steady decrease of A_z which always remains positive, and its use might be considered further. The Table also shows the effects on θ_z of incorrect assumptions about the values of D, or a change in the value of the linear and inverse decrements.

Inspection of the range of calculated temperatures at a depth of 7 km for all four models shows that the temperature is relatively insensitive to either the type of heat production distribution function used or indeed the value of D. Of more importance is the accurate observation of the heat flow, thermal conductivity and heat production data.

MEASUREMENT TECHNIQUES

Heat flow is measured by an indirect technique based on the solution of Equation 2.7 and requiring independent measurements of the geothermal gradient and the thermal conductivity of the corresponding rock unit. Seventy-one of the measurements in mainland UK made in the period up to 1978 were summarised by Bloomer and others (1979). Contributions by Oxburgh and others (1972), Pugh (1977), Richardson and Oxburgh (1979), Lee and others (1984) and

Area	Depth km	Heat production distribution used						
		Constant Eq 2.5 and Eq 2.6	Exponential Eq 2.5 and Eq 2.10		Linear Eq 2.5 and Eq 2.12		Inverse Eq 2.5 and Eq 2.14	
			D = 16.6	D = 8	u = 0.33	u = 0.6	w = 0.1	w = 0.2
Dartmoor	1	43.91	43.93	43.95	43.93	43.95	43.94	43.96
q_0 = 105.5	2	77.50	77.63	77.76	77.65	77.78	77.70	77.88
A_0 = 4.90	3	110.61	111.07	111.52	111.15	111.59	111.30	111.85
λ_0 = 3.10	4	143.10	144.21	145.27	144.42	145.50	144.74	145.95
	5	174.83	177.04	179.09	177.49	179.67	178.02	180.23
	6	205.65	209.53	213.03	210.39	214.29	211.13	214.73
	7	235.41	241.67	247.16	243.17	249.56	244.06	249.50
West Wales	1	28.82	28.82	28.83	28.84	28.85	28.83	28.84
q_0 = 59.4	2	47.41	47.46	47.52	47.56	47.68	47.49	47.56
A_0 = 1.94	3	65.76	65.93	66.10	66.26	66.67	66.02	66.22
λ_0 = 3.14	4	83.81	84.22	84.61	85.03	86.04	84.41	84.86
	5	101.55	102.35	103.09	103.98	105.98	102.70	103.50
	6	118.94	120.33	121.57	123.22	126.75	120.89	122.18
	7	135.94	138.15	140.09	142.88	148.59	139.00	140.91

Table 2.1 Sub-surface temperatures (in °C) for Dartmoor and West Wales assuming a variety of heat production distributions in the crust

Wheildon and others (1980, 1984c) complete the coverage. Methods and techniques involved and assessment of data reliability were reviewed in detail by Bloomer and others (1979) and only essential points are repeated here.

More recently Jessop (1983) has reviewed the manner in which world heat flow data is and should be reported, stressing the need for a rational presentation in which no subjective classification of reliability is required because data quality is apparent from the presentation itself. This has to include the frequency of temperature and thermal conductivity measurements. Much of the early UK heat flow data is of low quality, but the emphasis in the last decade has been towards the acquisition of high quality data, particularly with regard to adequate sampling for measurements of the thermal conductivity of the rock units involved.

Temperature gradient measurements

In the pioneering studies by Benfield (1939), Anderson (1940), Bullard and Niblet (1951) and Mullins and Hinsley (1957) which comprised 22 heat flow observations, maximum mercury-in-glass thermometers were used. These were deficient in that they provided temperature observations of doubtful accuracy at widely spaced observation points. More importantly they denied the observers any indication of convective flow within the boreholes where the measurements were made. In all subsequent work (see Burley and others (1984) for details) thermistor resistance thermometers have been used after designs by Beck (1965) and Roy and others (1968b). With these systems a complete one-dimensional temperature log, of high accuracy, is obtained.

The introduction of thermistor thermometers soon revealed the full extent of the problem of water circulation between different levels within boreholes. To overcome this Roy and others (1968b), working in the United States, lined boreholes with steel casing and filled the surrounding annulus with cement; this procedure has become widely adopted in the UK. Preserving boreholes in this way has the additional benefit of ensuring that they remain open sufficiently long for the temperature disturbance due to drilling to subside, variously reported at between 3 and 10 times the drilling time (Bullard, 1947; Lachenbruch and Brewer, 1959). Where boreholes cannot be retained as permanent heat flow stations, bottom hole temperatures (BHTs), being those disturbed the least by drilling, may be measured during breaks in drilling or when drilling has been completed. Numerical solutions for the estimation of undisturbed BHTs are available (Oxburgh and others, 1972; Barelli and Palma, 1981). Given good quality transient BHT data, acquired over a period of several hours, reasonable estimates of undisturbed BHTs can be made.

Thermal conductivity

Thermal conductivity measuring procedures in themselves are sufficiently accurate, but there remains a problem with regard to sampling frequency. A heat flow measurement based on, say 30 conductivity samples, each 10 mm thick, from a 300 m borehole is regarded as of high quality. Yet only 0.1% of the borehole material has been used in the determination of thermal conductivity. With conductivities ranging over a factor of 3, or so, this is a serious problem. It is particularly so in the case of sedimentary and metamorphic rocks where compositional variations occur on scale lengths from millimetres to tens of metres. Here sampling density must be increased to the point where the involved strata are adequately and proportionally represented.

Table 2.2 shows mean thermal conductivities for some of

Table 2.2 Thermal conductivities of some lithostratigraphical formations in the UK

System	Formation	Lithology	Code	n	λ	s.e	Reference
Palaeogene	Barton Beds	Smst	109	10	2.12	0.06	1
		Mdst	109	2	1.46	0.05	1
	Bracklesham Beds	Smst	109	14	2.20	0.16	1
		Mdst	109	4	1.58	0.01	1
	London Clay	Smst	108	5	2.45	0.07	1
	Reading Beds	Smst	107	4	2.33	0.04	1
		Mdst	107	10	1.63	0.11	1
Cretaceous	Chalk	Chlk	106	41	1.79	0.54	2
	Upper Greensand	Sdst	105	18	2.66	0.19	3
	Gault	Smst	105	32	2.32	0.04	1
		Mdst	105	4	1.67	0.11	1
	Hastings Beds	Slst	102	2	2.01		4
		Slcl	102	3	1.26		4
Jurassic	Kimmeridge Clay	Mdst	99	58	1.51	0.09	1
	Ampthill Clay	Mdst	98	60	1.29	0.03	1
	Oxford Clay	Mdst	97	27	1.56	0.09	1
	Kellaways Beds	Mdst	97	21	1.52	0.03	1
	Cornbrash	Lmst	96	5	2.29	0.17	3
	Forest Marble	Mdst + Lmst	95	37	1.80	0.07	1
	Frome Clay	Mdst	95	15	1.72	1.10	3
	Fullers Earth	Mdst	95	47	1.95	0.05	1
	Upper Lias	Sdst	93	13	2.87	0.12	3
		Mdst	93	11	1.27	0.03	1
		Slmd	93	11	2.22	1.10	3
	Middle Lias	Mdst	92	3	1.66	0.15	1
	Lower Lias	Mdst	91	37	1.80	1.10	1
Rhaetic	Penarth Group	Slmd	90	19	2.53	0.11	3
Triassic	Mercia Mudstone Group	Mdst	90	225	1.88	0.03	4
	Mercia Mudstone Group	Mdst	90	41	2.28	0.33	1
	Sherwood Sand-stone	Sdst	89	64	3.41	0.09	4
	Group	Mdst	89	6	2.37	0.23	4
Permian	Permian Marls	Marl	87	6	2.12	0.28	4
		Anhy	87	8	5.40	0.13	4
		Hali	87	14	4.87	0.22	4
	Magnesian Limestone	Lmst	86	12	3.32	0.17	4
Carboniferous	Westphalian	Sdst	82	37	3.31	0.62	5
		Slst	82	12	2.22	0.29	5
		Mdst	82	25	1.49	0.41	5
		Coal	82	8	0.31	0.08	5
	Namurian	Sdst	81	7	3.75	0.16	4
	Tournasian	Lmst	80	14	3.14	0.13	4
		Sdst	80	76	4.19	0.08	4
Devonian	Old Red Sandstone	Sdst	75–76	76	3.51	0.41	2
Lower Palaeozoic		Variable	62–74	58	2.87	0.08	4

Code = Geological code number used for geological formations represented on the 1:625 000 geological maps of the UK

n = Number of samples   λ = Thermal conductivity W/mK   s.e. = Standard error

References 1 Bloomer (1981) 2 Richardson and Jones (1981) 3 Bloomer and others (1982) 4 BGS records 5 Richardson and Oxburgh (1978)

Lithology Smst: sandy mudstone. Mdst: mudstone. Lmst: limestone. Sdst: sandstone. Slst: siltstone Slcl: silty clay. Slmd: silty mudstone. Hali: halite. Anhy: anhydrite.

	n	λ	s.e.	Reference
HERCYNIAN GRANITES				
South-west England				
Land's End	63	3.36	0.02	6
Carnmenellis	153	3.34	0.02	6
St Austell	64	3.26	0.03	6
Bodmin	164	3.31	0.02	6
Dartmoor	170	3.23	0.02	6
CALEDONIAN GRANITES				
Lake District and Weardale				
Shap	46	2.88	0.03	7
Skiddaw	45	3.51	0.04	7
Weardale	21	2.94*	0.02	8,10
Eastern Highlands				
Cairngorm	44	3.52	0.03	9
Ballater	47	3.26	0.04	9
Mount Battock	42	3.08	0.02	9
Bennachie	45	3.50	0.07	9

n = Number of samples
λ = Thermal conductivity W/mK
s.e. = Standard error
* England and others (1980) give 3.1 ± 0.25
References
6 Wheildon and others (1981)
7 Wheildon and others (1984a)
8 Bott and others (1972)
9 Wheildon and others (1984b)
10 England and others (1980)

Table 2.3 Thermal conductivities of Hercynian and Caledonian granites in the UK

the main lithostratigraphical formations of the UK taken from published data and BGS records and Table 2.3 shows similarly derived data for granites. Table 2.2 indicates that the mudstone and clay formations of the Jurassic and younger systems generally have thermal conductivities below 2 W/mK. In contrast sandstone formations commonly have conductivities about 3 W/mK as do most granites. The numerical code used on the two sheets of the 1:625 000 Geological Map of the UK is also given in Table 2.2, which indicates that the same formation can contain several different lithologies each with a different conductivity. As an example of this, Figure 2.3 is a plot of the measured thermal conductivity for Permo-Triassic mudstones (mainly from the Mercia Mudstone Group) against sample depth. This formation includes some sandstone horizons and the general scatter of points between 1 and 3 W/mK reflects this; the few points with conductivities above 4 W/mK are due to the presence of salts within the sample, either halite or anhydrite. The diagram illustrates the difficulty of assigning a mean formation conductivity, although it is clear from the variation of conductivity in the Thornton Cleveleys heat flow borehole (Figure 2.4), that small scale variations have little effect on the local temperature gradient.

The most desirable method of measuring thermal conductivity is with the divided bar (Birch, 1950), but this high accuracy technique is restricted to rock material which is sufficiently coherent for the cutting of accurately ground discs from original drill core. For poorly consolidated clays, mudstones and other soft sediments the transient line source needle probe technique (von Herzen and Maxwell, 1959) is used with good effect. Where only rock chips are recovered from the borehole the 'pillbox' adaptation of the divided bar (Sass and others, 1971) is used. This yields an apparent conductivity for the fragments which must then be converted to the whole-rock conductivity, through correction for in-situ porosity. Evaluation of porosity is from geophysical well logs and it is this step that normally limits the accuracy of the result.

HEAT FLOW CALCULATION

Heat flow is determined through some combination of the vertical temperature gradient $d\theta/dz$ with the thermal conductivity λ to yield the heat flow q using Equation 2.7. Significant variations in conductivity are normally present making it unacceptable to combine the mean temperature gradient with the mean conductivity for the borehole.

Heat flow is commonly calculated following the thermal resistance integral procedure of Bullard (1939). This involves assigning conductivity values, λ, to intervals Δz on the basis of borehole geological logs. The heat flow q is determined by a linear regression of the measured temperatures θ_z over their thermal resistances $\sum_i \left[\Delta z_i/\lambda_i\right]$ to each depth.

An alternative approach is to calculate heat flow for each lithological unit within the borehole and determine a mean value (Richardson and Oxburgh, 1978). Both techniques when properly applied will reveal local zones of convective flow where these are present.

Past climate

Changes in the earth's climate during recent geological and historical times affect sub-surface temperatures to depths of several hundred metres. The most severe cooling effects are those caused by the Pleistocene glaciations. The effect of past climates can be calculated by approximating the surface temperature backwards in time from the present, as a series of step functions (Birch, 1948). Most authors have, however, been inclined to neglect this effect on underground temperatures, presumably because of uncertainties in the magnitudes and durations of the perturbing surface temperature field.

For boreholes deeper than 300 m the effect is reasonably uniform so that for comparative regional studies this neglect is of no great significance. Most of the UK data are presented without climatic correction.

Nineteen of the boreholes drilled into the Cornubian granite batholith are limited in depth to 100 m only, in which range climatic fluctuations in the last 500 years cause significant perturbation of underground temperatures (Wheildon and others, 1980). A climatic correction was routinely applied to these shallow holes to bring them into agreement with uncorrected values at the 300 m level, so as to be consistent with the remainder of the UK heat flow dataset.

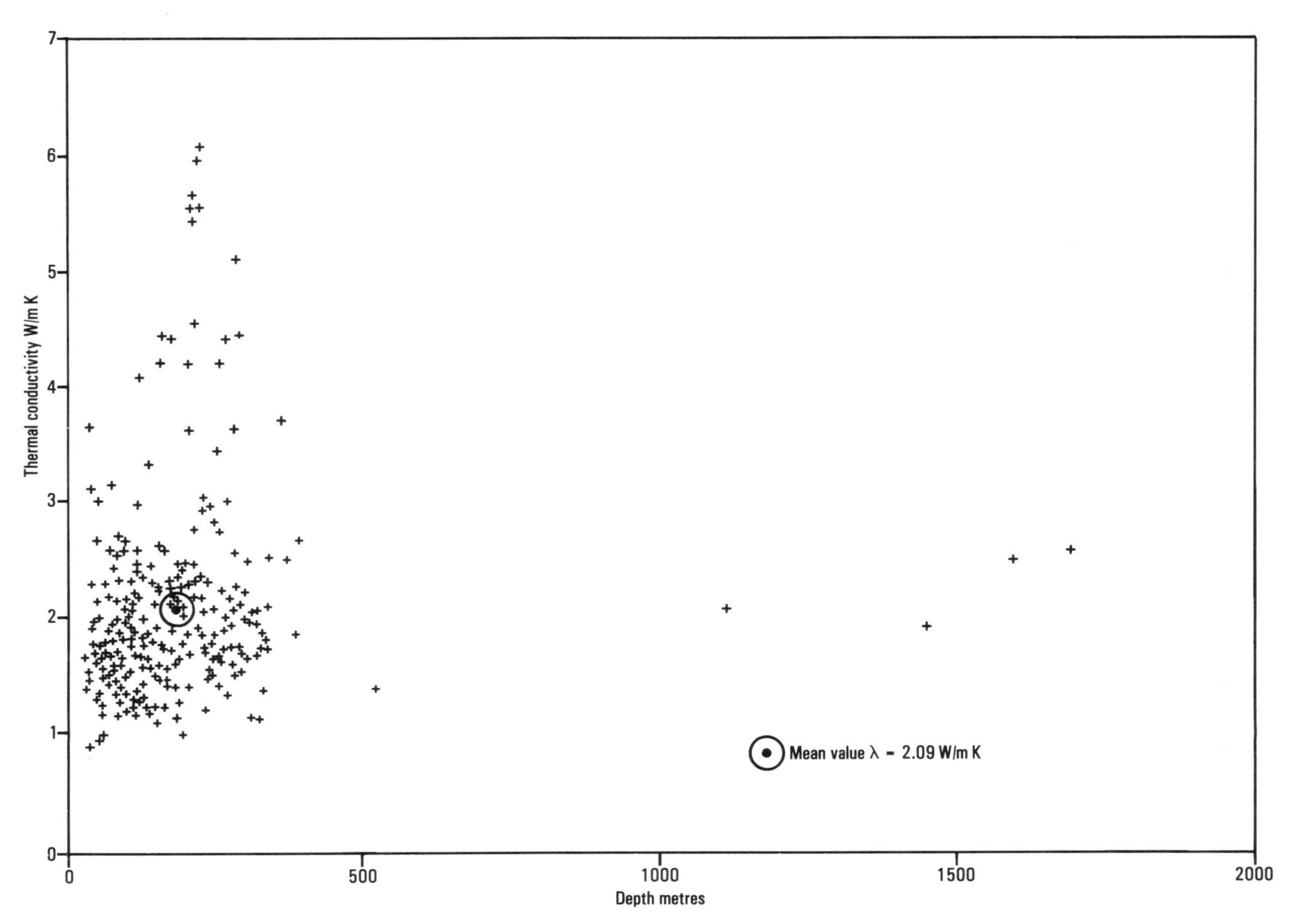

2.3 Thermal conductivity of Permo-Triassic mudstones plotted against depth of sample.

The fact that climatic corrections are generally ignored is acceptable in comparative regional studies but less so in the present context of geothermal exploration, where accurate values of θ_z are sought. The consequence of ignoring the most significant effect, that due to the Pleistocene glaciations, leads to an underestimate of heat flow, probably of between 5 and 10 mW/m^2, and a consequential modest underestimate of temperatures at depth.

The one-dimensional assumption

The equations developed earlier are based on the assumption of one-dimensional flow, directed vertically. However, thermal conductivity and/or heat production variations that depart from this condition will lead to horizontal components of flow. The case of granitic intrusions is particularly relevant to the case of the UK heat flow field. These bodies generally contrast positively with respect to thermal conductivity and heat production compared with their enclosing country rocks. England and others (1980) have reviewed the extent to which surface heat flow is affected by variations in these parameters and pluton geometry, by solving the heat conduction equation in two dimensions by a finite difference technique. Surface heat flow is augumented above a body of higher conductivity even when the body is deeply buried. But the positive heat production contrast in the granite pluton works in the reverse sense, transferring heat horizontally to the country rocks. England and others (1980) investigated the interplay between these contrasting effects for a range of pluton geometries, ranging between the situations in which maximum reduction of surface heat flow occurs over the centre of a deep, high heat production and low thermal conductivity structure, through to maximum enhancement in the case of a shallow, low heat production and high conductivity structure. If the involved parameters are known, then the heat flow can be adjusted to the one-dimensional situation as is required, for example, when considering the linear situation between the flow and heat production. Finite element models have been extensively used to take account of thermal conductivity and heat production contrasts in the interpretation of the heat flow anomalies associated with the granites in the UK (Wheildon and others, 1980; Lee and others, 1984).

Irregular surface topography around a heat flow borehole introduces shallow three-dimensional effects which can be corrected using the method proposed by Jeffreys (1938).

Convection effects

Convection of heat by groundwater flowing as a result of differences in fluid potential or horizontal temperatures can produce severe perturbations in the outward conductive heat flux. Furthermore, the recognition of these flows is extremely difficult. Even very slow flows and low volumes of mass movement can transfer large amounts of heat, equivalent to the conductive heat flow. For example Kappelmeyer (1979) has shown that seepage velocities of the order of 31.5 cm per year across a layer with a temperature difference of 1°C, will produce a variation in the measured heat flux of 42mW/m^2. The sensitivity of the heat flow to convection is such that heat flow observations in boreholes have recently been cited as a method of determining groundwater flow (Drury and others, 1984). Obvious convection effects can easily be seen in anomalous heat flow results, but the implication has to be that all shallow heat flow measurements could contain a convective effect, the size of which is generally unknown. Convection is usually assumed to be less of a problem in heat flow observations made in low-permeability basement rocks, although Drury (1984) has shown that even crystalline rocks can exhibit significant convective flows. In Eastern England, the low heat flow recorded at Eyam at the margin of the Pennines anticline, is generally assumed to be caused by the downward convection by groundwater flow. Enhanced heat flows can be observed further east over buried Carboniferous anticlinal structures, and are attributed to upward convective effects (Bullard and Niblett, 1951).

In the absence of detailed information of groundwater flow, the heat flow observations in mainland UK, listed in Appendix 2, are usually assumed to represent the conductive flux of heat through the earth's surface. However, in at least some areas the values are likely to be influenced by groundwater flows, possibly on a regional scale in the deep sedimentary basins. Recent work in the North Sea (Andrews-Speed and others, 1984) suggests that there is a significant variation of heat flow with depth of the order of 30 mW/m^2. This variation is on a very large scale and is presumed to be caused by groundwater circulation.

In view of the possible convection effects, heat flow observations made in the sedimentary cover rocks of the UK, including Carboniferous and younger strata, should be regarded as only apparent conductive heat flows.

THE HEAT FLOW FIELD IN THE UK

Data acquisition

Heat flow measurements

The manner in which research effort has been directed towards heat flow data acquisition is reflected in the Catalogue of Geothermal Data for the UK (Burley and others, 1984). In 1975 the total coverage was limited to just 32 values. Between 1975 and 1980, largely as a result of the first Commission of the European Communities' Energy Research and Development Programme and support from the Department of Energy, 121 new values were added to the coverage, substantially through the efforts of the Oxford University and Imperial College heat flow groups. The Oxford group had continued their programme (started earlier with support from the Natural Environment Research Council) of improving the national coverage, with strong emphasis on the acquisition of high quality data from specially preserved boreholes (Richardson and Oxburgh, 1978, 1979; Oxburgh, 1982). The Imperial College effort was directed towards a detailed definition of the south-west England heat flow anomaly using specially drilled boreholes (Wheildon and others, 1980).

Between 1980 and 1984, during the second European Communities' programme, again with support from the Department of Energy, the Imperial College group worked in collaboration with the British Geological Survey and the regional coverage was extended significantly (Wheildon and others, 1984c). Measurements were mainly in boreholes in sedimentary sequences but also in the newer Caledonian granites in northern England and Scotland (Wheildon and others, 1984a and b) as part of an investigation into their HDR potential (Lee and others, 1984). These programmes added 35 new heat flow values to the data set.

The observations were made in five types of borehole (the numbers for each are given in brackets):

Specially drilled heat flow boreholes, cement lined (17)

Deep geothermal exploration wells (2)

Exploration and stratigraphical test boreholes, taken over and cement lined (6)

Exploration and stratigraphical test boreholes, taken over but not lined (6)

Hydrocarbon exploration wells with no access for temperature measurements except for BHT data (4)

In all cases, comprehensive thermal conductivity sampling was achieved, which in a typical 300 m borehole was at 3 m intervals. Rather wider sampling intervals were used for deeper boreholes. In all boreholes except the hydrocarbon wells, a series of temperature logs, with observations at depth intervals of 3 m, was run in each borehole over a period of months after drilling had been completed, until equilibrium temperatures were observed.

Figure 2.4 shows an example of the results for one of the specially drilled heat flow boreholes at Thornton Cleveleys in west Lancashire. Attention is drawn to the significant variation in the measured conductivity of the Mercia Mudstone, a variation that is typical of other formations. In order to calculate the heat flow the integrated thermal resistance to each depth is plotted against the corrected temperature at that depth. The plot (Figure 2.5) should be a straight line, the slope of which is the heat flow.

Details of the heat flow dataset are contained in the Catalogue of Geothermal Data for the UK (Burley and others, 1984) and a summary is reproduced in Appendix 2 for easy reference. There are now about 188 known heat flow observations of varying reliability with a range of 17 to 136 mW/m^2 and a mean of about 69 mW/m^2. Figure 2.6 shows how this sample compares with the European heat flow data. The higher mean of the UK data and the bi-modal distribution is a consequence of the large number of measurements made over the high heat flow zone in south-west England.

Heat flow estimates

'The Catalogue of Geothermal Data for the UK' (Burley and others, 1984) contains an extensive list of temperature-depth observations. Where these are supported by geological

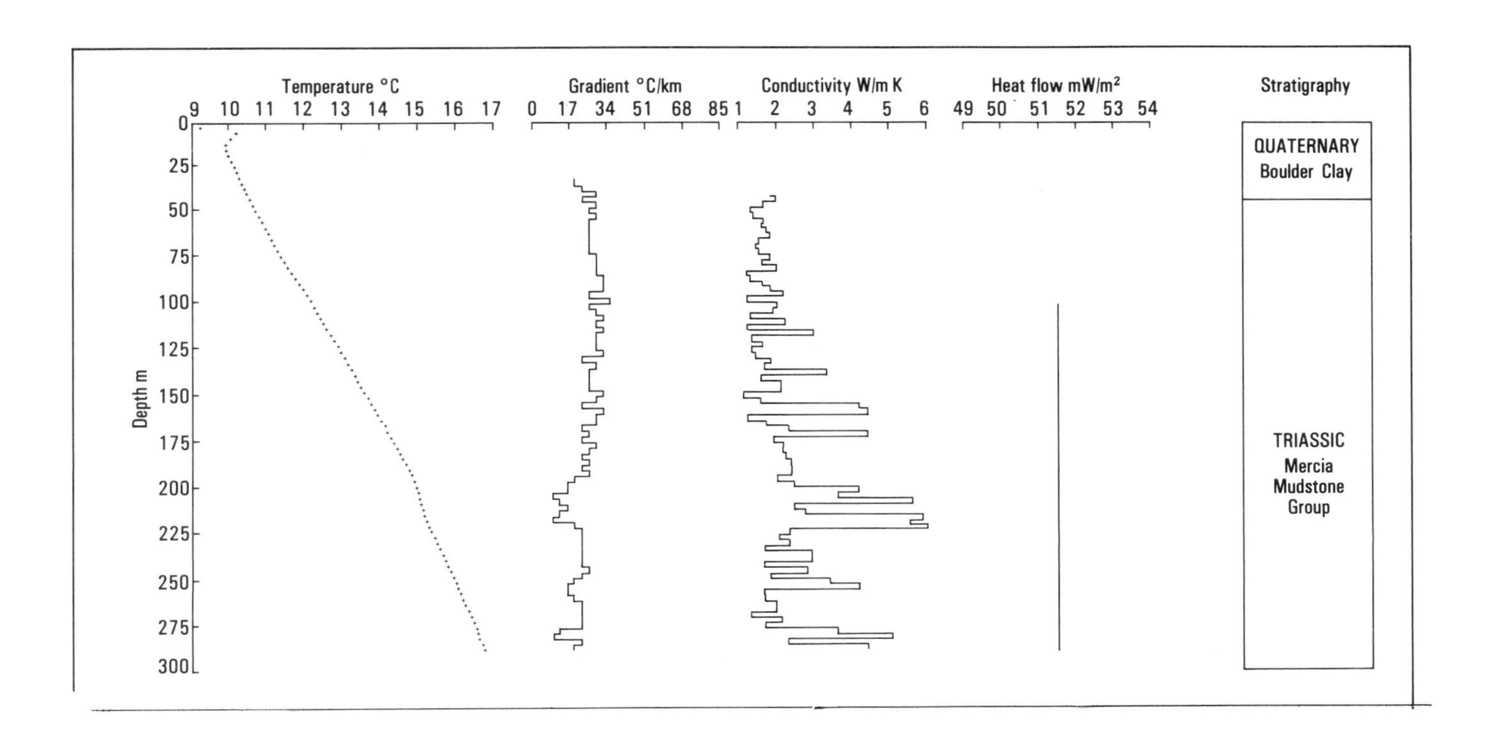

2.4 Equilibrium temperature log, thermal gradients, thermal conductivities and heat flow interpretation for the heat flow borehole at Thornton Cleveleys in west Lancashire.

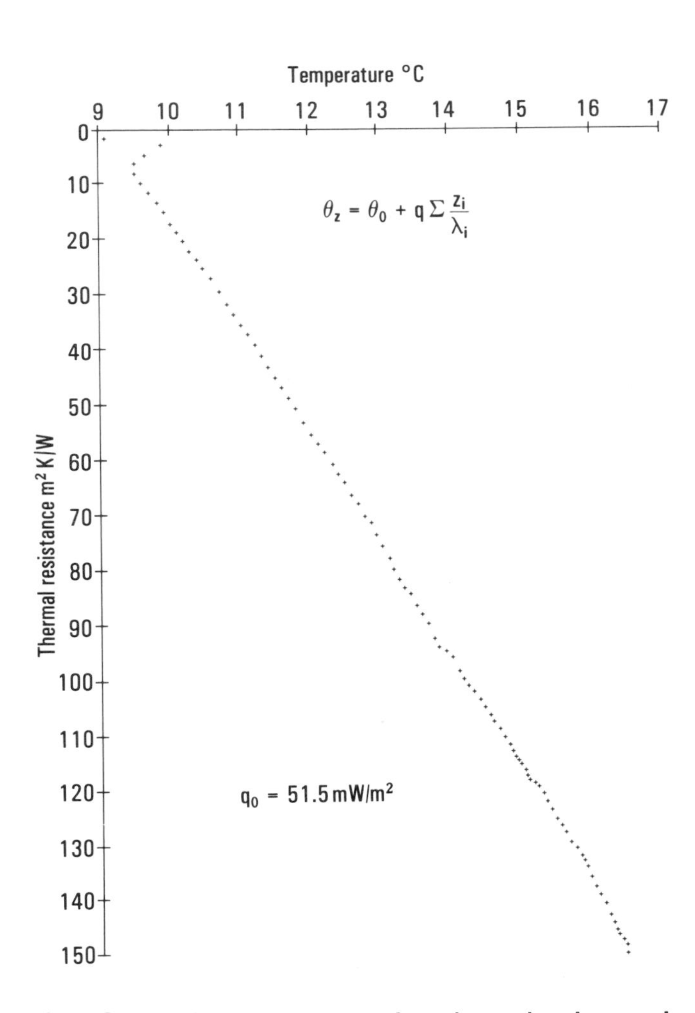

2.5 Graph of temperature against thermal resistance in the Thornton Cleveleys Borehole.

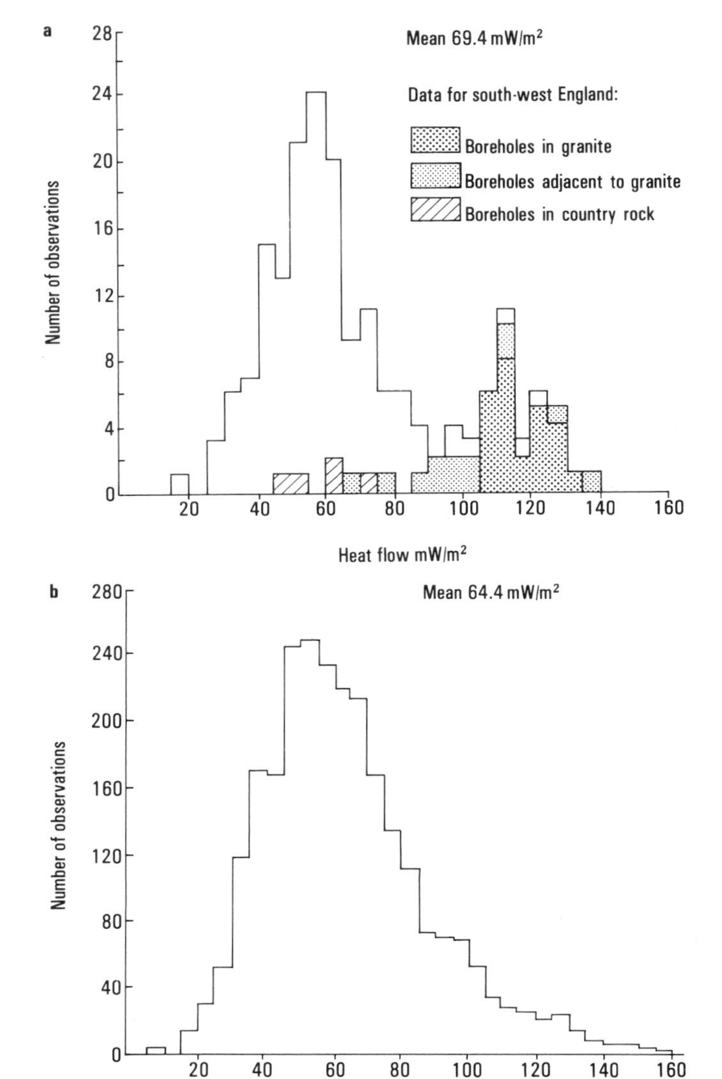

2.6 Histogram of heat flow data for the United Kingdom (a) contrasted with the coverage for continental Europe (b) European data are after Cermak (1979).

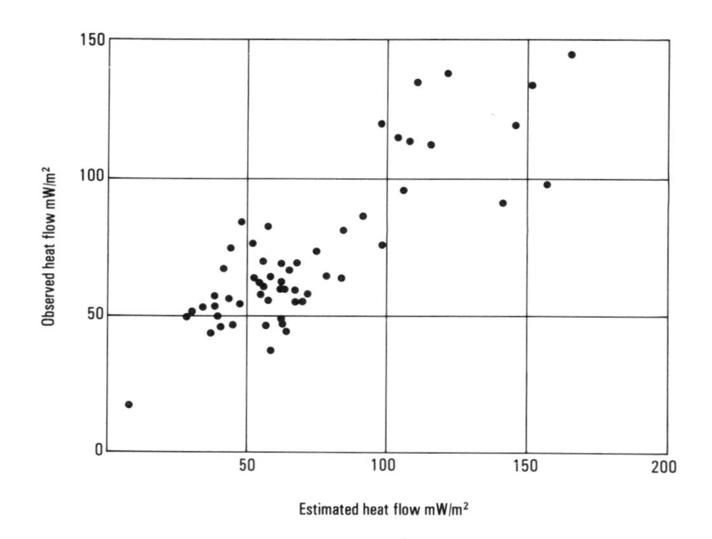

2.7 Plot of observed against estimated heat flow for 60 boreholes.

logs the opportunity exists to combine the observed temperatures with a datafile of mean thermal conductivities for the formations represented on the 1:625 000 Geological Map of the UK and thereby produce heat flow estimates. For each borehole in the catalogue for which a geological log is available, the procedure has been to calculate thermal resistance to each depth where a temperature has been observed by reference to a file of thermal conductivities and to the geological sequence penetrated by the borehole. Heat flow was then estimated by the application of Equation 2.7. The method assumes that each geological formation can be assigned a mean thermal conductivity. This seems reasonable remembering the intrinsic variation of conductivity measurements seen in the Thornton Cleveleys Borehole (Figure 2.4)

Using this method an estimate of heat flow has been made at about 580 borehole sites. Most of the estimates are located in lowland Britain at boreholes drilled for hydrocarbon exploration. The reliability of the estimates varies considerably, depending on the depth and type of temperature measurements and on the proximity to thermal conductivity observations. A single BHT observation made in a shallow borehole immediately after drilling, in an area where the formation conductivity has not been measured, is obviously less reliable than several temperatures measured in a deep borehole several days after drilling ceased, and in an area close to measurements of thermal conductivities. However, since in principle BHTs are under-estimates of the equilibrium values, heat flow estimates are unlikely to be over-estimates of the true heat flow. A sample of the estimates derived using bottom-hole temperatures, geological sections in boreholes and mean thermal conductivities, at 60 heat flow sites correlated well with the observed heat flow values at these sites (Figure 2.7). The product moment correlation coefficient is 0.90.

To improve the reliability of the dataset of heat flow estimates, data for shallow boreholes (less than 500 m deep) and for boreholes with measured geothermal gradients above 40°C/km were ignored. This reduced the number of acceptable estimates to about 200. Furthermore, to give preferential weighting to the observed data, any estimate closer than 20 km to any observation was also ignored. This filtering reduced the number of usable estimates still further, to about 100 sites.

The mean of these estimates is about 57 mW/m^2; this is the same as the mean of the heat flow observations in the UK outside south-west England.

The heat flow map

The first serious attempt to put form to the heat flow field for the UK was by Richardson and Oxburgh (1979) whose map was based on just over 100 observations. The broad regional patterns were clear. A general background of values in the range 40 to 60 mW/m^2 was crossed by two belts of variable but higher flux, one extending north-westwards across central and northern England and the other running from Cornwall eastwards along the south coast of the country. Much of the subsequent work, cited above, has been directed towards a better definition of these belts. The culmination of this effort has been the production of a revised heat flow map (Figure 2.8) which is based on the 188 heat flow observations supported by about 100 heat flow estimates derived as just described. The distribution of data is not uniform and the final contouring has been by hand, but it was derived from a machine-contoured version produced by Rollin (1983).

The new heat flow map is stark in its simplicity. The earlier identified high heat flow belts have shrunk into more localised features. Superimposed on what appear to be a fairly uniform background field are local anomalies, the most significant of which are clearly related to granite batholiths with high heat productions. The extent to which convective flow is causing undulations in the background field will only emerge as further high quality results become available. However, there is growing evidence that several of the major anomalous features that have no obvious connection with a heat source in the upper crust, do indeed have a convective origin.

The three most extensive anomalies, all of which are convincingly related to high heat production granites are as follows:

(i) South-west England (maximum amplitude 136 mW/m^2). This unusually high heat flow zone, with values rising to more than twice the background value, is constrained by over 40 observations and bears a close relationship with the space form of the Hercynian granite batholith, as Figure 2.9 illustrates. Model studies suggest that the granite is relatively unfractionated (Wheildon and others, 1980). Surface heat production is in the range 4.0 to 5.3 μW/m^3.

(ii) Lake District and Weardale (maximum amplitude 101 mW/m^2). This is less well constrained by data but a clear relationship with the newer Caledonian (Lower Devonian) granite batholiths of northern England exists. Model studies to reconcile heat flow with heat production require that in different areas the heat production is weakly or moderately fractionated towards the surface (Chapter 3 and Lee and others, 1984). Surface heat production is in the range 3.3 to 5.2 μW/m^3). This anomaly apparently extends southwards to embrace the Bowland Forest but evidence of a buried granite is lacking in the southerly area and this extension could be caused by convective groundwater circulation.

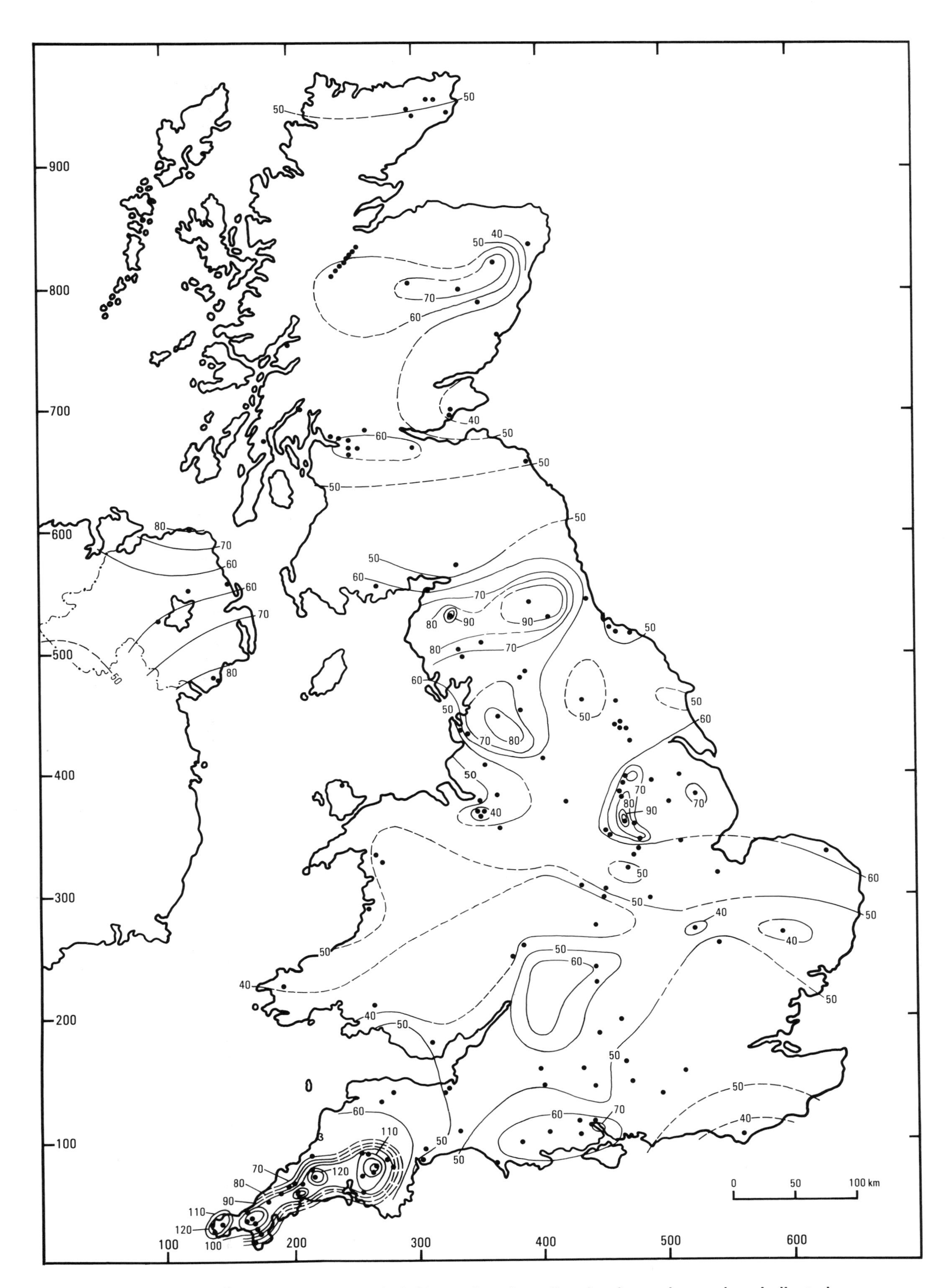

2.8 Heat flow map of the UK. Units are mW/m². Sites where heat flow has been observed are indicated.

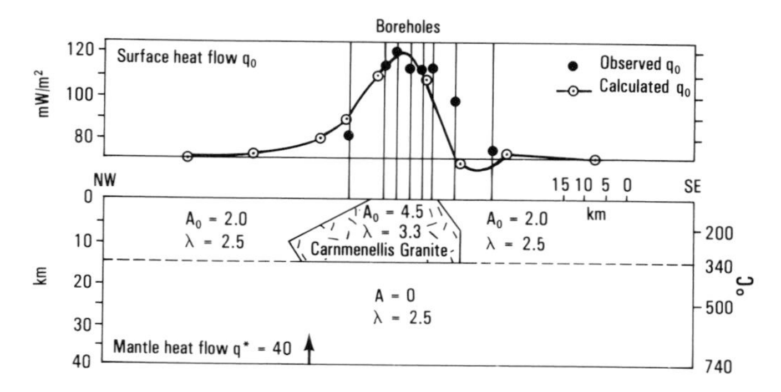

2.9 The relationship of heat flow to the form of the Carnmenellis Granite. Heat production is expressed in $\mu W/m^3$, thermal conductivity in W/m K and heat flow in mW/m^2. (After Wheildon and others, 1980).

(iii) Eastern Highlands of Scotland (maximum amplitude 76 mW/m^2). This anomaly is only supported by a limited number of observations, but again a clear relationship exists with high heat production of the newer Caledonian granites. To reconcile heat flow and heat production, relatively strong fractionation of heat producing elements is required (Lee and others, 1984). Surface heat production is in the range 4.8 to 7.3 $\mu W/m^3$.

Of the remaining features the most significant is that centred on east Nottinghamshire and extending eastwards. There is strong evidence that the east Nottinghamshire anomaly arises from convective eastward and upward movement of groundwater from the Pennine outcrop of Carboniferous strata (Bullard and Niblett, 1951). Richardson and Oxburgh (1978) suggested that the eastern extension of the anomaly into Lincolnshire may reflect higher than average flux from the basement, although regional upward groundwater circulation may be the cause.

A modest heat flow anomaly embraces the Wessex Basin. It is well constrained and there is no evidence of water circulation. The source is unknown, and although Wills (1973) proposed a buried granite between Southampton and Crewkerne the evidence for this is equivocal. The negative gravity anomaly could be accounted for by a sequence of relatively low density Palaeozoic rocks or possibly a deep Moho beneath the basin (Downing and others, 1984). The cause of the heat flow anomaly might be slow movement of groundwater in the basin.

The remaining minor undulations in the form of the heat flow field possibly arise from similar causes; only as new data becomes available will the origins of these minor features become apparent.

It is important to establish the amplitude of the background heat flow field against which the anomalous field may be judged. Referring back to Figure 2.6, the value of the mean heat flow depends very much on how it is calculated. The arithmetic mean of all the UK observations is 69.4 ± 28 mW/m^2. Averaging values on a 10 km grid, using values derived from the heat flow map (Figure 2.8 and the Plate), leads to an area-weighted mean of 54 ± 12 mW/m^2. Omitting those areas containing Caledonian and Hercynian granites reduces this area-weighted mean to 52 ± 9 mW/m^2, and this is probably a reasonable representation of the amplitude of the average background field. The mean of 10 high quality observations determined in specially drilled boreholes in sedimentary environments throughout the country, and completed in the recent programme, was 51 ± 6 mW/m^2, which lends further support to the amplitude of the average background field being about 52 mW/m^2. However, all these mean values of heat flow include unknown convective effects as noted earlier.

Knowledge of the heat flow distribution has application in fields other than geothermal energy, including studies of the nature of the crust, petroleum generation and migration, and the genesis of mineral deposits. But there are still large areas of the country with few or no observed values. There would be significant value in extending the data coverage, and in the geothermal context additional measurements would be useful to:

—predict accurately sub-surface temperatures in target areas for possible HDR development,

—examine other geophysical anomalies which might have associated geothermal anomalies,

—examine further the variations in background heat flow across the UK, identify anomalies and relate these variations to the tectonic framework.

Chapter 3 Hot Dry Rock

M. K. Lee

THE HOT DRY ROCK CONCEPT

The possibility of extracting geothermal energy from low permeability rocks, generally referred to as the Hot Dry Rock concept, has been described by numerous authors (e.g. Smith, 1975; Batchelor, 1982). The technique most commonly proposed is that of creating a heat exchange region between two boreholes by the stimulation of naturally occurring joints or the creation of an artificial fracture system. Cold water is pumped into the injection borehole, circulated through the fracture system and recovered as hot water from the production borehole (Figure 3.1). The heat is extracted with a heat exchanger at the surface and, depending on the temperature of the water, can be used for electricity generation, space heating, industrial processes or a combination of these applications. Several variations on the basic two borehole concept have been proposed which involve additional boreholes and different techniques for creating the circulation loop (e.g. International Energy Agency, 1980).

Research into the practical aspects of HDR exploitation originated at the Los Alamos National Laboratory in the United States in the early 1970s. Since then a number of exploration, injection and production boreholes have been drilled into Precambrian granitic basement at the Laboratory's test site at Fenton Hill in New Mexico. An artificial fracture system was created at a depth of some 3000 m, where the rock temperature is about 200°C, and the feasibility of the HDR concept was established by creating a closed circulation loop between the injection and production wells. More recently the research has been extended to a depth of around 4500 m where the rock temperature is about 320°C. The work at Fenton Hill is fully described in a series of reports from the Los Alamos National Laboratory (e.g. Smith and others, 1983).

In Britain an extensive research programme is being carried out by the Camborne School of Mines at its Rosemanowes test site on the Carnmenellis Granite in Cornwall. The first tests were conducted in a series of boreholes about 30 m deep followed by extensive testing at 300 m. The work has recently been extended to a depth of 2000 m where the rock temperature is 79°C. The objective is to investigate the rock mechanics and engineering aspects of creating a reservoir and extracting heat, including the concept of using small explosive charges to stimulate and gain access to the natural joint system within the granite (Batchelor, 1983). Long-term plans envisage the creation of a system at a depth of 5000 to 6000 m extracting heat at approximately 200°C.

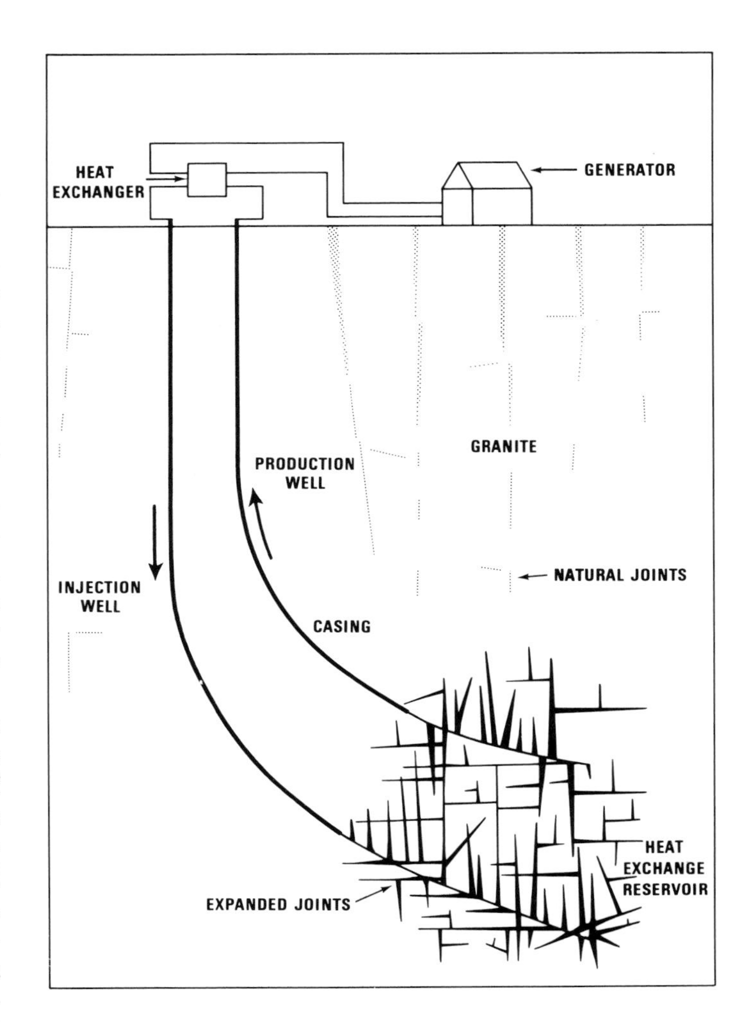

3.1 Conceptual diagram of a two borehole hot dry rock system.

BACKGROUND TO HDR PROSPECTS IN THE UK

The target temperature for an HDR system depends on the end use of the extracted heat. Most applications envisage temperatures in excess of 150°C, and usually 200°C, for electricity generation, but lower temperatures may be acceptable for other applications. Temperatures of 150°C can be reached at depths of less than 10 km over much of the earth's surface but the economic viability of an HDR system depends critically on the depth of, and thus the cost of drilling to, the target temperature.

The area-weighted mean heat flow in Britain (based on 10

km grid squares) is 54 ± 12 mW/m^2 (see Chapter 2) and the average geothermal gradient as determined from corrected bottom-hole temperatures, is 26.4°C/km (see Plate). As already pointed out, the latter value is biased towards high gradients because a disproportionate number of boreholes were drilled into low conductivity sedimentary rocks or on positive heat flow anomalies; a more likely value in basement rock is less than 20°C/km. In parts of the world with relatively recent volcanic activity the heat flow, and thus the temperature gradient, is often significantly above average and useful temperatures can be found at relatively shallow depths (e.g. 197°C at a depth of 2929 m at Fenton Hill; Geothermics, 1982). In Britain there are no heat flow anomalies related to recent igneous activity. However, individual values vary between 34 and 132 mW/m^2 (Class A observations in Burley and others, 1984) and rock conductivities vary considerably between lithologies; they are generally less than 2 W/mK in Jurassic and younger sediments and greater than 3 W/mK in metasedimentary basement rocks and granites (Tables 2.2 and 2.3). These factors result in wide variations of the geothermal gradient and thus the depth to any particular temperature.

Above average gradients in Britain tend to be associated with two types of geological terrain:

(1) Areas where the heat flow is close to or above average and the thermal conductivity of the upper few kilometres of sedimentary rock is low. A gradient in excess of 30°C/km would be expected in a formation with an effective conductivity of less than 2 W/mK for a heat flow of 60 mW/m^2 or greater.

(2) Major granite plutons where the regional heat flow is enhanced by the presence of above average concentrations of the heat-producing radioelements uranium, thorium and potassium. The highest heat flow values in the UK are observed over these 'radiothermal' granites; for example values greater than 110 mW/m^2 over the Cornubian batholith in south-west England (Wheildon and others, 1980). The thermal conductivity of granite is high, generally greater than 3 W/mK, and this reduces the beneficial effects of the high heat flow. However, if the heat flow is sufficiently enhanced then gradients in excess of 30°C/km may be observed.

The recognition that the highest heat flow values in the UK are observed over radiothermal granites, together with the considerable research effort into the extraction technology in granites, has resulted in priority being given during the present programme to the identification of granites as targets for investigation.

The Cornubian granite batholith in south-west England was the first area in the UK to be recognised as a potential HDR target (Dunham, 1974a) and the heat flow anomaly associated with the batholith has been thoroughly studied by Wheildon and others (1980 and 1981). The batholith is the only major granite of Hercynian age in the UK. However, there is a large number of Caledonian granites, mainly in northern Britain, occupying a total outcrop area of around 5500 km^2, many with a considerable subsurface extent. Prior to 1980, heat flow observations had been made on only two major Caledonian intrusions, namely the Weardale and Wensleydale granites, both in northern England, where the values were 95 and 65 mW/m^2 respectively (England and others, 1980). The high value at Weardale lent support to the view that at least some of the Caledonian granites might have significant potential for HDR exploitation.

IDENTIFICATION OF GRANITIC TARGETS

General criteria

The criteria for selecting the most promising granites in the UK were outlined in an earlier phase of the programme (Lee, 1978) and have since been discussed by several authors (e.g. Brown and others, 1979; Brown and others, 1982; Lee and others, 1983). The starting point for the selection process is the recognition that in plutonic rocks the surface heat flow (q_0) generally increases with increasing surface heat production (A_0). In many parts of the world an empirical linear correlation has been demonstrated between the two, and several different models have been proposed for the distribution of radioactive elements with depth to satisfy this relationship (see Chapter 2 and the discussion on the relationship between heat flow and heat production in this Chapter). Whichever model is finally accepted it is clear that the two main prerequisites for high heat flow over igneous intrusions are high heat production and large intrusive volume with a vertical extension of many kilometres.

Practical criteria

Both the heat production and the density of igneous intrusive rocks vary widely with composition. In very general terms the relatively dense basic and intermediate rocks are comparatively low in heat producing radioelements while granites, which are less dense and geochemically more evolved, tend to be relatively rich. The mean surface heat production of Caledonian granites in Britain is about 2.7 μW/m^3 (Table 3.9) but there is considerable variation between intrusions and values of twice this amount are sometimes observed. The abundances of primary radioelements depend on such factors as the derivation and physico-chemical evolution of the granite magma. However, in general the highest concentrations of heat producing elements tend to be found in the most evolved granites which have the highest silica contents and thus the lowest densities. Such granites are generally less dense than the basement rocks into which they are intruded and those with a combination of low density and large intrusive volume can be recognised from their characteristic negative gravity anomalies.

3.2 Bouguer gravity anomaly map of the UK. Generated from the BGS National Gravity Data Bank by numerical approximation on a 2 km grid smoothed over the eight nearest grid points. Contour interval is 5 mGal. The three areas emphasised are referred to in Figures 3.4, 3.9 and 3.10.

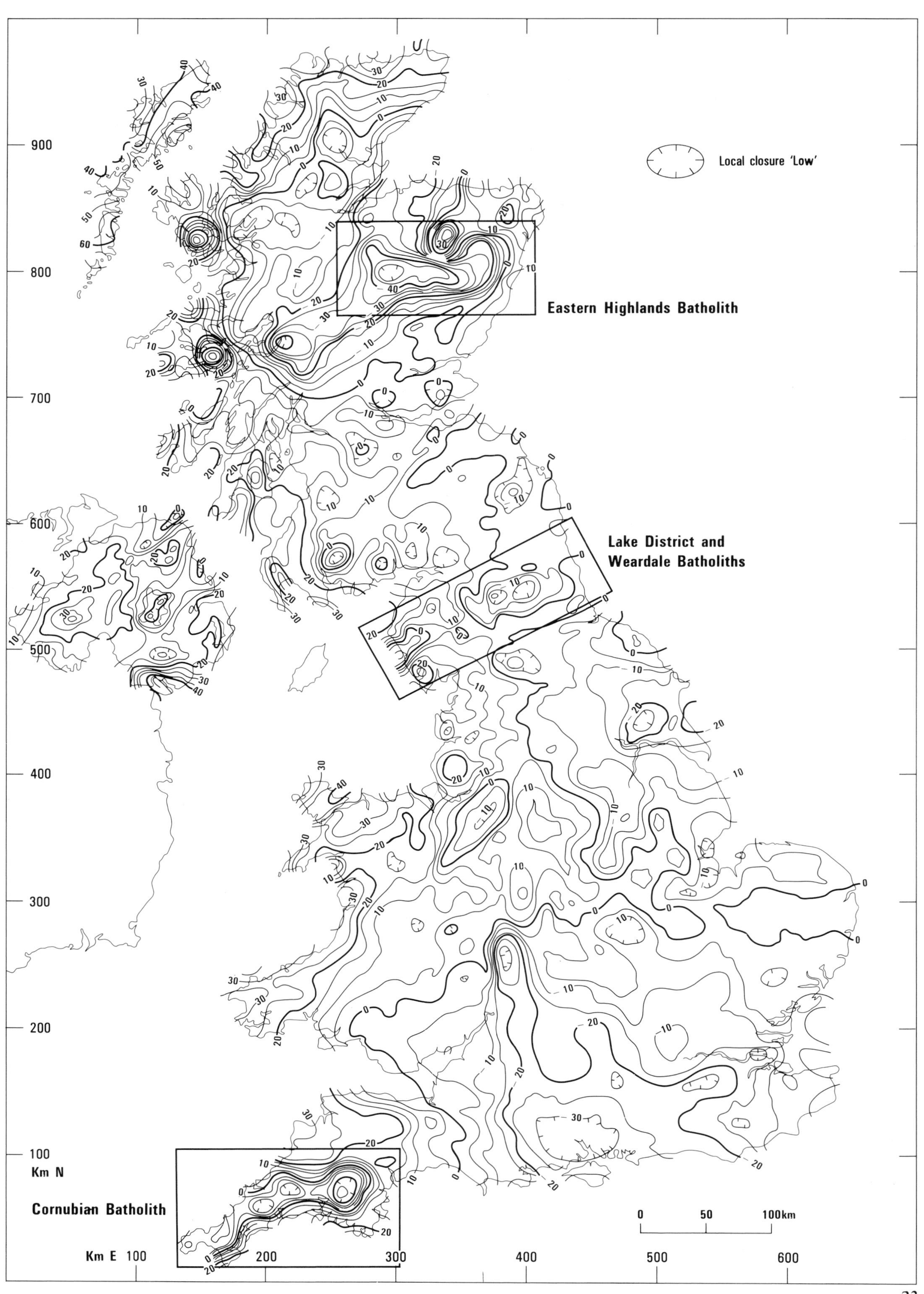
Local closure 'Low'
Eastern Highlands Batholith
Lake District and
Weardale Batholiths
Cornubian Batholith
Km N
Km E
0
50
100km

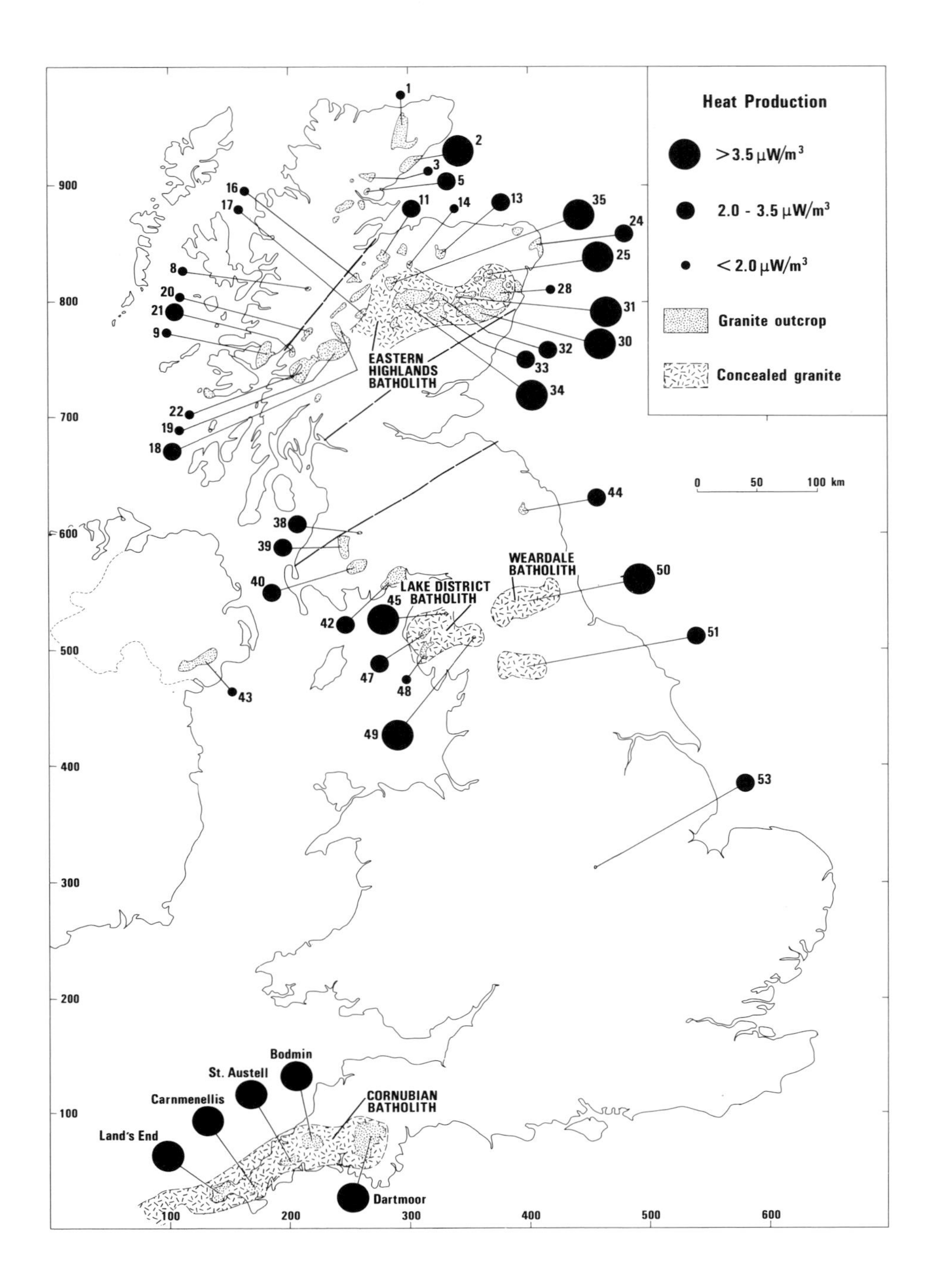

3.3 Summary of heat production data for selected British granites. From a compilation of published and unpublished data by P.C. Webb of the Open University, and listed by Lee and others (1984) and shown in Tables 3.1, 3.9 and 3.10. Values are given only for intrusions with five or more samples and are based on borehole and surface data. Where both types of data exist they are in agreement, within the limits of the categories shown on the Figure, for all intrusions except the Skiddaw Granite (No. 45) where the higher value derived from borehole core is shown. Caledonian granites are numbered as in Tables 3.9 and 3.10.

In practical terms, therefore, the highest heat flow values are likely to be observed over granites characterised by:

(1) high concentrations of primary radioelements (i.e. high evenly distributed heat production)

(2) large negative gravity anomalies indicating large intrusive volume and implying that relatively evolved, potentially radioelement-rich, low density crystalline rocks extend to a depth of many kilometres.

Two additional factors which also affect the HDR potential of an area are a high background heat flow and a cover of low conductivity sedimentary rocks.

Target areas in the UK

Gravity and heat production maps for the UK are shown in Figures 3.2 and 3.3 respectively. Three areas stand out clearly as satisfying criteria 1 and 2 just referred to; these are

(1) The Cornubian batholith in south-west England

(2) Northern England (the Lake District and Weardale batholiths)

(3) The Eastern Highlands of Scotland.

By the early 1980s, the Cornubian batholith had been confirmed as a possible HDR resource but the potential in northern England was less well established and that in the Eastern Highlands almost unknown. This paucity of heat flow data over Caledonian granites led to the formulation of a collaborative project between BGS, the Open University and Imperial College to investigate the HDR potential of northern England and the Eastern Highlands.

Detailed studies were carried out on selected intrusions to measure the heat flow and heat production, study the geochemical variations and the distribution of radioactive elements, interpret the sub-surface shape from gravity data and develop thermal models in order to predict sub-surface temperatures and assess the HDR potential.

In view of the proven high heat flow over the Weardale Granite (England and others, 1980), efforts in northern England were concentrated on the adjoining Lake District batholith, in particular on the Shap and Skiddaw granites. In the Eastern Highlands, the Cairngorm, Ballater, Mount Battock and Bennachie granites were the subjects of detailed study. The results of these recent investigations are discussed below but first the principal features of the Cornubian batholith, as established from the earlier work, are reviewed.

CORNUBIAN BATHOLITH

The Cornubian granite batholith is exposed in five main outcrops in south-west England and also in the Scilly Isles. Interpretation of the associated large negative gravity anomaly (Tombs, 1977) suggests that the batholith extends to depths of between 10 and 20 km. The thermal anomaly has been studied thoroughly and the results fully reported by Wheildon and others (1980 and 1981). The sub-surface form of the batholith, as interpreted from the gravity data, together with the heat flow results are shown in Figure 3.4 and the available thermal data are summarised in Table 3.1.

Table 3.1 Summary of thermal data for granites in south-west England

Granite Borehole or mine	National Grid reference	Depth logged (metres)	Heat flow* (mW/m^2)	Heat production ($\mu W/m^3$)	Thermal conductivity (W/mK)	References
Carnmenellis						
Grillis Farm	SW 6795 3846	100	112.9	4.5 ± 0.5	3.31 ± 0.16	1
Polgear Beacon	SW 6927 3663	100	121.7	3.8 ± 0.9	3.57 ± 0.35	1
Medlyn Farm	SW 7083 3404	100	113.4	3.4 ± 0.8	3.32 ± 0.23	1
Trevease Farm	SW 7185 3180	100	111.9	3.8 ± 0.5	3.27 ± 0.12	1
Trerghan Farm	SW 7353 3033	100	112.9	4.4 ± 1.1	3.21 ± 0.17	1
Troon	SW 6570 3677	122	122.7	—	3.40 ± 0.16†	1
Rosemanowes A	SW 7352 3456	303	105.5	—	3.30 ± 0.20	1
Rosemanowes D	SW 7352 3460	292	106.4	—	3.09 ± 0.21	1
Longdowns	SW 7368 3462	182	111.7	3.6 ± 0.7	3.09 ± 0.34	1
South Crofty Mine	SW 6680 4105	650	128.9	5.3 ± 0.4	3.60 ± 0.22	2
Bodmin						
Bray Down	SX 1907 8177	100	113.4	5.2 ± 1.8	3.37 ± 0.17	1
Blackhill	SX 1835 7820	100	119.0	3.1 ± 0.5	3.42 ± 0.18	1
Pinnockshill	SX 1892 7450	100	120.7	3.5 ± 0.8	3.09 ± 0.13	1
Browngelly	SX 1924 7247	100	108.4	5.0 ± 0.8	3.41 ± 0.17	1
Gt Hammett Farm	SX 1885 6986	100	118.8	4.3 ± 1.0	3.28 ± 0.16	1
Land's End						
Newmill	SW 4608 3435	100	123.8	4.9 ± 0.8	3.36 ± 0.15	1
Bunker's Hill	SW 4022 2726	100	123.9	5.2 ± 0.7	3.36 ± 0.21	1
Geevor Mine	SW 3750 3450	403	128.6	6.5 ± 0.2	3.39 ± 0.29	2
St. Austell						
Tregarden Farm	SX 0553 5945	100	125.8	3.5 ± 0.6	3.14 ± 0.14	1
Colcerrow Farm	SX 0679 5763	100	126.5	4.8 ± 0.7	3.38 ± 0.25	1
Dartmoor						
Winter Tor	SX 6117 9156	100	107.4	5.7 ± 1.3	3.23 ± 0.12	1
Blackingstone	SX 7850 8593	100	105.5	4.9 ± 0.6	3.09 ± 0.32	1
Soussons Wood	SX 6733 7971	100	132.2	5.0 ± 2.0	3.12 ± 0.28	1
Laughter Tor	SX 6562 7549	100	114.2	5.9 ± 1.0	3.28 ± 0.19	1
Foggin Tor	SX 5663 7334	100	110.9	4.9 ± 1.1	3.41 ± 0.17	1
Hemerdon	SX 5733 5849	128	107.9	—	3.45 ± 0.50†	1

References
1 Wheildon and others (1981)
2 Tammemagi and Wheildon (1974)

Notes
* A correction for recent palaeoclimate changes has been applied to heat flow values in shallow boreholes (less than 200 m deep) to be consistent with values from deeper boreholes.
† Thermal conductivity values determined from disc samples. All other values determined from borehole chippings.

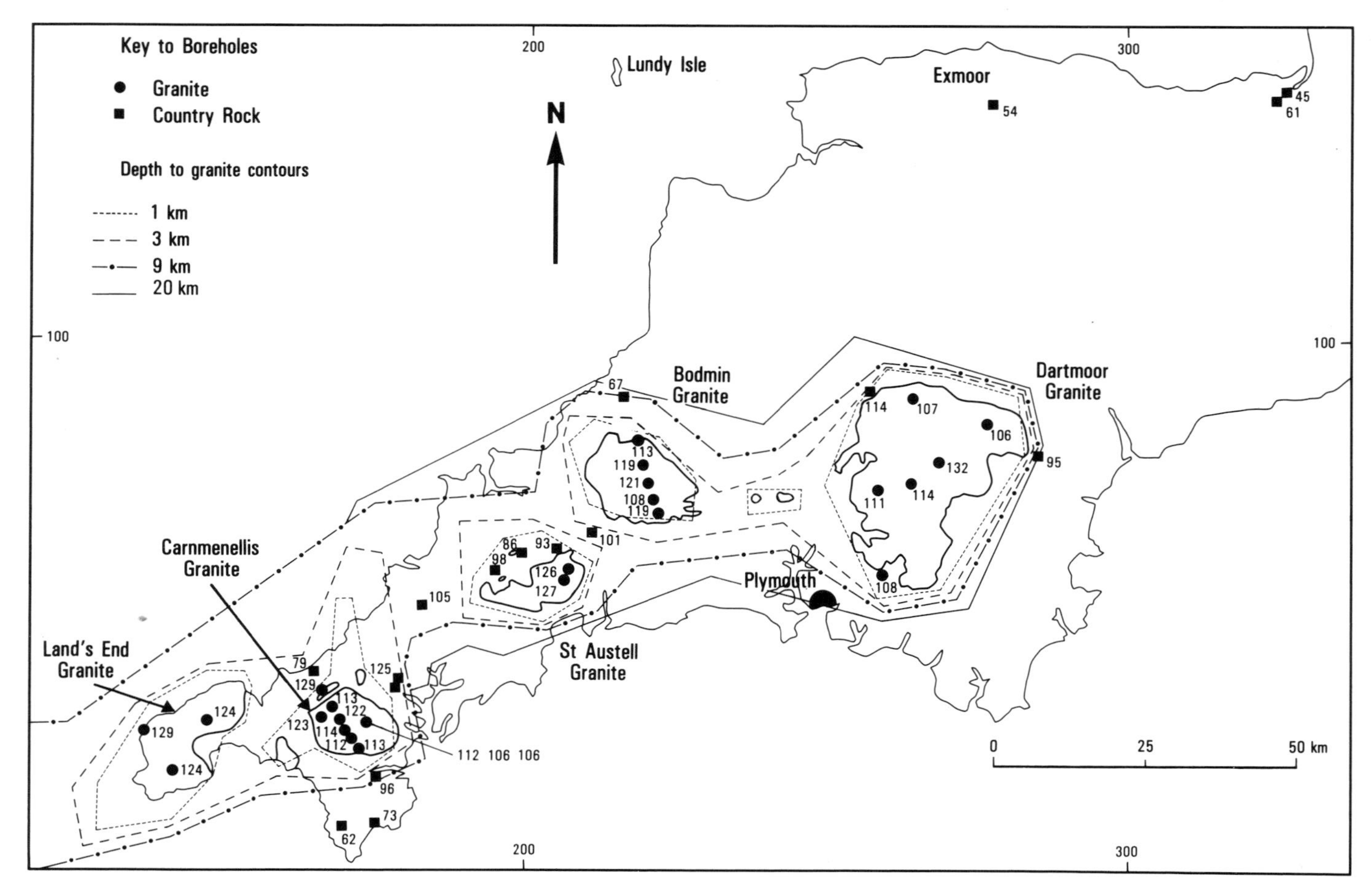

3.4 Three-dimensional model of the granite batholith of south-west England showing the distribution and values of heat flow measurements. Heat flow values in mW/m² have been corrected for topography and recent climate only (after Wheildon and others, 1980). Geometry of the granite batholith as interpreted from regional gravity data by Tombs (1977).

The heat flow is equally high, and generally greater than 100 mW/m², over all five major outcrops. Heat production values are similarly uniformly high, with average values from boreholes in each outcrop in the range 4.0 to 5.3 μW/m³. The uniformity of the values and the positive correlation between the space form of the batholith and the surface heat flow pattern (Figure 3.4) tend to suggest a basically conductive heat transfer mechanism. This is supported by a two-dimensional finite element model of the Carnmenellis Granite (Figure 2.9) which explains the observed high surface heat flow in terms of uniformly radiothermal granite extending to mid-crustal depths. The observed pattern of heat flow values across the granite is accounted for by refraction and lateral heat transfer due to thermal conductivity and heat production contrasts between the batholith and the country rock.

A study of groundwaters in the Carnmenellis Granite (Chapter 6) shows that saline groundwater has circulated on a geological timescale in response to hydraulic head differentials. This system has mixed with a more recent circulation system, extending to a depth of about 1.2 km, which has been induced or greatly accelerated by mining activity. The results show that natural circulation of groundwater occurs to considerable depths and imply that fracture permeability is an important feature in the granite. However, there is no evidence for a deep convection cell which might seriously distort the surface heat flow pattern, although some local perturbations occur due to the mining activity.

Predicted sub-surface temperatures within the Cornubian batholith are illustrated by the calculated profile for the Carnmenellis Granite shown in Figure 3.5. The profile was calculated according to the one-dimensional heat transfer equation (Lee and others, 1984, based on Richardson and Oxburgh, 1978), assuming a surface temperature of 10°C, a heat flow of 115 mW/m², a heat production of 4 μW/m³, decreasing exponentially with depth with a scale length of 15 km*, and a temperature dependent conductivity of 3.3 W/mK. The model predicts a temperature of 79°C at a depth of 2 km which is in accord with that observed at the same depth in the boreholes drilled into the Carnmenellis Granite by the Camborne School of Mines (Batchelor, 1983). The profile suggests that a temperature of 150°C might be reached at a depth of about 4 km and 200°C at 5.4 km. The model is most sensitive to changes in the assumed heat flow and thermal conductivity values, and relatively insensitive to values of the surface heat production and scale length. The agreement between observed and predicted temperatures at a depth of 2 km enhances the credibility of the model, and thus the predicted temperatures at greater depth, and lends support to the contention that the heat transfer mechanism is mostly conductive.

* As indicated in Chapter 2, there are a number of plausible models. For the vertical distribution of heat-producing elements. However, temperatures within the upper few kilometres of the earth's crust are relatively insensitive to the type of model (see Table 2.1). In this and other examples given later, where sub-surface temperature profiles have been calculated from the one-dimensional heat transfer equation, an exponential distribution has been assumed with a scale length of 15 km (that is the heat production decreases to A_0/e at a depth of 15 km).

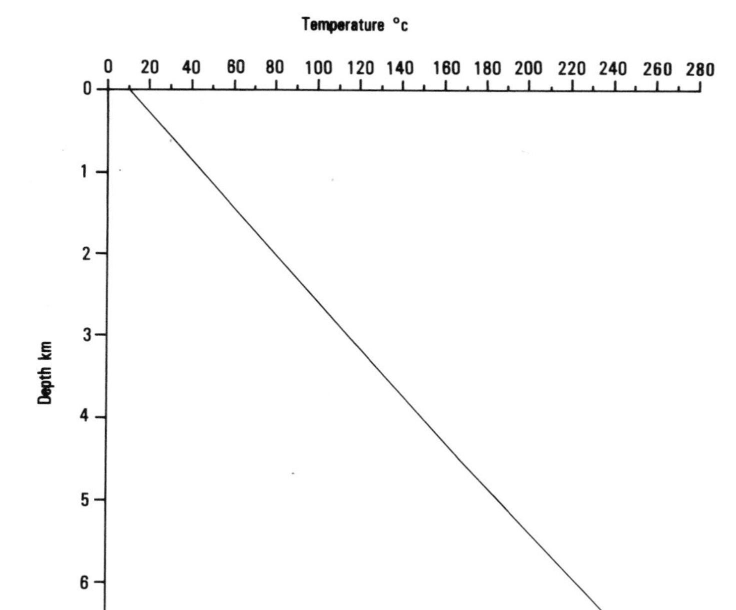

3.5 Predicted temperature-depth profile for the Carnmenellis Granite calculated from a one-dimensional heat transfer model. This assumes a surface temperature of 10°C, surface heat flow of 115 mW/m², surface heat production of 4 μW/m³ decreasing exponentialy with depth with a scale length of 15 km, and a temperature dependent thermal conductivity of 3.3 W/m K at the surface.

NORTHERN ENGLAND

Lake District

The Lake District batholith covers an area of approximately 1500 km². It is a composite body which was emplaced during two main intrusive phases. The Eskdale and Ennerdale granites and probably much of the concealed part of the batholith were emplaced around 430 to 420 Ma (Rundle, 1979) whereas the Shap and Skiddaw granites are early Devonian in age (394 ± 3 Ma and 399 ± 8 Ma according to Wadge and others (1978) and Rundle (1981), respectively). A simplified three-dimensional model of the batholith, as interpreted from the regional gravity data (Lee 1984a) is shown in Figure 3.6. The central part of the Lake District is underlain by a granite ridge at a relatively shallow depth, which appears to be an extension of the Eskdale Granite. The later Shap and Skiddaw granites appear to be somewhat separate bodies with steeper sides than the main batholith. In practice, the concealed part of the batholith is itself likely to be a composite intrusion with a range of granite types and density values. Overall the batholith extends to a depth of at least 9 km.

Heat flow boreholes were drilled to a depth of approximately 300 m into the Shap and Skiddaw granites during 1982 (Wheildon and others, 1984a). The Skiddaw Borehole was sited at the north-eastern edge of the central (Caldew) outcrop and the Shap Borehole as near to the centre of the

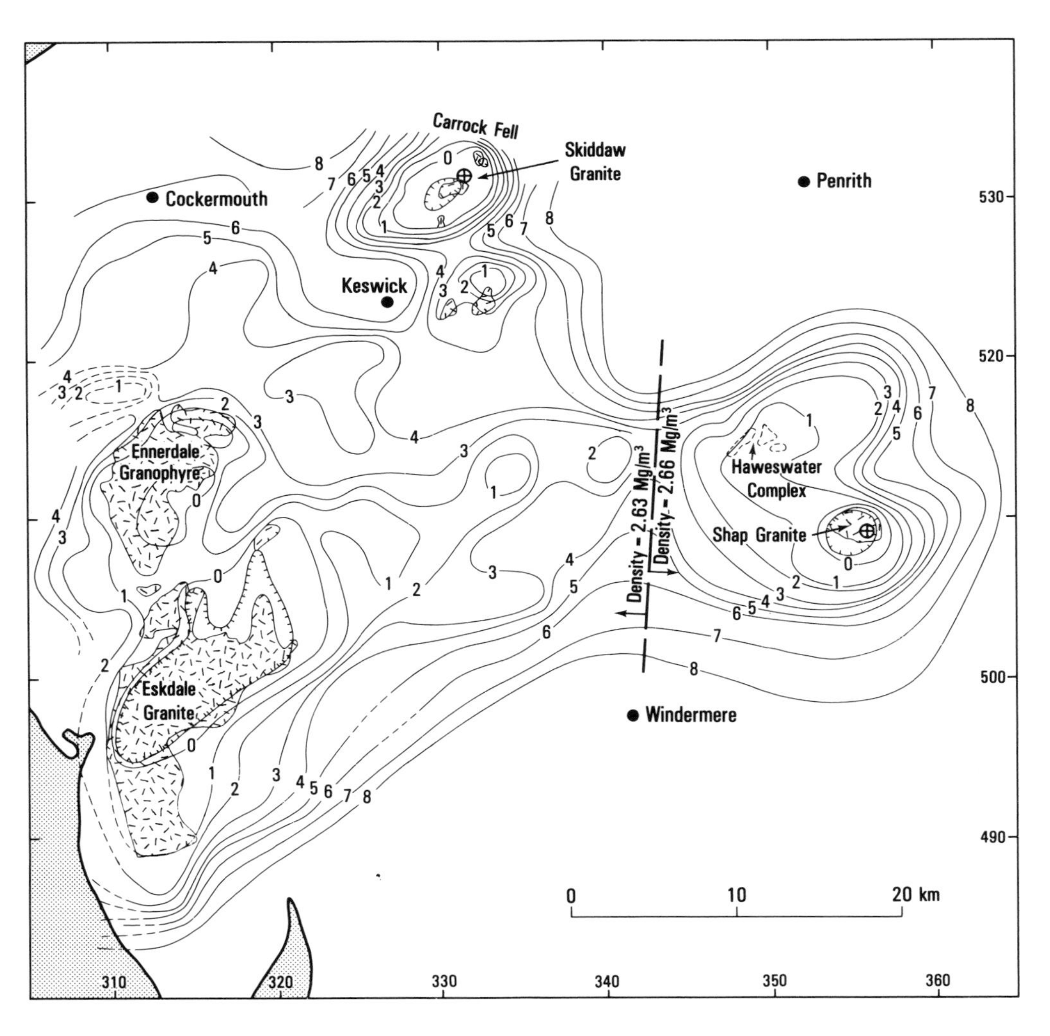

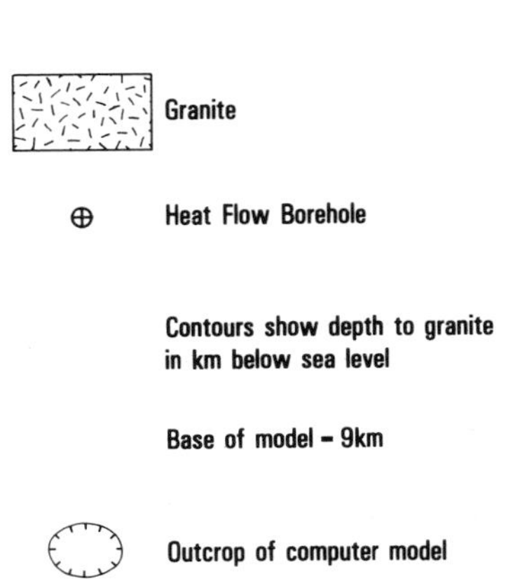

3.6 Three-dimensional model of the Lake District granite batholith (after Lee, 1984a) showing location of heat flow boreholes. In this simplified model the concealed part of the batholith has been interpreted assuming a density of 2.63 Mg/m³. In practice there is likely to be a range of granite types with a range of density values.

outcrop as access would allow (Figure 3.6). The boreholes were drilled mainly by the downhole rotary percussion methods with short coring runs at approximately 100 m, 200 m and at total depth to recover samples for heat production, physical property and thermal conductivity measurements as well as for geochemical analysis. The heat flow values were determined as described in Chapter 2 and the results are given in Table 3.2. The values at both sites are well above the UK average although the value of 77.8 mW/m^2 at Shap is not as high as might have been expected considering the high heat production of the Shap Granite (see below).

Table 3.2 Heat flow results from the Shap and Skiddaw boreholes

Borehole and Grid reference	Uncorrected heat flow (mW/m^2)	Topographic correction (mW/m^2)	Corrected heat flow (mW/m^2)
Shap NY 559 087 (Pink Quarry)	72.8	+5.0	77.8
Skiddaw NY 314 314 (Burdell Gill)	118.5	−17.6	100.9

Heat production determinations on borehole core and outcrop samples have been reported and discussed in detail by Webb and Brown (1984a). Their data are summarised in Tables 3.3 and 3.4 and in Figure 3.7. A critical factor in evaluating the HDR potential of the Lake District is the assessment of how representative are the samples of the individual intrusions and of the batholith as a whole.

Fresh samples from both the Shap and Skiddaw intrusions contain radioelements dominantly in primary accessory minerals and have heat production values generally greater than 4 $\mu W/m^3$. For the Shap Granite a 'preferred' value of 5.2 $\mu W/m^3$ for mean 'surface' heat production was proposed by Webb and Brown (1984a). This is based largely on the borehole data (Table 3.3) for the 'pink' granite in the upper and lower core sections. Account was also taken of the proportion of haematised and slightly radioelement-depleted 'red' granite visible at outcrop. A previous value of 4.4 $\mu W/m^3$ (Brown and others, 1982), based on surface sampling alone (Figure 3.7), was an underestimate because the uranium contents of outcrop samples are depleted relative to fresh borehole core. This loss is attributable to leaching of uranium by oxidising groundwaters in the weathered zone; leachable sites being

Table 3.3 Average radioelement content and heat production of core from the Shap and Skiddaw boreholes (from Webb and Brown 1984a)

Core section	No.of samples	Uranium (ppm)	Thorium (ppm)	Potassium (%)	Heat production ($\mu W/m^3$)
Shap					
Upper	3	11.7 ± 0.1	31.0 ± 2.2	4.15	5.4
Middle	3	8.3 ± 1.2	27.0 ± 1.2	3.82	4.3
Lower	3	10.7 ± 2.7	27.0 ± 2.9	4.13	4.9
Skiddaw					
Upper	6	13.0 ± 3.6	13.0 ± 0.8	4.14	4.5
Middle	4	9.6 ± 3.6	20.0 ± 1.0	4.04	4.1
Lower	4	7.8 ± 1.6	23.1 ± 1.9	4.04	3.8

Notes
1. U and Th data by Instrumental Neutron Activation Analysis (INAA), K data by X-ray fluorescence (XRF).
2. Density values used for calculating heat production were 2.66 Mg/m^3 and 2.61 Mg/m^3 for the Shap and Skiddaw granites respectively.

Table 3.4 Summary of properties relating to the radiothermal potential of Lake District granites (from Webb and Brown 1984a)

	Surface area (km^2)	Age (Ma)	Range of Uranium (ppm)	Range of Thorium (ppm)	Range of SiO_2 (%)	Surface heat production ($\mu W/m^3$)
Shap	6	R394± 3	6.1-14.8	23.5-37.1	67-71	4.3(5.2)
Skiddaw	3	K399± 9	2.8-17.6	3.3-22.4	69-77	3.5(4.2)
Ennerdale	58	R420± 4	4.0-6.4	17.2-22.5	72-77	2.8
Eskdale	82	R429± 4	1.1- 6.2	6.1-25.1	63-77	1.9*
Threlkeld	4	R438± 6	4.0-4.4	8.5-10.9	67-73	1.9

Notes
1. R = Rb/Sr dating: K = K/Ar dating
2. Data in brackets are preferred values for heat production based on borehole and surface sampling
3. *Expected to be substantially underestimated by surface data

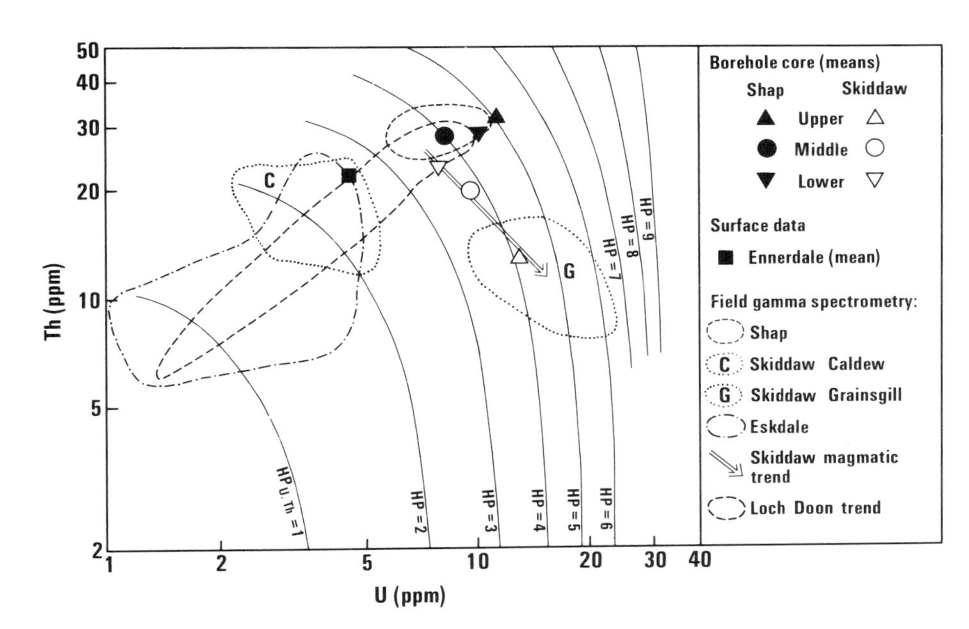

3.7 Summary of uranium and thorium contents and their contribution to heat production of Lake District granites (from Webb and Brown, 1984a). Heat production contours in μW/m³ relate to the combined contribution of uranium and thorium based on a rock density of 2.65 Mg/m³. Total rock heat production requires the addition of a contribution from potassium equivalent to 0.4 μW/m³ for 5% K_2O.

uraninite, which accounts for 35% of the uranium, and grain-boundary sites, which may account for up to 20% of the uranium in haematized rocks. An extensive enclave of 'grey' granite encountered during drilling the middle core section has a lower mean heat production of 4.3 μW/m³. If it represents part of a disrupted early-crystallising part of the Shap intrusion, as the geochemistry suggests, it is unlikely that the mean 'surface' heat production value persists throughout the body.

For the Skiddaw Granite Webb and Brown (1984a) proposed a 'preferred' value of 4.2 μW/m³ for the mean 'surface' heat production. It is a compromise between the values of 4.5 μW/m³ for white biotite granite, which comprises the upper core section and much of the Grainsgill (northern) outcrop, and 3.8 to 4.1μ W/m³ for grey biotite granite, which forms the middle and lower core sections and much of the Caldew (central) outcrop. A value of 2.6 μW/m³ reported earlier (Brown and others, 1982) was derived from measurements on surface samples which are now known to have been seriously affected by up to 75% depletion of uranium in the Caldew outcrop (Figure 3.7). This occurs because a high proportion of the uranium is present as leachable uraninite; hydrothermal sericitic alteration, identified in the borehole and developed extensively in the Caldew outcrop, has produced porous rock susceptible to weathering and leaching of uranium. In borehole and Grainsgill samples, U/Th ratios increase regularly with magmatic evolution (Figure 3.7), and their heat production values range from 3.8 to 5.5 μW/m³, apparently unaffected by hydrothermal alteration. The present level of outcrop is close to the original roof of the Skiddaw Granite, so it is strictly only the outer skin of the intrusion that has been examined. If this represents an early-crystallising outer layer of the highest level intrusive phase, then progressively more fractionated, and more highly radiothermal rock with heat production greater than 4.2 μW/m³ might be present at depth.

Looking at the Lake District batholith as a whole, the Shap and Skiddaw granites appear to be the most highly radiothermal but they are small in surface area and marginal to the main batholith. The surface geochemistry of the northern Eskdale Granite (Webb and Brown, 1984a) suggests that uranium is strongly depleted, but the granite is at least as evolved as that at Skiddaw and may have had similarly high primary uranium contents. Either weathering or oxidation by hydrothermal fluids could have released much of the original uranium, especially if it had been hosted by uraninite. Therefore the surface heat production value of 1.9 μW/m³ for the Eskdale Granite is probably a substantial underestimate.

This uncertainty weakens the assessment of the radiothermal potential of the concealed Lake District batholith because, in view of the gravity interpretation, the Eskdale Granite is most likely to be representative of the central part. The occurrence of metalliferous mineralisation in exposed parts of the batholith roof may indicate that hydrothermal circulation was focused by buried radiothermal parts of the batholith. However, the presence of haematitic alteration overlying its southern flanks suggests that Eskdale-type alteration, with its possible uranium depletion, could be extensive. Thus there may be considerable variation in heat production and heat flow throughout the batholith and further studies are required before a definitive answer can be given.

The Shap and Skiddaw intrusions are relatively limited in extent and surface heat flow values are affected by lateral heat transfer, both by the refraction of heat into the intrusions and by lateral conduction of radiogenic heat away from the intrusions. The magnitude of these effects and the resultant total perturbation of the one-dimensional heat flow depends on the granite geometry and on the heat production and thermal conductivity contrasts between the granite and the surrounding country rock.

A computer program based on the finite element method was developed by Wheildon and others (1984a) to take account of these effects and predict sub-surface temperatures within the Shap and Skiddaw granites. A series of two-dimensional and three-dimensional models were generated based on the granite geometries as interpreted from the gravity data (Figure 3.6) and the observed heat flow, heat production and thermal conductivity values. The temperature profiles from the preferred models are shown in Figure

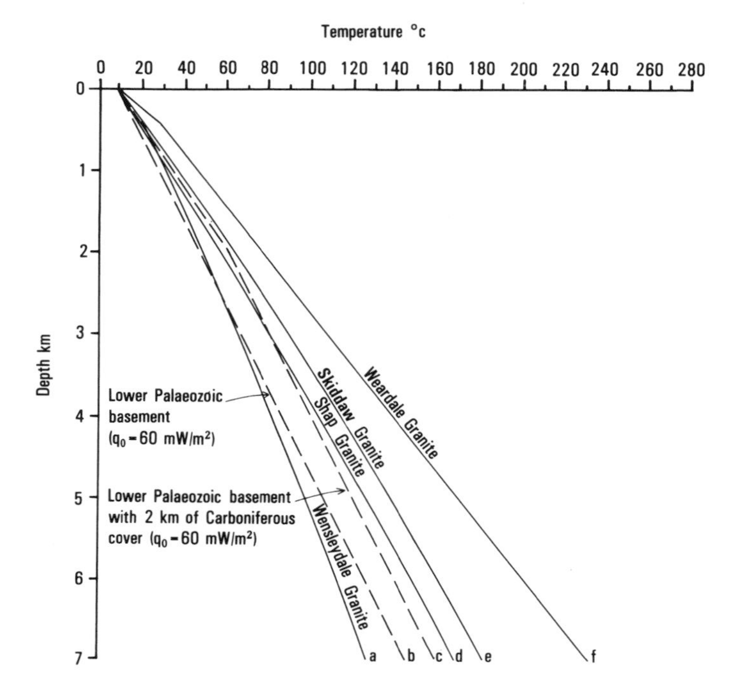

3.8 Predicted temperature-depth profiles for granites and basement rocks in northern England. Profiles for the Shap and Skiddaw granites are from the preferred models of Wheildon and others (1984a). All other profiles were calculated as described in the text assuming one-dimensional heat transfer.

3.8, together with other profiles from northern England discussed below. The models predict temperatures of 57 and 63°C at a depth of 2 km in the Shap and Skiddaw intrusions respectively, compared with 79°C measured at the same depth in the Carnmenellis Granite. Predicted temperatures at a depth of 6 km are 148 and 160°C respectively. In neither case are these values particularly high from the HDR point of view, although they compare very favourably with temperatures predicted for basement rocks in many other parts of Britain (see below). The reason for the relatively low value at Shap is simply that the heat flow, although above average, is not sufficient to produce a high geothermal gradient in a rock with a relatively high conductivity of 2.9 W/mK. At Skiddaw where the heat flow is higher, the granite has a particularly high conductivity of 3.5 W/mK and the limited size of the intrusion allows axial transfer of radiogenic heat into the surrounding rock. The thermal models suggest the difference in heat flow between the two sites may be partially accounted for by a more rapid decrease of heat production with depth in the Shap Granite than in the Skiddaw Granite.

It is difficult to estimate the HDR potential of the rest of the Lake District batholith from the observations on the Shap and Skiddaw intrusions alone. The heat production data from the other granites are equivocal and the heat flow values from the Shap and Skiddaw granites are almost certainly not representative of the rest of the batholith. On the positive side, the larger intrusive volume of the Eskdale/Ennerdale granites (Figure 3.6) would tend to reduce lateral heat transfer and thus boost sub-surface temperatures for a given heat flow, and their surface heat production values may under-estimate their true heat production at depth. It is possible, therefore, that the heat flow over the western and central Lake District may be significantly above average, but additional heat flow boreholes would be required to confirm this suggestion.

Weardale Granite

The Weardale Granite was intruded into the Lower Palaeozoic basement rocks of the Alston Block and lies concealed beneath Carboniferous sedimentary rocks. It occupies an area of approximately 1500 km² (Figure 3.9) and according to the interpretation of gravity data by Bott (1967) extends to a depth of about 9 km. The existence of the granite was proved by the Rookhope Borehole which penetrated almost 400 m of Carboniferous strata before entering the granite (Dunham and others, 1961, 1965). A heat flow of 92 mW/m² was reported by Bott and others (1972) but England and others (1980) have since revised this to 95 mW/m² after remeasuring the geothermal gradient and rock conductivities.

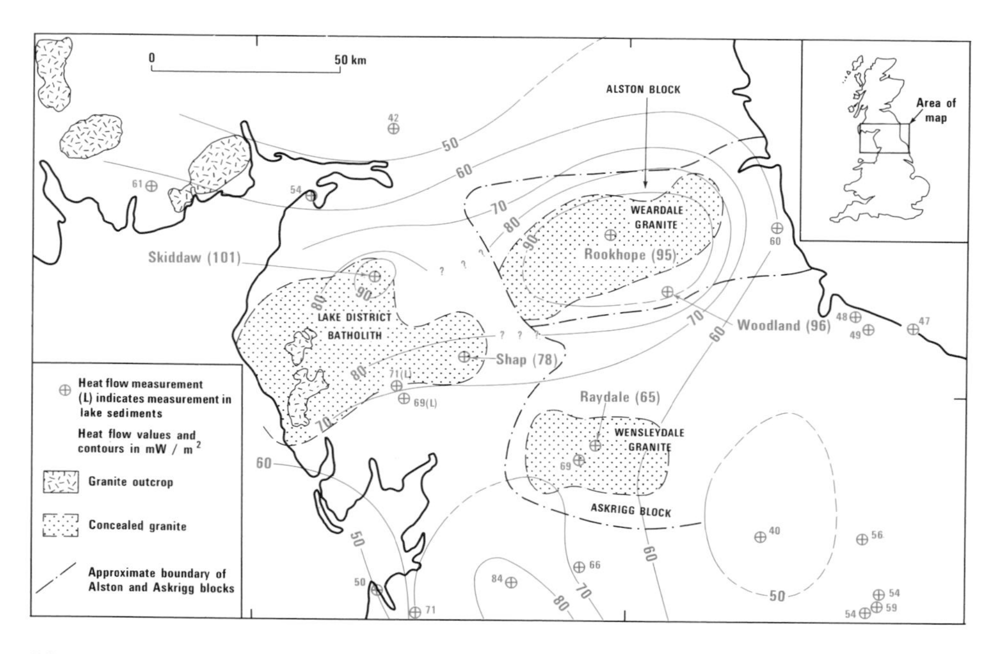

3.9 Location of granite batholiths and heat flow values in northern England. Heat flow contours are as defined in Figure 2.8 and on the Geothermal Map of the UK (in pocket).

The age of Weardale Granite is 394 ± 34 Ma (Dunham (1974b), recalculated from Holland and Lambert (1970)), which is similar to that for the Shap and Skiddaw granites. Bott and others (1972) reported a heat production for the Weardale granite of 4.5 $\mu W/m^3$ but this has since been revised to 3.7 $\mu W/m^3$ on the basis of new data from the Open University, reported by Lee and others (1984). In the absence of further boreholes it is not possible to be certain whether other parts of the batholith are of the same age and similarly radiothermal. The batholith may be a composite intrusion with a range of heat production values as in the Lake District.

The high heat flow at Rookhope suggests the presence of a large volume of highly radiothermal granite in the central part of the batholith. A heat flow of 96 mW/m^2 from the Woodland Borehole (Bott and others, 1972) on the southern margin of the Alston Block (Figure 3.9) also tends to suggest that large parts of the block may be characterised by high heat flow. However, Bott and others (1972) pointed out that heat production within the Weardale Granite does not contribute significantly to the heat flow at Woodland, which lies outside the margin of the batholith, and they suggested water movement along the nearby Butterknowle Fault as a possible explanation.

Additional heat flow observations are clearly required before the HDR potential of the Weardale batholith as a whole can be fully assessed. However, the potential in the central part of the intrusion can be evaluated by assuming that data from the Rookhope Borehole are representative. England and others (1980) carried out a study of heat refraction in the Pennine granites and concluded that refraction effects in the Weardale batholith are relatively insignificant. The geothermal gradient beneath the Rookhope Borehole has therefore been predicted on the basis of one-dimensional heat transfer, assuming a heat flow of 95 mW/m^2, a thermal conductivity for the granite of 3.1 W/mK and a heat production at the top of the granite of 3.7 $\mu W/m^3$, decreasing exponentially with depth with a scale length of 15 km. The profile (Figure 3.8) predicts an average geothermal gradient between the roof of the granite and a depth of 7 km of about 31°C/km and indicates that temperatures of 150 and 200°C should be encountered at depths of about 4.4 and 6 km respectively. It should be noted that the 400 m of low conductivity Carboniferous sedimentary rocks have the effect of raising temperatures within the granite by about 7°C with respect to an equivalent granite at outcrop.

Comparison of sub-surface temperatures in northern England granites and Lower Palaeozoic basement rocks

The thermal data for granites in northern England are summarised in Table 3.5. The new heat flow values of 101 mW/m^2 for Skiddaw and 78 mW/m^2 for Shap, together with the previously reported value of 95 mW/m^2 for Weardale, confirm that the Lake District-Weardale line of granites forms a high heat flow zone within northern England (Figure 3.9). The value of 101 mW/m^2 for Skiddaw is the highest recorded over a granite outside south-west England. The Wensleydale Granite, to the south of the Weardale Granite, is less radiothermal with a heat flow of 65 mW/m^2, and lies outside the main high heat flow zone.

Predicted temperature profiles for northern England granites and basement rocks are compared in Figure 3.8. In addition to the temperature profiles previously discussed, the figure also includes profiles for the Wensleydale Granite and the Lower Palaeozoic basement (from Lee and others, 1984). The profile for the Wensleydale Granite is based on data from the Raydale Borehole (Table 3.5); the model assumes a surface temperature of 8°C, a temperature at the top of the granite (at a depth of approximately 500 m) of 22°C, and an exponential decrease in heat production with a scale length of 15 km.

Temperatures within the Lower Palaeozoic basement are more difficult to predict. For a given heat flow, sub-surface temperatures depend critically on the conductivity of the basement rocks and the thickness and thermal resistance of the sedimentary cover. Profile b (Figure 3.8) represents outcropping Lower Palaeozoic basement rocks assuming a heat flow of 60 mW/m^2. The model assumes a conductivity of 2.9 W/mK, and a surface heat production of 2.2 $\mu W/m^3$ which is in line with values quoted for the Skiddaw Slates (Wheildon and others, 1984a).

Profile c (Figure 3.8) predicts the temperatures within Lower Palaeozoic basement covered by an arbitrary 2 km of Carboniferous sedimentary rocks. The thermal

Table 3.5 Summary of thermal data for granites in northern England

Granite (Borehole)	National Grid reference	Depth logged (metres)	Heat flow (mW/m^2)	Heat production ($\mu W/m^3$)	Thermal conductivity (W/mK)
Weardale (Rookhope)	NY 938 428	799	95.4 (a)	3.7 (c)	3.10 ± 0.25 (a)
Wensleydale (Raydale)	SD 903 847	593	65.0 (a)	3.3 (c)	3.65 ± 0.08 (a)
Shap (Pink Quarry)	NY 559 087	300	77.8 (b)	5.2 (d)	2.9 ± 0.2 (b)
Skiddaw (Burdell Gill)	NY 314 314	281	100.9 (b)	4.2 (d)	3.5 ± 0.3 (b)

References
(a) England and others (1980)
(b) Wheildon and others (1984a)
(c) Heat production data provided by the Open University and reported by Lee and others (1984)
(d) Preferred values, Webb and Brown (1984a)

Borehole	National Grid reference	Uncorrected heat flow (mW/m²)	Topographic correction (mW/m²)	Corrected heat flow (mW/m²)
Cairngorm (Coire Cas)	NH 989 062	72.2	−2.7	69.5
Mount Battock (Auchabrak)	NO 542 905	65.6	−6.9	58.7
Ballater (Tomnakeist)	NO 401 986	75.6	−4.2	71.4
Bennachie (Birks Burn)	NJ 669 210	81.4	−5.6	75.8

Table 3.6 Heat flow results from the boreholes in the Eastern Highlands.

resistance of the Carboniferous cover depends critically on the relative proportions of the various rock types. Sequences with a high proportion of sandstones and limestones will have a relatively low thermal resistance (i.e. a high effective conductivity) and sequences with a high proportion of shales a relatively high thermal resistance. The relative proportions of the three main lithologies vary considerably from area to area within the Upper and Lower Carboniferous. For the purpose of the present calculation an effective conductivity of 2.2 W/mK has been adopted. This is approximately equivalent to the values measured in the Rookhope Borehole for the Carboniferous Limestone Series between the Great Whin Sill and the granite (England and others, 1980), and in the Woodland Borehole for the Millstone Grit and Carboniferous Limestone Series (Bott and others, 1972). An effective conductivity of 2.2 W/mK results in a temperature gradient of 27°C/km within the Carboniferous rocks and raises the temperature beneath by an additional 7°C for each 1 km of cover, for a heat flow of 60 mW/m².

Predicted temperatures within the Weardale and Skiddaw granites are greater than those expected in the Lower Palaeozoic basement (with or without Carboniferous cover) at all depths (Figure 3.8). Temperatures in the Shap Granite are comparable with those expected in the Carboniferous but slightly higher than those expected in the basement in either situation. Temperatures within the Wensleydale Granite are lower than those expected within the basement for depths greater than about 2 km.

Although predicted temperatures within the Shap and Skiddaw granites are generally higher than those expected in adjacent basement rocks (Figure 3.8) in neither case is a temperature of 150°C likely to be reached at a depth of less than about 5.5 km. On the evidence of the heat flow value from the Rookhope Borehole, the Weardale Granite remains the principal HDR prospect in northern England. However, further heat flow boreholes are required to confirm its potential and also to investigate the potential in the western part of the Lake District.

EASTERN HIGHLANDS OF SCOTLAND

The heat flow data for the Eastern Highlands region are shown in Figure 3.10. The granite batholith is a composite body comprising a series of intrusions emplaced into the Moine/Dalradian basement between about 400 and 415 Ma. Interpretation of the large negative Bouguer gravity anomaly suggests that the batholith extends at least to a depth of 13 km and may continue beneath the Moine to link with the Etive batholith in the south-west (Rollin, 1984).

Reconnaissance heat production surveys indicated that five of the major granites (Cairngorm, Mount Battock, Ballater, Bennachie and Monadhliath) are characterised by heat production greater than 4 $\mu W/m^3$ (Webb and Brown, 1984b). Heat flow boreholes to a depth of about 300 m were subsequently drilled on the Cairngorm, Ballater, Mount Battock and Bennachie granites during 1982. The heat flow values were determined as described in Chapter 2 and are listed in Table 3.6. Although the values are lower than might have been expected in view of the high heat

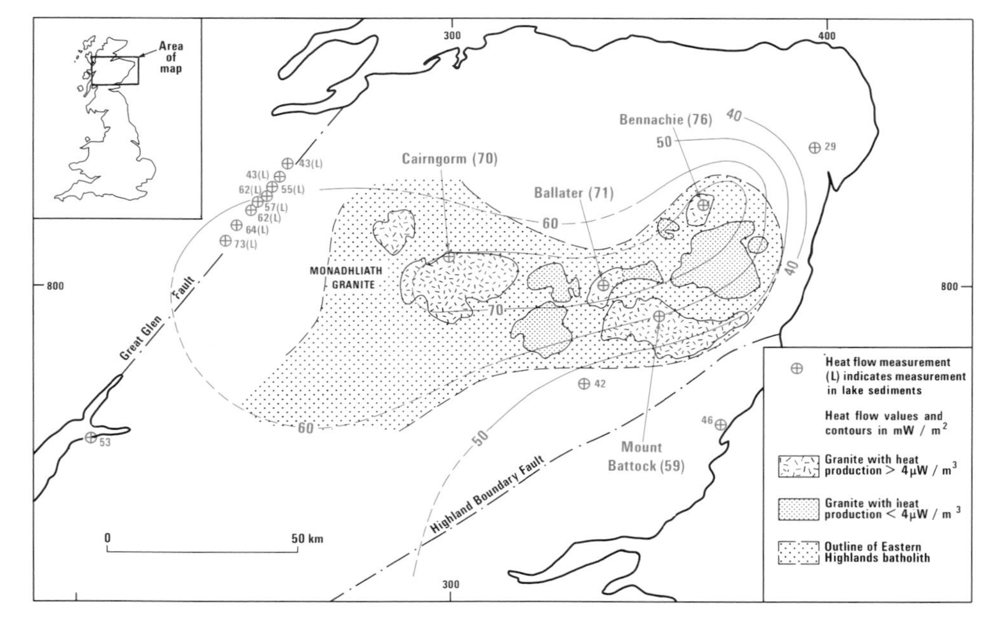

3.10 Location of granites and heat flow values in the Eastern Highlands of Scotland. Heat flow contours are as defined in Figure 2.8 and on the Geothermal Map of the UK (see Plate).

production (see below), those on the Cairngorm (70 mW/m^2), Ballater (71 mW/m^2) and Bennachie (76 mW/m^2) granites are still well above the average for the UK. They are the highest values recorded in Scotland and show that the Cairngorm-Bennachie line of granites form an above average heat flow zone within Scotland (Figure 3.10).

As in the Lake District the availability of borehole core afforded the opportunity to study the heat production and geochemistry of the granites in some detail (Webb and Brown, 1984b). Because of the larger surface area of these intrusions compared with the accessible areas of the Shap and Skiddaw granites, more extensive surface sampling was necessary to identify intra-intrusion variations in geochemistry and heat production. More emphasis was placed on studying geochemical relationships between borehole core and representative surface rocks over a substantial area (about 20 km^2) in the vicinity of each borehole in order to derive 'preferred' values of heat production for each intrusion. This was essential because the limited and intermittent sections of core sampled a large proportion of unrepresentative material, and the geophysical logs (Lee 1984b) showed evidence of considerable downhole variation in heat production and rock condition between the cored sections. The heat production values derived from borehole core are given in Table 3.7. The radiothermal properties of all the principal Eastern Highlands granites, including the 'preferred' values for the intrusions studied in detail are shown in Table 3.8.

Table 3.7 Average radioelement content and heat production of core from the boreholes in the Eastern Highlands (from Webb and Brown, 1984b)

Borehole/core section	No. of samples	Uranium (ppm)	Thorium (ppm)	Potassium (%)	Heat production (μW/m^3)
Cairngorm					
Upper	3	10.9 ± 1.0	40.2 ± 2.6	4.2	5.8
Middle	3	21.8 ± 1.9	41.3 ± 2.5	4.1	8.5
Lower	3	23.4 ± 2.2	46.2 ± 3.6	4.1	9.2
Mount Battock					
Upper	2	4.6 ± 0.2	36.2 ± 1.3	4.4	4.0
Middle	2	6.1 ± 0.2	38.7 ± 0.8	4.5	4.5
Lower	2	9.6 ± 0.2	41.2 ± 0.8	4.6	5.5
Ballater					
Upper	3	18.1 ± 6.4	29.6 ± 1.7	4.8	6.9
Middle	3	15.0 ± 1.9	38.6 ± 2.0	4.5	6.7
Lower	3	16.4 ± 3.9	40.6 ± 1.5	4.6	7.2
Bennachie					
Upper	3	14.7 ± 6.8	34.8 ± 2.8	4.0	6.3
Middle	3	28.6 ± 12.7	21.7 ± 5.2	4.1	8.9
Lower	2	7.0 ± 0.0	33.9 ± 1.0	4.5	4.6

Notes
1. U and Th data by Instrumental Neutron Activation Analysis (INAA)
2. K data by X-ray fluorescence (XRF)
3. Density values for heat production calculation: 2.60 Mg/m^3 (Cairngorm and Bennachie), 2.61 Mg/m^3 (Mount Battock and Ballater)

Table 3.8 Summary of radiothermal properties of granites in the Eastern Highlands (from Webb and Brown, 1984b)

Intrusion	Area	Age	Range of			Mean 'surface' heat production
	(km^2)	(Ma)	U (ppm)	Th (ppm)	SiO_2 (wt%)	(μW/m^3)
Monadhliath	100		7.8–21.7	30.9–41.6	74–78	5.7
Cairngorm	380	R408 ± 3	2.3–25.0	13.9–55.3	70–78	5.0 (7.3)
Glen Gairn	90		1.0–18.0	5.0–27.0	56–76	2.8
Lochnagar	180	R415 ± 5	1.0–12.1	8.8–31.0	60–75	2.7
Ballater	50		4.0–22.0	17.0–72.0	72–77	5.7 (6.8)
Mount Battock	350		3.0–19.0	13.0–55.0	71–78	5.0 (4.8)
Bennachie	50	K404 ± 5	4.0–31.0	21.0–52.0	70–77	5.7 (7.0)
Hill of Fare	55	R413 ± 3	5.1–8.0	22.6–35.1	75–77	3.9
Skene	280		<5.0	3.0–14.0	50–75	1.6

Preferred heat production values derived by Webb and Brown (1984b) are given in brackets
R = Rb/Sr dating: Brook and others (1982), Halliday and others (1979)
K = K/A dating: Brown and others (1968).

The results show that the granitic intrusions chosen for detailed study are the most highly radiothermal, both in terms of widespread surface sampling and borehole sampling, of any voluminous granites in the UK. In general, the Eastern Highlands granites may be divided into two groups, those containing high primary abundances of radioelements, and those with lower abundances. The former comprises Cairngorm, Monadhliath, Ballater, Mount Battock, Bennachie and probably the Hill of Fare (*sensu stricto*) granites, and is distinguished also by high SiO_2 contents (Table 3.8) and generally evolved geochemical characteristics. They are all coarse-to medium-grained biotite granites, and are variably affected by haematisation which imparts a reddish-orange to brown colouration to all the rocks, hence the collective term Eastern Highlands 'Red Granites'. Examples of the less radiothermal granites are Lochnagar, Glen Gairn and the Skene Complex. Although parts of these bodies are highly radiothermal, such as the central portion of the concentrically-zoned Lochnagar intrusion and the eastern part of the Glen Gairn Granite, these are outweighed by the abundance of less radiothermal dioritic and granodioritic rocks at the level of outcrop.

The Eastern Highlands granites are believed to have been intruded over a relatively narrow time span, between 400 and 415 Ma, but the contrasting geochemistry between the 'Red Granites' and the others suggests that their evolution and possibly even their sources were different (Brown and others, 1984b). How much of the Eastern Highlands granite batholith is in geochemical continuity with the more radiothermal 'Red Granites' is, therefore, open to speculation. However, in view of the extent of the outcrop comprising the Monadhliath, Cairngorm, Ballater and Mount Battock granites and the more northerly but isolated Bennachie Granite, it must be a very substantial amount.

Of the four intrusions studied, the lowest heat flow was recorded over the Mount Battock Granite (59 mW/m^2) which also has the lowest heat production (4.8 $\mu W/m^3$). It would be expected therefore, that those other parts of the batholith with a heat production of less than 4 $\mu W/m^3$ will also have a heat flow of less than 60 mW/m^2.

As in the Lake District, computer models of each intrusion were developed to explain the observed heat flow and predict sub-surface temperatures (Wheildon and others, 1984b). In all four cases the results demonstrate that the combination of very high heat production and only moderately elevated heat flow require a relatively rapid decrease in the concentration of radioactive elements with depth (see also the discussion on heat flow – heat production relationships below).

With regard to sub-surface temperatures, the thermal conductivity values of the Eastern Highlands granites are generally high, in the range 3.1 to 3.5 W/mK (Wheildon and others, 1984b) and the heat flow values, although above average, are not sufficiently elevated to give high geothermal gradients. The thermal conductivity of the Moine and Dalradian country rocks is, however, also high (approximately 3.5 W/mK), and even lower gradients can be expected in surrounding areas of lower heat flow. The predicted temperature profile in the Ballater Granite, which has marginally the highest gradient of the four Eastern Highlands borehole sites, is compared in Figure 3.11 with that in the outcropping Moine/Dalradian basement. Two profiles are shown for the basement, representing probable upper and lower limits of surface heat flow of 55 and 40 mW/m^2 respectively. Both assume a conductivity of 3.5 W/mK, a heat production of 1.7 $\mu W/m^3$ at the surface (decreasing exponentially with a scale length of 15 km), and a surface temperature of 7°C. The profiles show that below a depth of about 3 km equivalent temperatures are found about 1 to 3 km deeper in the basement than in the granite. In neither case are suitably high temperatures encountered at reasonable depths.

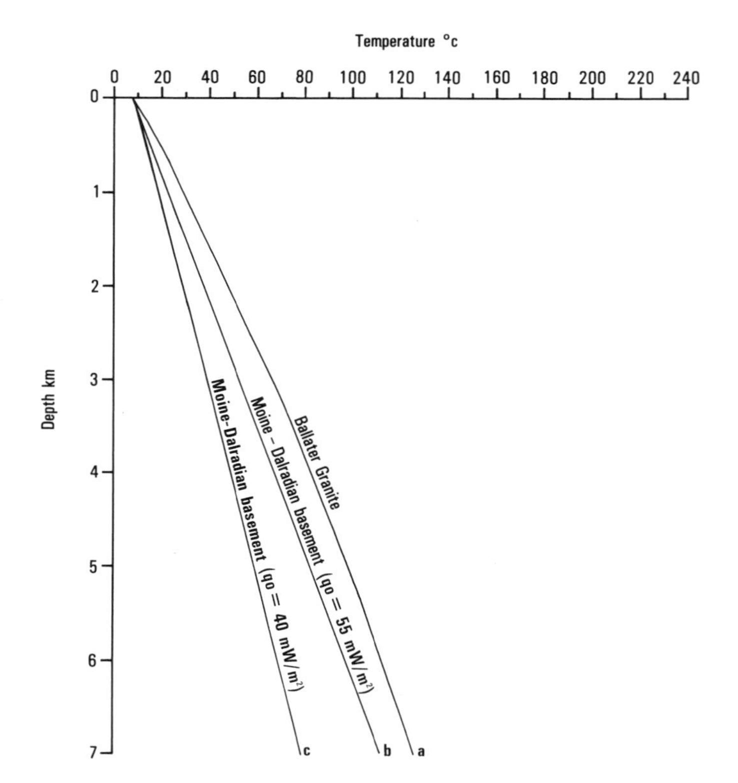

3.11 Predicted temperature-depth profiles for the Ballater Granite and the Moine/Dalradian basement. The profile for the Ballater Granite is from the preferred model of Wheildon and others (1984b). Profiles b and c are for the Moine/Dalradian basement representing possible upper and lower limits for the surface heat flow of 55 and 40 W/m² respectively; they were calculated as described in the text assuming one-dimensional heat transfer.

To summarise, the HDR potential of the Eastern Highlands granites is poor but it is worth noting that sub-surface temperatures within the granites are somewhat higher than those expected in the surrounding basement.

Data sources and methods referred to in Table 3.9 opposite

1. Webb and Brown (1984a and b): INAA and XRF
2. Barritt (1983): INAA and field gamma-spectrometry calibrated by INAA
3. Tindle (1982): INAA
4. Cassidy (1979): Field gamma-spectrometry calibrated by INAA
5. Hennessy (1979): INAA
6. Gollop (pers. comm. to P.C. Webb of the Open University): Laboratory gamma-spectrometry
7. Plant and others (1980)
8. Simpson and others, (1979): Delayed neutron analysis for U.
9. Tweedie (1979): XRF
10. Bowie and others (1973)
12. Cassidy (pers. comm. to P.C. Webb)
14. Tindle (pers. comm. to P.C. Webb): INAA

Table 3.9 Heat production and radioelement data from surface sampling of Caledonian granites

Granite intrusion	Heat production ($\mu W/m^3$)	No. of samples	U (ppm)	Th (ppm)	K (%)	SiO_2(%) range	Data source
A. Northern and Grampian Highlands of Scotland							
2 Helmsdale	4.1	97	8.6	23.0	4.4	69–74	7,8,9
3 Lairg-Rogart	1.4	17	1.9	9.8	2.6	60–72	1,6
4 Grudie	3.8*	3	6.9	26.5	4.0	76	1
5 Migdale	2.4	9	3.1	19.0	3.7	72–73	1
6 Fearn	5.1*	4	9.2	37.3	3.5	76–78	1
7 Abriachan	4.0*	2	6.0	33.5	2.1		1
8 Cluanie	1.1	10	2.2	3.9	2.6		6
9 Strontian	1.9	37	3.6	11.5	2.5	58–72	5,6
10 Ross of Mull	1.5*	4	2.7	6.3	3.8	66	5,6
11 Moy	2.1*	6	2.5	17.2	3.8	72–76	1
12 Ardclach	1.4*	4	2.0	7.8	3.5	73	1
13 Ben Rinnes	3.2*	6	4.3	25.5	4.4	73	1
14 Grantown	1.3	6	2.3	5.5	3.6	73	1
15 Tomatin	1.4	10	2.9	4.1	3.5		1
16 Foyers	1.1	72	1.5	5.2	3.3	53–76	5,6,7
17 Laggan	0.6	5	0.7	2.7	2.3		1
18 Strath Ossian	2.2	6	4.1	13.0	2.7		6
19 Rannoch Moor	1.7	34	2.7	11.4	2.8	56–75	5,6
20 Ben Nevis	1.8	18	2.8	11.7	3.2		5,6
21 Ballachulish	3.2	17	4.8	24.0	3.4	55–70	5
22 Etive	1.9	236	3.0	11.6	3.4	56–58	2,6
23 Strichen	2.2*	2	3.3	15.2	4.3	68–72	1,10
24 Peterhead	2.2	18	3.3	15.4	3.6		1,5,6
25 Bennachie	5.7	32	11.9	36.2	4.1	70–77	1
26 Coull	2.6*	2	4.5	16.0	4.4	74–77	1
27 Hill of Fare	3.9*	2	6.5	28.9	4.1	75–77	1
28 Skene	1.6	17	2.5	9.3	3.2	50–75	1,5,8
29 Aberdeen	2.2	3	3.0	17.0	4.0		1
30 Mount Battock	5.0	48	8.6	37.3	4.2	71–78	1
31 Ballater	5.7	34	10.5	40.8	4.4	72–77	1
32 Glen Gairn	2.8*	8	6.7	12.3	3.4	56–76	1
33 Lochnagar	2.7*	14	4.4	18.7	3.8	60–75	1
34 Cairngorm	5.0	233	10.8	29.6	4.2	70–78	1,2,7
35 Monadhliath	5.7*	6	11.6	36.7	4.1	74–78	1
B. Southern Uplands of Scotland, Northern Ireland and Northern/Central England							
36 Priestlaw	1.4*	2	2.6	8.4	2.1		1
37 Distinkhorn	2.0*	4	3.7	11.6	2.8	55–66	1
38 Carsphairn	2.2*	17	3.6	15.0	2.5	58–78	1,14
39 Loch Doon	2.5	164	4.6	15.7	2.9	50–72	3,4,5
40 Fleet	3.0	146	4.6	22.8	4.3	68–75	4,5,14
41 Galloway	2.1	4	3.1	15.6	2.6		5
42 Criffell	2.2	18	3.7	15.0	3.2	60–71	5,14
43 Newry	1.9	18	3.3	12.1	2.5	53–70	1
44 Cheviot	3.0*	6	5.7	18.2	3.6		12
45 Skiddaw	3.5	31	8.6	15.2	4.1	69–77	1,4,8
46 Threlkeld	1.9	3	4.3	10.1	2.0	67–73	1,8
47 Ennerdale	2.8*	8	5.0	20.7	1.7	72–77	1,12
48 Eskdale	1.9	81	3.3	11.2	4.0	63–77	4
49 Shap	4.3	36	7.7	29.2	4.0	68–71	1,4
53 Mountsorrel	2.5	8	3.8	19.2	3.1		12

Notes
Data compiled by P. C. Webb of the Open University
The intrusions are numbered according to the locations shown Figure 3.3.
The heat production values quoted in the table are the means of surface data.
The number of samples quoted is the minimum number of values for U *or* Th contributing to the averages.
* Distribution of samples and variability of data inadequate to derive a satisfactory heat production value.

OTHER CALEDONIAN GRANITES

A summary of surface and borehole heat production data for all Caledonian granites is given in Tables 3.9 and 3.10, from a compilation by P. C. Webb, reported by Lee and others (1984). The location of intrusions with five or more samples is shown in Figure 3.3. Although intrusions with a high heat production are mainly confined to the major batholiths discussed above, a few smaller or zoned intrusions reach comparably high values.

Southern Uplands of Scotland

The mean heat production values of the four Southern Uplands granites (Fleet, Loch Doon, Criffell and Carsphairn) are not high (Table 3.9). However, they are zoned intrusions with centres more radiothermal than their margins. The Fleet, Loch Doon and Criffell granites are characterised by large negative gravity anomalies which suggest the presence of large volumes of evolved granite at depth. Interpretations of the gravity data imply that they extend to a depth of about 10 to 12 km (Allsop, 1977; Bott and Masson Smith, 1960).

The heat flow, and thus the sub-surface temperature, will depend critically on the actual vertical distribution of heat-producing radioelements and the background heat flow. If the heat production decreases relatively rapidly with depth, as has been suggested in the Eastern Highlands, then the geothermal gradients will be low. However, there is much evidence to suggest that the crustal structure and tectonic setting of the Southern Uplands granites has more in common with northern England than the Eastern Highlands of Scotland. The geothermal potential of the Loch Doon and Fleet granites, in particular, might therefore be similar to that of the Lake District and Weardale granites.

Highlands of Scotland

The Fearn, Abriachan and Helmsdale granites all have heat productions of at least 4 $\mu W/m^3$. Fearn and Abriachan are unlikely to be of great interest as they are small and the limited exposures casts some doubt on how representative the sampling has been. The Helmsdale Granite is enriched in uranium due to secondary processes associated with metalliferous mineralisation; how deep the enrichment extends is uncertain.

Parts of the zoned Etive, Strontian and Ballachulish intrusions have heat production values exceeding $4 \mu W/m^3$. Of these only the Etive Granite is associated with a large negative gravity anomaly (Figure 3.2). The Ballachulish Granite has no obvious gravity expression and the anomaly due to the Strontian Granite is strongly distorted by the adjacent positive field over the Mull Tertiary centre. An interpretation of the Etive anomaly by Rollin (1980) suggests that the granite extends to a depth of 13 km and is connected to the Eastern Highlands batholith by a concealed granite ridge. The size of the negative gravity anomaly clearly indicates that a large volume of low density and, therefore, potentially radioelement-rich granite, underlies the Etive intrusion. The degree of radiothermal enhancement of background heat flow depends critically on the vertical distribution of heat producing elements. If their concentration decreases relatively rapidly with depth, as has been suggested in the Eastern Highlands, then the heat flow should be similarly low. However, if uniformly radioelement-rich granite continues at depth, as appears to be the case in south-west England, the heat flow may be higher.

Concealed granites

A review of national gravity data (Lee and others, 1984) suggests that there is no evidence for additional major

Table 3.10 Heat production and radioelement data from boreholes in Caledonian granites

Granite intrusion		Heat production ($\mu W/m^3$)	No. of samples	U (ppm)	Th (ppm)	K (%)	SiO_2 range (%)	Data source
1	Strath Halladale	1.4	22	1.8	10.3	2.8	66–73	13
25	Bennachie	6.8 (7.0)	8	18.0	29.7	4.2	75–77	1
30	Mount Battock	4.7 (4.8)	6	6.8	38.7	4.5	71–73	1
31	Ballater	6.9 (6.8)	9	16.5	36.3	4.6	73–75	1
33	Cairngorm	7.8 (7.3)	9	18.7	42.6	4.1	75–77	1
45	Skiddaw	4.2 (4.2)	14	10.5	17.9	4.1	68–75	1
49	Shap	4.9 (5.2)	9	10.2	28.3	4.0	67–70	1
50	Weardale	3.7	16	10.5	11.1	4.0	71–74	5,12
51	Wensleydale	3.3	19	6.5	19.6	4.2	73–77	12
52	Melton Mowbray	1.7	1	1.4	14.0	4.1	69	12

Data compiled by P. C. Webb of the Open University

Heat production values quoted in the table are the means of the borehole data.

Values in brackets are the preferred values taking into account various geological and geochemical criteria (see text)

Data sources and methods
1. Webb and Brown (1984a and b): INAA
5. Hennessy (1977): INAA
12. Cassidy (pers. comm. to P. C. Webb)
13. Storey and Lintern (1981)

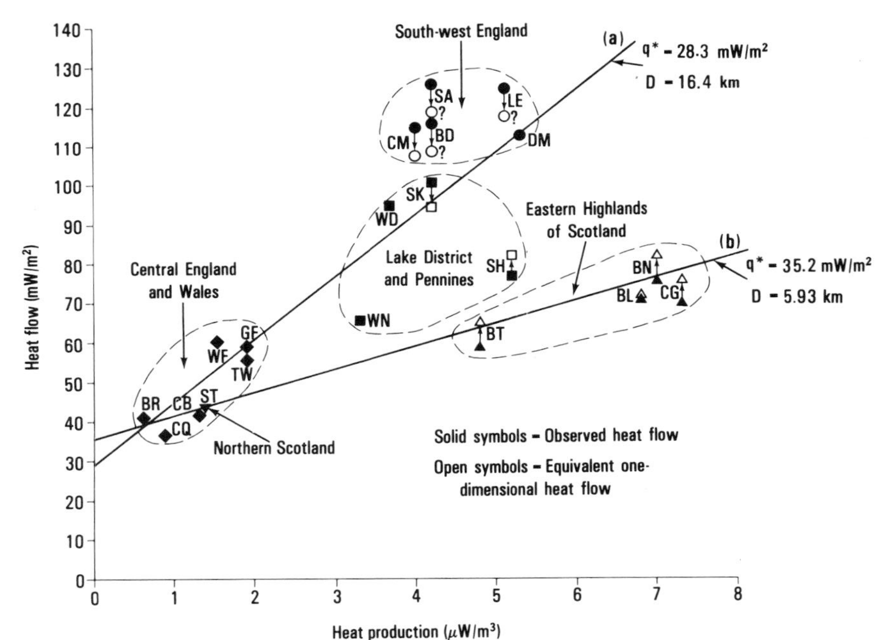

3.12 Relationship between heat flow and heat production for British granites and basement rocks. Compiled from data listed in Table 3.11. Line (a) is a linear regression of data from south-west England, the Lake District, the Pennines, Central England and Wales. Line (b) is a linear regression of data from the Eastern Highlands and northern Scotland. Both are based on the equivalent one-dimensional heat flow values listed in Table 3.11. Codes identify granites or boreholes as follows: CM = Carnmenellis Granite, BD = Bodmin Granite, LE = Lands End Granite, SA = St Austell Granite, DM = Dartmoor Granite, WD = Weardale Granite, WN = Wensleydale Granite, SH = Shap Granite, SK = Skiddaw Granite, CG = Cairngorm Granite, BT = Mount Battock Granite, BL = Ballater Granite, BN = Bennachie Granite, ST = Strath Halladale Granite, CQ = Croft Quarry Borehole, BR = Bryn Teg Borehole, CB = Coed-y-Brenin Borehole, GF = Glanfred Borehole, TW = Thorpe-by-Water Borehole, WF = Withycombe Farm Borehole.

radiothermal batholiths similar to the Cornubian, northern England and Eastern Highlands batholiths anywhere in the UK. However, there is some evidence, often equivocal, for a number of smaller granites in southern Britain, as for example near Market Weighton and around the Wash. These may be of some geothermal interest if their heat production is sufficiently high to raise the heat flow, and thus sub-surface temperatures within and beneath the overlying sedimentary strata, to values above those in surrounding areas.

THE RELATIONSHIP BETWEEN HEAT FLOW AND HEAT PRODUCTION IN BRITISH GRANITES

The existence of an empirical linear correlation between surface heat flow (q_0) and heat production (A_0) in granite and high-grade metamorphic terrain is discussed in Chapter 2. The relationship takes the form:

$$q_0 = q^* + A_0 D \qquad 3.1$$

where q^* is the heat flow from the lower crust and mantle, and D is a parameter with the dimension of length which is a function of the thickness of the upper crustal radioactive layer. Several different models for the distribution of radioactive elements with depth satisfy the linear relationship, the most favoured being an exponential decrease (extending at least to a depth of $3D$) of the form

$$A_z = A_0 e^{-z/D} \qquad 3.2$$

where z is the depth and A_z is the heat production at depth z.

The recent investigations in the Lake District and Eastern Highlands have added significantly to the number of heat flow and heat production measurements for British granites. The data, together with those for metasedimentary and metavolcanic basement given by Richardson and Oxburgh (1978), are summarised in Table 3.11. The equivalent one-dimensional heat flow (i.e. the observed heat flow corrected for refraction and lateral heat transfer) is given where this can be deduced from thermal models.

The data shown in Table 3.11 are plotted on a q_0-A_0 diagram in Figure 3.12. It is immediately apparent that no single linear correlation is appropriate to the UK as a whole and the data imply that, as a first approximation, there are two heat flow provinces in the UK, a southern province covering England and Wales and a northern 'Scottish' province (however, see further discussion below). Regression lines for the two 'provinces' have values of 28.3 mW/m² for q^* and 16.4 km for D for England and Wales, and 35.2 mW/m² and 5.9 km for the Scottish Highlands (based on the equivalent one-dimensional heat flow values shown in Table 3.11). The parameters for the southern 'province' are similar to those given by Richardson and Oxburgh (1978 and 1979). They pointed out that the value of D for southern Britain is considerably higher than values for heat flow provinces studied elsewhere in the world, most of which lie within the internal zones of orogenic belts. The relatively unfractionated heat production of the basement rocks was taken to reflect the low grade of metamorphism and relative paucity of igneous activity during the late Precambrian and Palaeozoic in England and Wales. In the case of the granite plutons, their emplacement into relatively cool country rocks, which limited the length of time in which fractionation could take place, was suggested as a possible cause of the relatively unfractionated heat production compared with petrographically similar granites elsewhere. The low value of 5.9 km for D for the Scottish Highlands contrasts sharply with the value for southern Britain and implies relatively strong fractionation of radioelements. The Scottish granites are emplaced within the internal zone of the Caledonian orogenic belt and the q_0-A_0 data are not inconsistent with those for such environments elsewhere (see for example the summary of worldwide data by Jaupart, 1983).

In recent years there has been considerable discussion about the meaning and validity of the empirical linear relationship between heat flow and heat production. In particular, England and others (1980) showed that when data from a series of isolated plutons of different heat

Table 3.11 Heat flow and heat production data for granites and basement rocks in the UK.

Granite	No. of heat flow measurements	Mean observed heat flow (a) (mW/m²)	Equivalent one-dimensional heat flow (b) (mW/m²)	Heat production (a) (μW/m³)
Southwest England				
Carnmenellis	−10	115 ± 7	108	4.0 ± 0.5 (d)
Bodmin	−5	116 ± 5	109?	4.2 ± 0.9 (d)
Land's End	−3	125 ± 3	118?	5.1 ± 0.2 (d)
St. Austell	−2	126 ± 0.5	118?	4.2 ± 0.9 (d)
Dartmoor	−6	113 ± 9	113	5.3 ± 0.5 (d)
Northern England				
Weardale	−1	95	95	3.7 (e)
Wensleydale	−1	65	65?	3.3 (e)
Shap	−1	78	82	5.2 (f)
Skiddaw	−1	101	95	4.2 (f)
Scotland				
Cairngorm	−1	70	76	7.3 (f)
Mount Battock	−1	59	65	4.8 (f)
Ballater	−1	71	72	6.8 (f)
Bennachie	−1	76	82	7.0 (f)
Strath Halladale (c)	−1	43	43	1.4 (e)
Basement in Central England and Wales (see Richardson and Oxburgh (1978) for source data)				
Croft Quarry (Quartz monzonite)	−1	37	37	0.9
Bryn Teg (Late Precambrian)	−1	41	41	0.6
Coed-y-Brenin (Ordovician)	−1	42	42	1.3
Glanfred (Silurian)	−1	59	59	1.9
Thorpe-by-Water (Lower Palaeozoic)	−1	57	57	1.9
Withy Combe Farm (Lower Palaeozoic)	−1	60	60	1.5

Notes

(a) Standard deviations refer to the mean of mean values for each borehole

(b) Equivalent one-dimensional heat flow values were derived as follows:

(i) The Shap, Skiddaw, Cairngorm, Mount Battock, Ballater and Benachie granites from the thermal models given by Wheildon and others (1984a and b)

(ii) The Carnmenellis Granite from the model given by Wheildon and others (1980). The Bodmin, Land's End and St Austell granites are of similar size and the observed heat flow is assumed to be enhanced by the same amount as at Carnmenellis (6%). The Dartmoor Granite is larger and the average observed value is assumed to be equivalent to the one-dimensional heat flow.

(iii) The Weardale and Wensleydale granites from models by England and others (1980).

(iv) All other observed values are assumed to represent one-dimensional heat flow.

(c) The Strath Halladale Granite is in northeast Scotland.

(d) Heat production values are the mean of the data from the 100 m heat flow boreholes only. Values were derived from borehole chippings and probably underestimate the true heat production.

(e) Heat production values from borehole core.

(f) Preferred heat production value for the intrusion based on borehole and surface material (Webb and Brown, 1984a and b)

production are plotted on a q_0-A_0 diagram, positive heat production and conductivity contrasts have the effect of elevating the apparent value of q^* and reducing the apparent value of D. Jaupart (1983) suggested that the linear relationship itself may be partially a function of horizontal heat transfer between blocks of contrasting heat production, with the deepest radioactivity variations smoothed out and included in the background heat flow. The net result is a line whose slope is some average of the various thicknesses involved and which appears as a series of relatively thin plutons on a relatively high background.

In addition to these general doubts about the interpretation of the q_0-A_0 relationship, it is instructive to look at the distribution of q_0-A_0 data in Britain. In the southern province the data relate to three discrete areas: the Cornubian batholith of Hercynian age; the Lake District and Pennine (Caledonian) granites; and the metasediments, metavolcanics and early Caledonian intrusions of the English Midlands and Wales. Each of these areas is characterised by its own distinct tectonic and thermal history; in none of them is there a sufficient range of data points to define a reliable individual linear relationship between q_0 and A_0, and none of the regional data subsets overlap. Although they may have certain characteristics in common, it is questionable whether a single regression line should be applied to all three areas. In the northern province the data relate mainly to the Eastern Highlands batholith, with a single point from the Strath Halladale Granite in northern Scotland.

Although a distinction between granites which are fractionated and relatively unfractionated in heat production on either side of the Iapetus Suture is apparent from Figure 3.12, it is probably invalid to define two British 'heat flow provinces' with characteristic values of q^* and D. Rather the q_0-A_0 data should be explained in terms of the characteristics of the *tectonic* provinces represented by the four regional subsets, or more especially in terms of the characteristics of the individual granite batholiths within those tectonic provinces.

In a geological context the surface heat flow (q_0) can be thought of as the sum of three components: (1) the heat flow from beneath the crust which might vary from tectonic province to tectonic province due to factors such as age since the last thermal event and lithospheric thickness (see for example Sclater and others, 1980), (2) the heat flow contribution from heat production within the lower crust, which might vary on a regional scale, and (3) the heat flow contribution from heat production within the upper crust, which might vary on a scale of a few kilometres. The revised heat flow map of Britain (Figure 2.8 and Plate) reflects the variations in these components.

Each of the regional q_0-A_0 subsets can be explained in terms of a number of plausible models with different values of mantle heat flow and different vertical distributions of heat production within the upper and lower crust. The likely range of models in the UK can be considered by comparing the Eastern Highlands and the Cornubian batholiths. The published models of the Cornubian batholith (Wheildon and others, 1980; and Figure 2.9) and the Eastern Highlands batholith (Wheildon and others, 1984b) explain the contrasting q_0-A_0 data mainly in terms of different vertical distributions of heat production within the granites themselves. However, they were not designed to investigate regional differences in mantle heat flow and lower-crustal heat production, both of which could possibly be less in the Eastern Highlands than in south-west England. A reduced mantle heat flow would be consistent with the greater (Caledonian) age of the Scottish province, and a radioelement depleted lower crust would be consistent with seismic evidence (Bamford and others, 1978) which shows a mid-/lower-crustal layer beneath the Scottish Highlands, identified as granulite facies Lewisian basement.

The extent of possible regional differences in mantle heat flow and/or lower-crustal heat production between the Eastern Highlands and south-west England are illustrated by examining the consequences of assuming the *same* vertical distribution of heat production in both batholiths. Models based on an exponential decrease in heat production to a depth of 15 km, with a scale length of 10 km in both cases, imply components of heat flow from beneath 15 km of 23 mW/m² in the Eastern Highlands and 74 mW/m² in south-west England (assuming representative surface heat flow values of 77 and 113 mW/m² respectively, and representative surface heat production values of 7 and 5 μW/m³ respectively). Regional differences of this magnitude would be expected to feature on the heat flow map (Figure 2.8 and Plate) as areas of contrasting background heat flow away from the influence of the batholiths. The heat flow field in the Scottish Highlands is poorly defined and the background may be up to 20 mW/m² less than in south-west England. However, there is no evidence for a difference of over 50 mW/m².

An alternative explanation in terms of a large difference in depth extent between the two batholiths is not supported by the gravity evidence. Interpretations suggest that the Eastern Highlands batholith extends to a depth of about 13 km (Rollin, 1984), while the Cornubian batholith extends to a depth of 9 km in the west and 20 km in the east (Tombs, 1977, and Figure 3.4). The variation in depth extent along the Cornubian batholith is not matched by any corresponding variation in heat flow (see Figure 3.4) and an average depth extent of about 14 or 15 km is probably appropriate. Given the inherent uncertainty in gravity interpretations, both batholiths can be said to merge geophysically with the basement at mid-crustal depths.

In geological terms it must be concluded that, provided there is not some additional mechanism responsible for the elevated heat flow over the Cornubian batholith, such as convective fluid flow within or beneath the batholith, the contrast between the Eastern Highlands and Cornubian batholiths must be accounted for by somewhat more fractionated heat production within the Eastern Highlands batholith and perhaps slightly reduced values of mantle heat flow and lower-crustal heat production in the Scottish Highlands.

In northern England the q_0-A_0 plot (Figure 3.12) shows that for a given heat production the heat flow is considerably higher in the Skiddaw and Weardale granites than in the Shap and Wensleydale granites. It is difficult to argue for any significant variation in either lower-crustal heat production or mantle heat flow within the area covered by the four sites. Gravity interpretations suggest that the Weardale and Lake District granites merge with the basement at similar depths of about 9 km (Bott, 1967 and Lee, 1984a) while the Wensleydale Granite may extend only to 4 km (Wilson and Cornwell, 1982). A shallow depth may account for the relatively low heat flow over the Wensleydale Granite but a difference in the vertical distribution of heat producing elements seems a more likely explanation for the contrast between the Shap and Skiddaw granites (Webb and Brown, 1984a; Wheildon and others, 1984a; and Lee and others, 1984). In short, the northern England q_0-A_0 data can be said to reflect the geological, geochemical and geophysical diversity of the Weardale, Wensleydale and Lake District batholiths.

In central England and Wales there are no high heat flow or high heat production values (Table 3.11). The data fit a regression line with values of 26.8 mW/m² for q^* and 16.7 km for D, which are consistent with the view put forward by Richardson and Oxburgh (1978) that heat production within the upper part of the crust in central England and Wales is less fractionated than in axial zones of orogenic belts characterised by high grade metamorphism and magmatism. A single step interpretation of the linear relationship between q_0 and A_0 gives a minimum value for the heat flow at a depth of 16.7 km of 26.8 mW/m².

The relevance of this discussion to HDR potential lies in the use of the q_0-A_0 relationship to predict heat flow in possible target areas. The assumption of a linear correlation with an inappropriate value of D can lead to serious errors; for example a value for D of 16.4 km would imply heat flow values in the range 107 to 148 mW/m² in the Eastern Highlands, whereas the measured values lie in the range 59 to 76 mW/m².

In south-west England the heat flow anomaly has been thoroughly studied and the problem of predicting heat flow values no longer exists. In the Eastern Highlands of Scotland the four new measurements of q_0 and A_0 should give a reasonably reliable guide to heat flow values elsewhere within the batholith. However, in northern England the range of possible heat flow values for a given heat production is large and it would still be difficult to predict with any certainty the heat flow elsewhere in the region; for instance, in the western part of the Lake District batholith. The prediction of heat flow from heat production values in other parts of Britain is even more uncertain. The diversity of the existing q_0-A_0 data suggests that granites in other areas, such as the Southern Uplands and Western Highlands of Scotland, might also be characterised by their own particular relationship between q_0 and A_0.

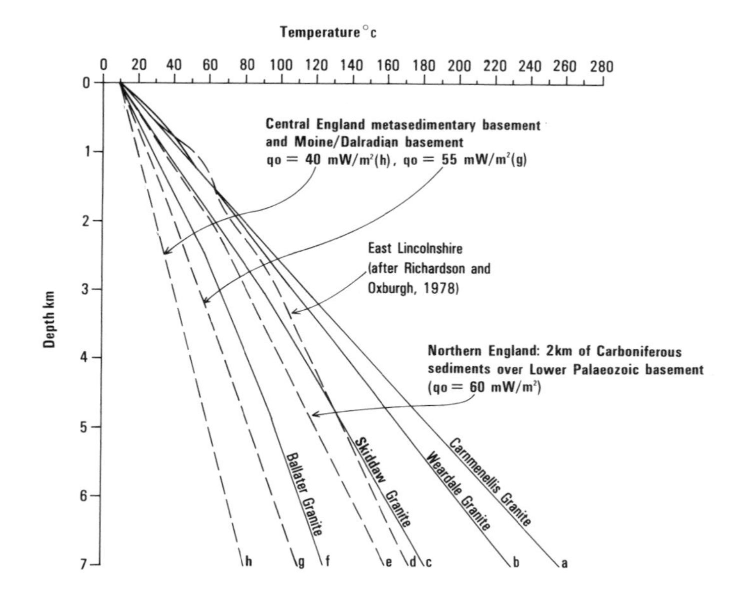

3.13 Comparison of predicted sub-surface temperature profiles for British granites and basement rocks. Profile d for East Lincolnshire is that given by Richardson and Oxburgh (1978). All other profiles are as previously shown in Figures 3.5, 3.8 and 3.11 calculated as described in the text.

OVERVIEW OF HDR POTENTIAL IN BRITAIN

Major granite batholiths

The heat flow boreholes in the Lake District and Eastern Highlands granites have shown that both areas are characterised by above average values of heat flow. In neither case are the values as high as those observed over the Cornubian batholith, and the HDR potential is consequently less than in south-west England. In both regions, however the heat flow is sufficiently elevated to raise sub-surface temperatures above those expected in the surrounding basement rocks.

The Cornubian batholith is characterised by the highest heat flow values in the UK and remains the area of greatest HDR potential. The heat flow is equally high, generally in excess of 110 mW/m², over all five major outcrops. The uniformity of the heat flow values, their correlation with the space form of the batholith, and the observed temperature of 79°C at a depth of 2 km in the Carnmenellis Granite, all tend to support a conductive heat transfer mechanism in which high surface heat flow results from the presence of fairly uniform radiothermal granite extending to mid-crustal depths. The predicted temperature profile for the Carnmenellis Granite suggests that a temperature of 150°C might be reached at a depth of about 4 km and a temperature of 200°C at about 5.4 km.

The Weardale batholith is the most favourable of the northern England granites and is the best site outside south-west England. Predicted sub-surface temperatures, based on a single heat flow value from the Rookhope Borehole, are lower than those for the Cornubian batholith but considerably higher than those elsewhere in the UK, in either granites or basement rocks (Figure 3.13). A temperature of 150°C might be reached at a depth of about 4.4 km and 200°C at about 6 km.

The evidence of the Shap and Skiddaw boreholes suggests that the HDR potential of the Lake District batholith is less than that of the Weardale batholith. However, sub-surface temperatures are well above those predicted for basement rocks over most of the UK and comparable with those in basement rocks under the most favourable circumstances (see below).

The thermal anomaly over the Weardale-Lake District region has not been studied in as much detail as that over the Cornubian batholith and the predicted sub-surface temperature profiles are therefore, less well constrained. Additional heat flow boreholes are required, in particular over the Weardale batholith and the western Lake District, before the HDR potential can be fully assessed.

Radiothermal enhancement of background heat flow through the Eastern Highlands granites is insufficient to raise the surface heat flow to values as high as those observed over the Cornubian and northern England granites. Predicted gradients are higher than those in the surrounding basement but in absolute terms the area has little potential for the application of HDR technology. Sub-surface temperatures are comparable with those predicted for much of the English basement.

Normal basement

The HDR programme has been concerned mainly with the potential of the major granites. However, it is useful to compare predicted sub-surface temperatures with those expected in the Lower Palaeozoic and Precambrian basement elsewhere in Britain (Figures 8.2 and 10.1). Where basement rocks are close to the surface, in areas of normal heat flow, their relatively high thermal conductivity results in low geothermal gradients. This is illustrated by profiles g and h (Figure 3.13) which represent outcropping metasedimentary basement in central England and Wales for heat flow values of 55 and 40 mW/m^2 respectively. In both cases a heat production of 2.22 $\mu W/m^3$ and a thermal conductivity of 3.52 W/mK are assumed (after Richardson and Oxburgh, 1978). These are similar to values of 1.7 $\mu W/m^3$ and 3.5 W/mK respectively quoted by Wheildon and others (1984b) for the Moine/Dalradian basement of Scotland, and the predicted temperature profiles are therefore almost identical in both areas.

In the major Mesozoic sedimentary basins of southern Britain the comparatively low thermal conductivity of the sedimentary rocks raises gradients within the sediments to values similar to those observed in the Cornubian and Weardale granites. This is illustrated by the predicted profile for east Lincolnshire shown in Figure 3.13. The profile is that given by Richardson and Oxburgh (1978) and represents the most favourable case for normal basement overlain by sedimentary rocks in central Britain. The model assumes a heat flow of 63 mW/m^2, basement of conductivity 3.52 W/mK and heat production 2.22 $\mu W/m^3$, overlain by 1000 m of Upper Coal Measures and 1750 m of Permian and younger sediments. Observed gradients in the Wessex basin are similarly high; the average gradient in the Marchwood Well, where the measured heat flow is 61 mW/m^2, is about 34°C/km to the base of the Mesozoic (at a depth of 1725 m). The gradient within the underlying Devonian rocks is lower (less than 20°C/km) due to their higher thermal conductivity; the temperature at the base of the borehole, at 2615 m, is about 86°C (Burgess and others, 1981).

From a purely thermal viewpoint, the relative merits of radiothermal granites and normal basement beneath thick low conductivity sedimentary rocks, as HDR targets, depend largely on the target temperature. Temperatures up to about 100°C can be reached within a depth of about 3 km in both geological environments (Figures 3.13 and 8.4), but the Mesozoic basins have the advantage of being located closer to the main population centres of Britain where the demand for low grade geothermal heat is greatest. For higher target temperatures of say 150 to 200°C for power generation, then the Cornubian and Weardale granites, and possibly the Lake District batholith, offer the best prospects in the UK.

Chapter 4 Mesozoic Basins

I. F. Smith

INTRODUCTION

Successful exploitation of low enthalpy geothermal energy requires the presence of permeable rocks at depths where temperatures are suitable for economic development. In the UK such favourable situations exist in the deeper parts of the post-Carboniferous sedimentary basins, which are often referred to as the Mesozoic basins even though the oldest sediments they contain are Permian in age. A general examination of the sediments preserved in these basins reveals that, although several significant aquifers occur, those with the greatest potential are in the Permo-Triassic sandstones, simply because these aquifers occur at the greatest depth and they tend to have favourable aquifer properties.

There are six Mesozoic basins (Figure 4.1) containing significant thicknesses of Permo-Triassic rocks, which have been investigated to assess their geothermal potential. They are the East Yorkshire and Lincolnshire Basin (Gale and others, 1983), the Wessex Basin (Allen and Holloway, 1984), the Worcester Basin (Smith and Burgess, 1984), the Cheshire and West Lancashire basins (Gale and others, 1984a), the Carlisle Basin (Gale and others, 1984b) and the basins in Northern Ireland (Bennett, 1983). The following review of the geology and hydrogeology of these basins is based on these studies. Because the Permo-Triassic sandstones represent the best prospects for development, attention is concentrated on these rocks, but other Mesozoic sequences are discussed where appropriate. Before reviewing the individual basins, the general geological history of the Mesozoic is summarised to emphasise the variability shown by some of the deposits. The factors which are important in identifying suitable aquifers are also discussed briefly.

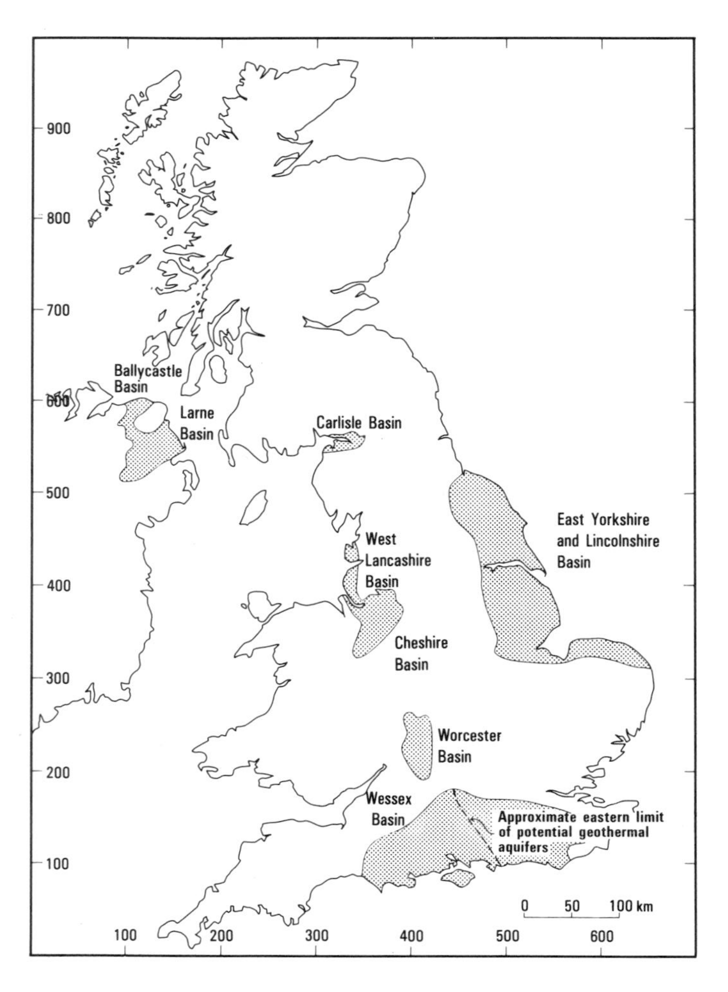

The principal objectives of the various basin studies were:

- to identify potential geothermal aquifers
- to define the three-dimensional structure of the aquifers
- to estimate the temperature of fluids within the aquifers
- to determine the properties of the aquifers and the fluids they contain, such as permeability, porosity, transmissivity, salinity and piezometric head

and hence to evaluate the geothermal resources.

GEOLOGICAL HISTORY OF THE POST-CARBONIFEROUS SEDIMENTARY BASINS

During the last stages of the Variscan (Hercynian) orogeny, in late Carboniferous times, a phase of uplift affected Britain, with consequent subaerial erosion in a desert-like environment. Simultaneously, in response to tensional stresses, local downwarps developed in many cases along pre-existing fault-lines, and these were separated by relatively elevated terrain. Other downwarps were caused by gentle flexuring, perhaps related to crustal thinning. These basins continued to develop throughout the Permian and Triassic periods and they gradually filled with sediments, including the thick sandstones at the base, which are the subject of much of this chapter.

The Permo-Triassic deposits in the various basins have been given different names for coeval rocks and the usage adopted here is shown in Figure 4.2.

The initial products of the erosion of the highlands were coarse breccias and sandstones, which were swept into the

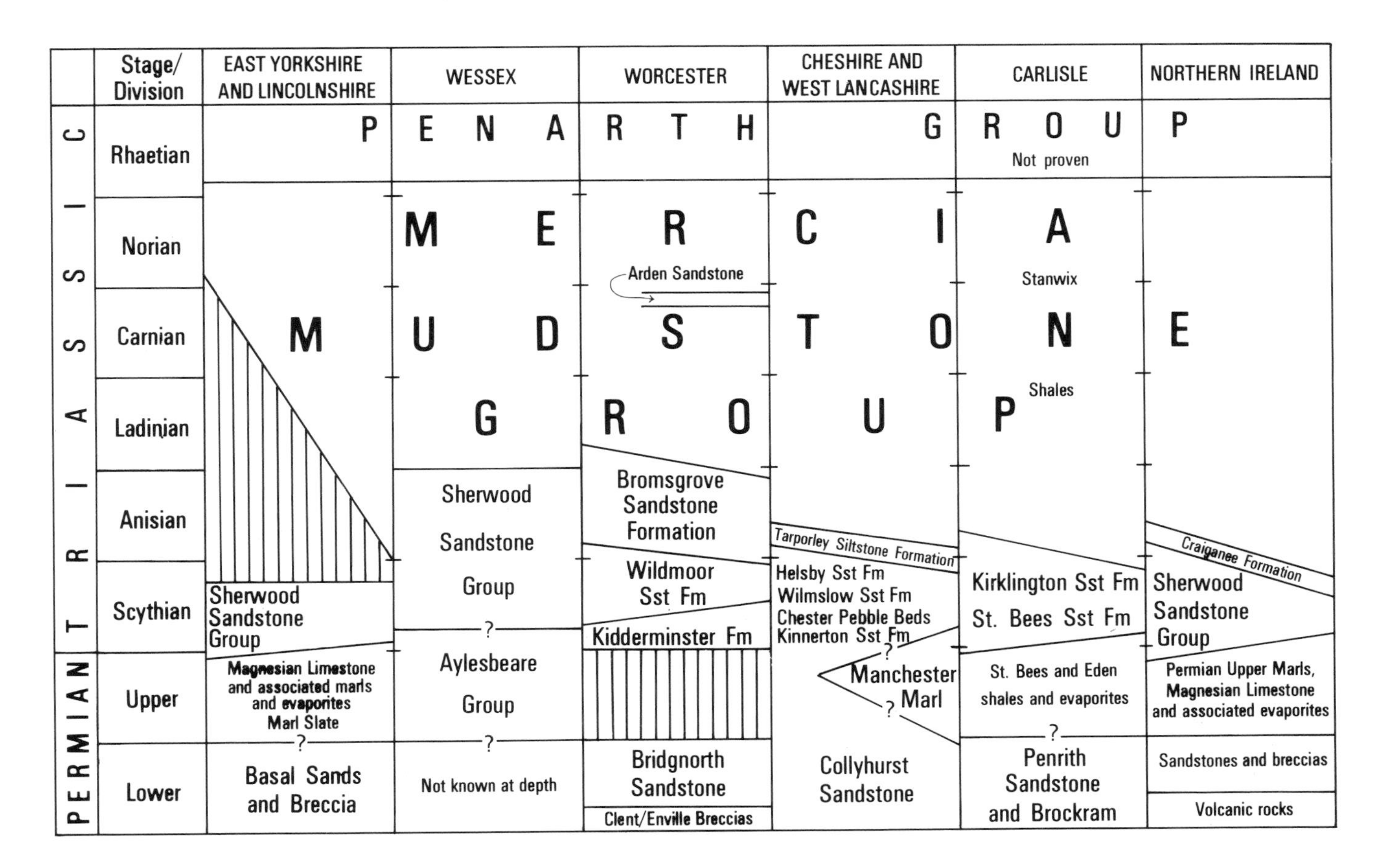

4.2 Generalised correlation of stratigraphical nomenclature for the major Permo-Triassic basins, based on G. Warrington (personal communication).

basins as flood deposits. They are concentrated along the basin margins but occur impersistently and variably over much of the land surface that existed at that time. They recur locally throughout the Permian and Triassic sequences. Volcanic rocks are interbedded locally with these breccias and presumably result from eruptions associated with the tensional stress regime. The breccias are overlain by coarse-grained, well-sorted, cross-bedded sandstones, which have been interpreted as aeolian dunes, and which merge laterally into water-laid deposits. Together these sediments attain thicknesses of several hundreds of metres and they tended to smooth out the pre-Permian topography.

Upper Permian rocks are represented commonly by marine deposits, laid down as a result of marine transgression across a Lower Permian landscape of low relief. The deposits comprise complex and variable formations of limestones, dolomites and evaporites which grade upwards and laterally towards the margins of the basins into continental marls and sandstones. Analysis of these sediments suggest that they result from at least five major periods of marine transgression and regression.

The end of the Permian and the early Triassic period heralded a return to a continental environment. The basins initiated in the Permian continued to subside and thick clastic deposits accumulated and spread gradationally and diachronously across the older rocks; these sandstones are of Scythian and Anisian age and are called the Sherwood Sandstone Group. They are largely of fluviatile origin, but locally wind-blown deposits, marls and breccias occur, suggesting different local depositional environments, typical of modern desert basins. The succession thins against older uplifted areas, such as the London Platform and the Pennines. A number of cycles of gradational grain-size occur within the sequence and as a whole the grain-size decreases upwards. During the late Scythian, a depositional break, which may be correlated with the Hardegsen Disconformity of Germany, can be identified in many parts of England. Following this break, to the east of the Pennines, thin conglomerates were overlain by red marls with evaporites, including dolomite, gypsum and halite. Elsewhere in England the fluviatile sandy facies continued for a period into the Anisian.

Subsequently the partly marine red mudstone and siltstone deposits of the Mercia Mudstone Group covered the sandstone. Halite beds and persistent sandstones (or skerries), such as the Arden Sandstone, are common and quite widespread in this group.

During the closing stages of the Triassic (the Rhaetian) open marine conditions were established, with a shallow shelf sea depositing mudstones and limestones. This led into the Jurassic period, during which time predominantly fine-grained clastic sediments were formed. Limestones occur widely but are relatively thin, and sandstones tend to be local in development. The rocks were formed in shallow subsiding basins, divided by broad 'swells' over which thin or condensed sequences occur.

The seas were warm, relatively shallow, and at times with restricted circulation, producing black mudstones in the Lower Jurassic. During this period there is evidence of tectonic warping, with the development of diachronous sandy formations in the south of England, such as the Bridport Sands Formation of the Wessex Basin.

The Middle Jurassic rocks are very variable across England. In the south limestones, such as the Inferior and Great Oolite, are well-developed, particularly in the Cotswold Hills, and are known at depth in the Wessex Basin, where they merge into an argillaceous succession. The limestones were probably deposited in shallow shoal or reef areas, whilst at greater sea-depths the clays of the Fuller's Earth were formed. Farther north, in the East Midlands and Yorkshire, the Middle Jurassic rocks contain mudstones, siltstones and sandstones interbedded with limestones which formed on coastal flats adjacent to the London Platform. In north Yorkshire rocks of this age mainly originated as fluvial deposits on coastal plains or as deltaic deposits.

The Upper Jurassic rocks are laterally more uniform, but include, as well as the thick clay deposits of the Oxford and Kimmeridge clays, limestones and sandstones of the Cornbrash, Corallian and Portland beds. The clays were deposited in broad, shallow seas adjacent to the London Platform, with subtle changes in environment allowing the development of the shelly limestones.

Similar types of conditions continued into the Lower Cretaceous, when mudstones and sandstones were deposited in several basins in coastal and deltaic environments, with periods of non-deposition and local emergence of low-lying landmasses. Such conditions produced laterally and vertically variable deposits. The periods of emergence are correlated with the sandy facies.

The final episode of Mesozoic deposition is the Upper Cretaceous, when the Chalk was deposited. This is a clean, white limestone, in which whole or broken coccoliths make up the bulk of the rock; debris of bivalve shells is also an important constituent. Much of Britain may have been covered by the Chalk, which has been removed subsequently by erosion. Minor flexuring during the Tertiary era moulded the Chalk into broad synclinal structures.

AQUIFERS WITH GEOTHERMAL POTENTIAL

The combination of an understanding of the nature of those formations at or near the surface which are water-bearing, and studies in deep boreholes drilled mainly for hydrocarbon exploration, has led to the identification of aquifers which might be suitable for the production of large quantities of hot water from depth. It has long been recognised that the porosity and permeability of a formation vary with depth of burial and distance from outcrop, so that an aquifer, although widely known and accepted as such near the surface, will not necessarily behave in a similar way at greater depths. Near the surface fissure permeability tends to be a major contributor to the transmissivity of aquifers, but because at depth fissures tend to close, intergranular permeability then becomes relatively more important. A consequence is that extrapolation of surface parameters to greater depths is not straightforward and the results of flow tests at depth show significant differences from those which might be expected by extrapolation of surface values.

Guidelines for the production rate from aquifers for viable geothermal schemes indicate that fluids must be produced at a rate of 25 to 50 l/s at temperatures of at least 40°C, although in the future even 20°C may be acceptable. An intrinsic transmissivity of at least 5 darcy metres (D m) and generally more than 10 D m is necessary (see chapters 9 and 10). The objective of geothermal investigations is to identify aquifers which have properties exceeding these minima; such aquifers are the Basal Permian Sands and Breccia, the Sherwood Sandstone Group, and also the Lower Greensand in a limited area along the south coast of England. Obviously the guidelines are flexible, for an aquifer with a large transmissivity at a lesser depth (and hence lower temperature) contains as much enthalpy as the converse deeper (higher temperature) aquifer with a lower transmissivity. However, an important factor is that heat energy cannot be extracted so efficiently from fluids at a lower temperature.

Clearly, effective development of hot groundwaters requires the presence of water-bearing beds at depths where appropriate temperatures occur. The aquifers should also be overlain by a confining bed of low thermal conductivity and the base of the confining bed should be at the depth where the minimum required temperature occurs. This avoids the risk of vertical downward movement of cold water into zones producing hot water in the immediate vicinity of a well. It is, of course, a risk that should also be guarded against on a regional scale so far as this can be anticipated.

EAST YORKSHIRE AND LINCOLNSHIRE BASIN

The East Yorkshire and Lincolnshire Basin is the on-shore extension of the Southern North Sea Basin. The structure is of post-Hercynian age and includes a sequence ranging in age from Permian to Cretaceous. Along the east Yorkshire coast, the base of the Permian has been buried to depths of more than 2 km. At such depths, with an average geothermal gradient, the temperature may be expected to be about 60°C, and any water held in permeable rocks would represent a geothermal resource. Aquifers at depths of more than 500 m, with temperatures of more than 20°C, could be of potential value as a resource and on this basis most of the basin is of interest from the point of view of geothermal potential (Gale and others, 1983).

General geology

A simplified geological map is shown in Figure 4.3 and the post-Carboniferous sequence in the basin is given in Table 4.1. The principal aquifers are the Permo-Triassic sandstones and limestones, the Lincolnshire Limestone, the Corallian in Yorkshire, the Lower Cretaceous/Upper Jurassic Spilsby Sandstone in Lincolnshire, and the Chalk.

The earliest Permian deposits are the continental Basal Permian Sands and Breccia. Near the outcrop these are discontinuous but towards the coast they become continuous and increase in thickness. Successive carbonate-

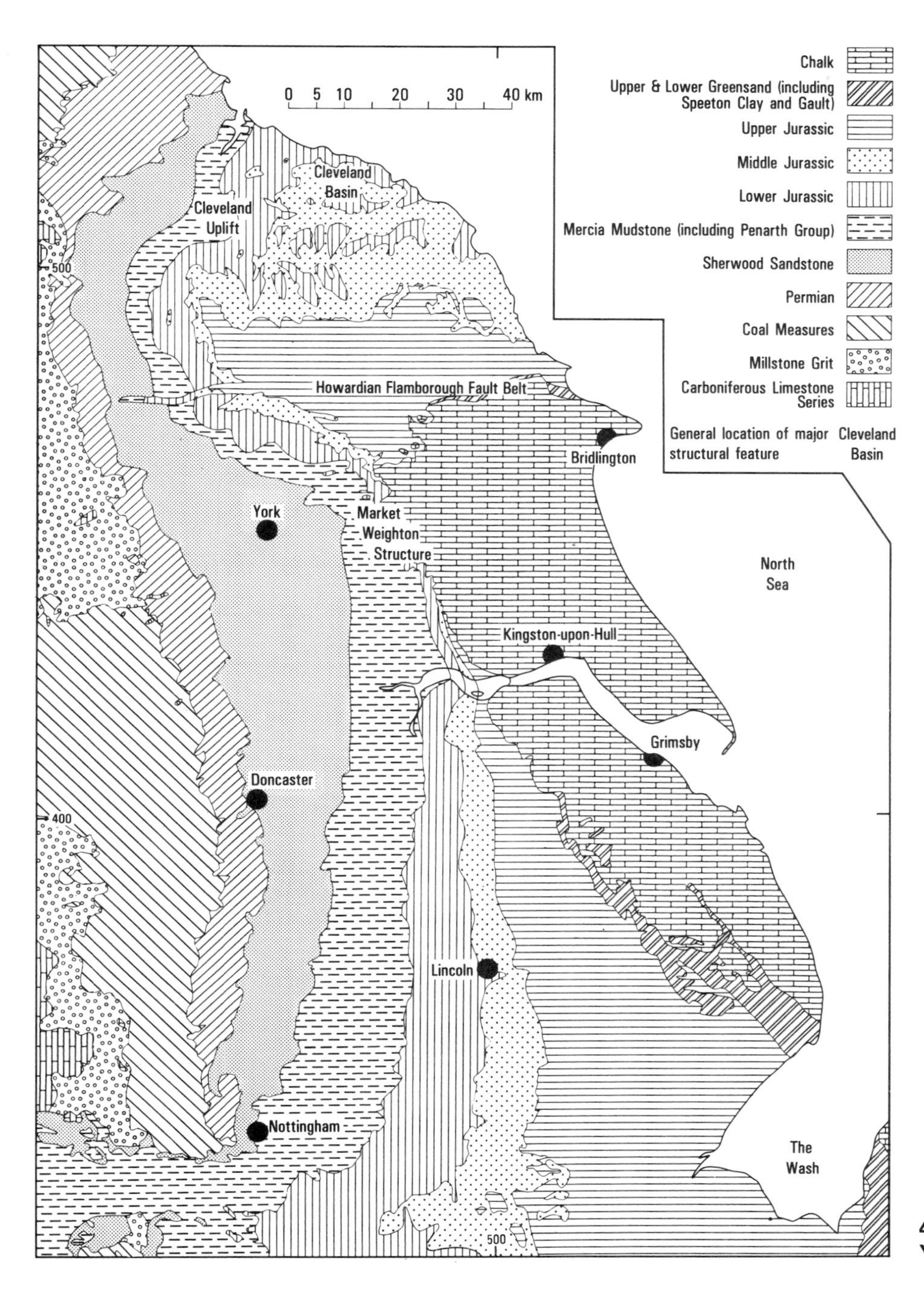

4.3 Geological map of the East Yorkshire and Lincolnshire Basin.

evaporite sequences have been recognised in the basin and arise from repeated transgressions and regressions of the Zechstein Sea. The carbonate bodies — the Lower Magnesian Limestone, the Kirkham Abbey Formation and the Upper Magnesian Limestone — are wedges of limestone and dolomite which thin towards, and extend progressively further into, the centre of the basin (Smith, 1974). They are interbedded with marls and evaporite deposits. Reef deposits also formed locally over much of the area when the Lower Magnesian Limestone and Kirkham Abbey Formation were forming.

On the southern margin of the basin, adjacent to the ancient landmass of the London Platform, continental redbeds were deposited continually throughout Permian times and this persisted into the Triassic (Smith, 1974), but there is little information available about these restricted deposits of sandstone.

The Zechstein Basin was filled by the beginning of the Triassic, and the Permian marine sediments were covered by a thick layer of red fluvial sandstones and conglomerates, known collectively as the Sherwood Sandstone Group. Near the basin edge, the 'Lower Mottled Sandstone', which is partly of Permian age, is covered by the 'Bunter Pebble Beds', which pass progressively basinwards into sandstone and argillaceous sandstone. The Sherwood Sandstone Group is overlain by the Mercia Mudstone Group, which consists largely of argillaceous and evaporite deposits; localised arenaceous deposits occur adjacent to ancient land areas (Warrington, 1974) and they are here incorporated with the Sherwood Sandstone Group.

The Penarth Group was deposited as a result of a marine transgression which gave rise to open marine conditions. This continued into the Jurassic period, and mainly argillaceous and calcareous rocks with little permeability were deposited until the end of the Lower Jurassic. During the Lower Jurassic, differential uplift occurred (particularly

Table 4.1 Geological succession in the East Yorkshire and Lincolnshire Basin

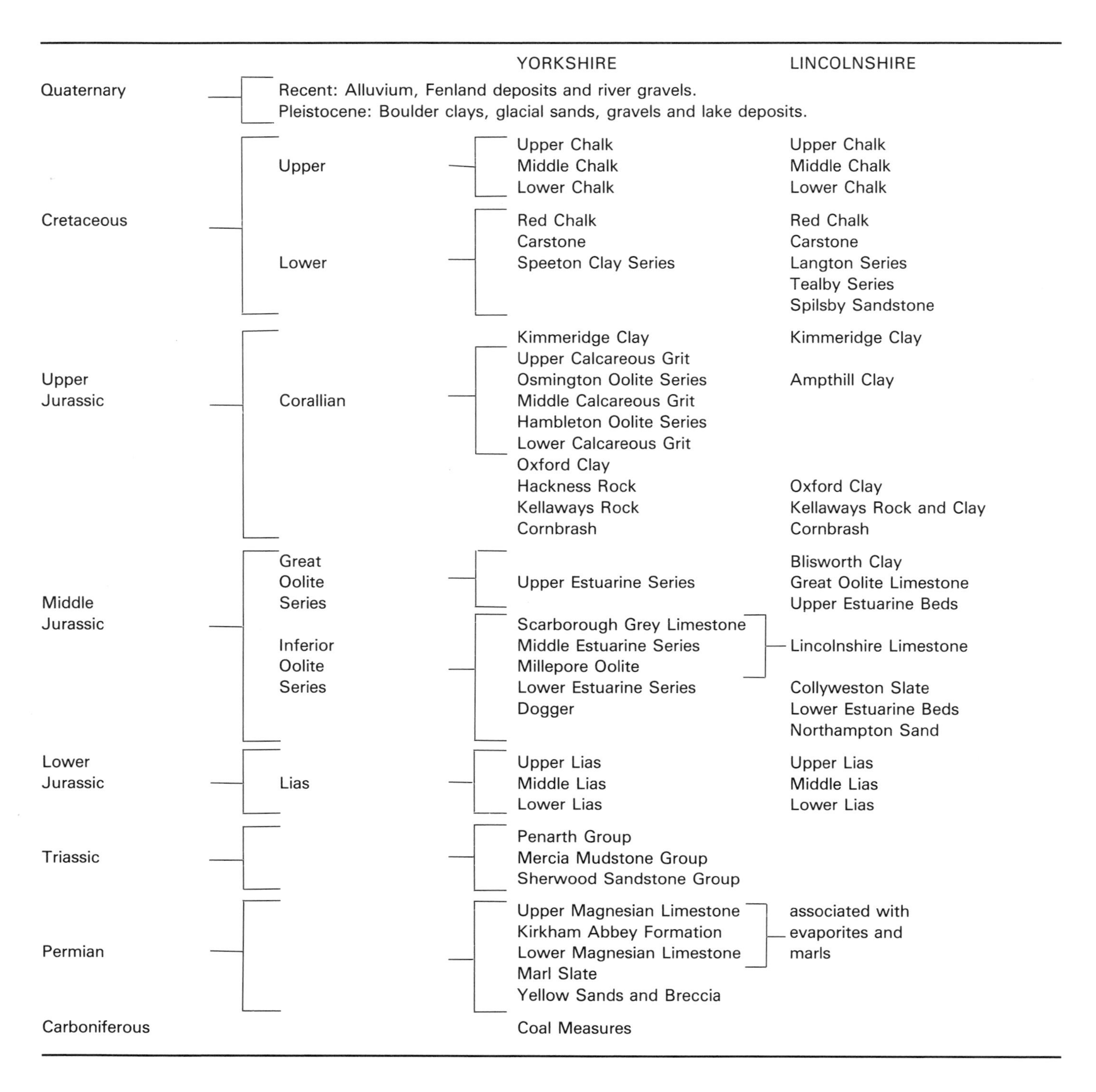

		YORKSHIRE	LINCOLNSHIRE
Quaternary	Recent: Alluvium, Fenland deposits and river gravels.		
	Pleistocene: Boulder clays, glacial sands, gravels and lake deposits.		
Cretaceous	Upper	Upper Chalk	Upper Chalk
		Middle Chalk	Middle Chalk
		Lower Chalk	Lower Chalk
	Lower	Red Chalk	Red Chalk
		Carstone	Carstone
		Speeton Clay Series	Langton Series
			Tealby Series
			Spilsby Sandstone
Upper Jurassic		Kimmeridge Clay	Kimmeridge Clay
	Corallian	Upper Calcareous Grit	
		Osmington Oolite Series	Ampthill Clay
		Middle Calcareous Grit	
		Hambleton Oolite Series	
		Lower Calcareous Grit	
		Oxford Clay	
		Hackness Rock	Oxford Clay
		Kellaways Rock	Kellaways Rock and Clay
		Cornbrash	Cornbrash
Middle Jurassic	Great Oolite Series		Blisworth Clay
		Upper Estuarine Series	Great Oolite Limestone
			Upper Estuarine Beds
	Inferior Oolite Series	Scarborough Grey Limestone	Lincolnshire Limestone (Scarborough Grey Limestone, Middle Estuarine Series, Millepore Oolite)
		Middle Estuarine Series	
		Millepore Oolite	
		Lower Estuarine Series	Collyweston Slate
		Dogger	Lower Estuarine Beds
			Northampton Sand
Lower Jurassic	Lias	Upper Lias	Upper Lias
		Middle Lias	Middle Lias
		Lower Lias	Lower Lias
Triassic		Penarth Group	
		Mercia Mudstone Group	
		Sherwood Sandstone Group	
Permian		Upper Magnesian Limestone	associated with evaporites and marls (Upper Magnesian Limestone, Kirkham Abbey Formation, Lower Magnesian Limestone)
		Kirkham Abbey Formation	
		Lower Magnesian Limestone	
		Marl Slate	
		Yellow Sands and Breccia	
Carboniferous		Coal Measures	

along the Market Weighton axis) and the shore-line moved towards the Yorkshire area. Hemingway (1974) has described how deltaic facies were developed between minor incursions of the sea in the area north of the Humber. The Middle Jurassic marine deposits south of the Humber are the Inferior and Great Oolite series, which consist of massive limestones and brackish and non-marine sandstones (Kent, 1980). They thin northwards against the Market Weighton Block. Their equivalent north of the uplift are deltaic deposits which thicken into the Cleveland Basin. The Upper Jurassic in Lincolnshire is largely argillaceous but the sandy Kellaways Beds, which are up to 17 m thick, occur at the base. The Corallian Series of oolitic limestones and calcareous grits to the north of the Market Weighton Block is equivalent to the Ampthill Clay to the south of the block. Gray (1955) considered that a sandy limestone found in Yorkshire immediately south of the block is a local equivalent of the Corallian.

The Lower Cretaceous rests unconformably on the Jurassic. South of the Market Weighton Block it is represented by the Spilsby Sandstone and an overlying sequence of clays, ironstones, limestones and sandstones. Thin sands which are probably of Lower Cretaceous age occur on the block while, in the southern part of the Cleveland Basin, the Speeton Clay is of this age. The sequence is overlain unconformably by the Red Chalk and the Upper Cretaceous

Chalk. Along the east coast the Chalk is obscured by Quaternary deposits.

A major fault system, referred to as the Howardian-Flamborough Fault Belt, and which continues below the North Sea as the Dowsing Fault Line, separates the Cleveland Basin from the more stable area to the south. It is partly of post-Cretaceous age, but the initial movements were in the Jurassic, if not pre-Permian (Kent, 1975).

Geothermal potential

The physical properties of all the principal aquifers in the basin were considered by Gale and others (1983) but only the Sherwood Sandstone Group and the Basal Permian Sands and Breccia were shown to have significant potential within the currently accepted criteria and with available technology. The potential of other aquifers may be worth reviewing if the economics and technical aspects alter, but at this stage only these two major aquifers are discussed.

Sherwood Sandstone Group

A maximum thickness of more than 500 m of Sherwood Sandstone is attained to the south of Bridlington (Figure 4.4). Conglomerates represent parts of the Sherwood Sandstone Group in the south west. In part of this area, a marginal sandstone facies (previously referred to as the Keuper Waterstones), which has a low permeability, overlies and is in hydraulic continuity with the conglomerates. In east and north Yorkshire, the sequence includes a shaly sandstone horizon (Figure 4.4). The Sherwood Sandstone dips gradually towards the east and lies at depths of over 1000 m in Holderness and north east Lincolnshire (Figure 4.5).

Over much of the area, the porosity of the Sherwood Sandstone exceeds 20%. A tongue within which values exceed 25% extends from Nottingham to Bridlington and embraces much of Lincolnshire and south-east Yorkshire. In north Yorkshire values are between 10 and 20%. From a

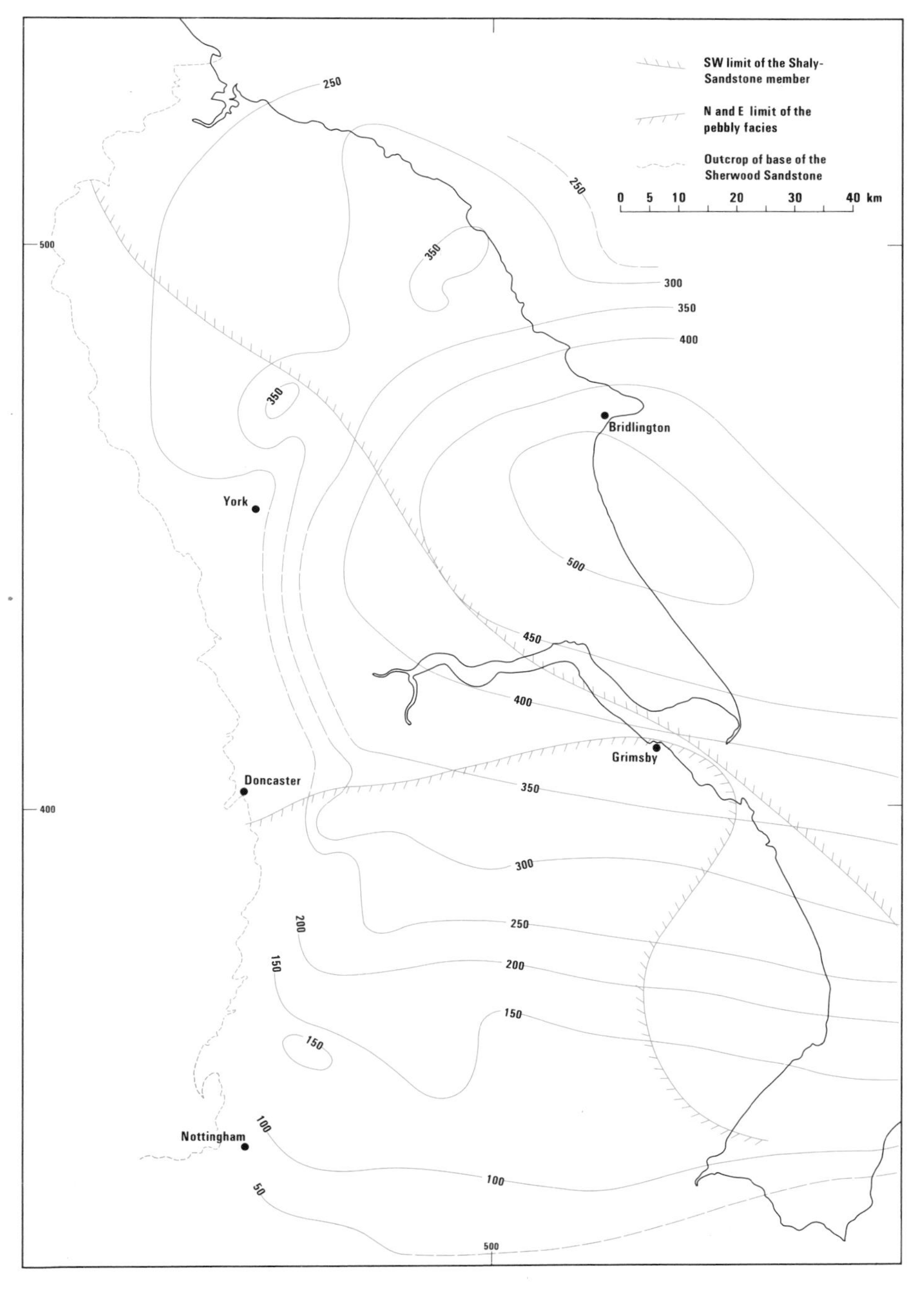

4.4 Isopachytes of the Sherwood Sandstone Group in the East Yorkshire and Lincolnshire Basin (metres).

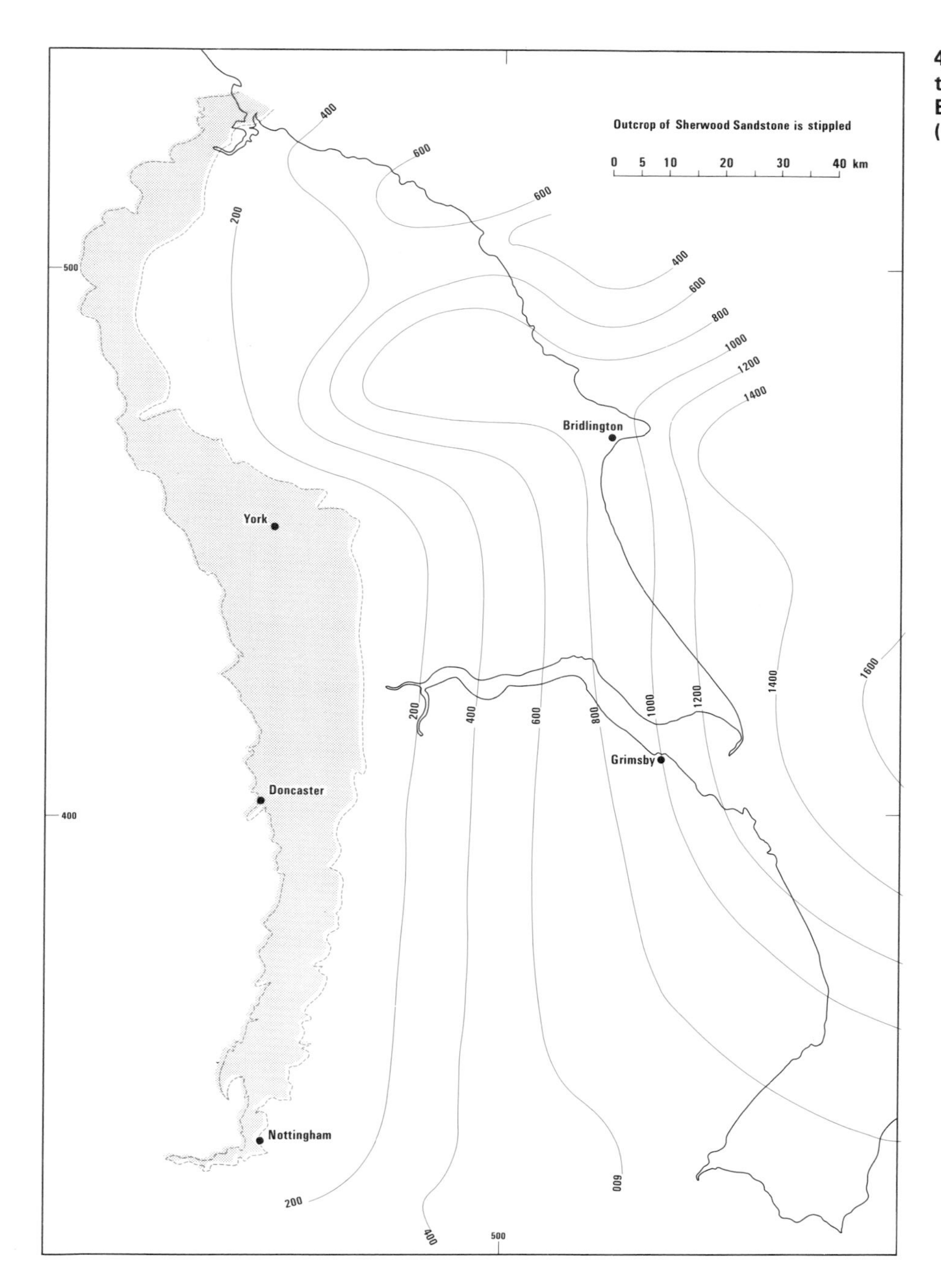

4.5 Structure contours on the top of the Sherwood Sandstone Group in the East Yorkshire and Lincolnshire Basin (metres below sea level).

relationship between porosity (derived from sonic velocity logs) and depth two groups of values can be recognised: those around the Cleveland Uplift have a lower porosity and the remainder, which includes outcrop samples and those derived from off-shore wells, plot close to a straight line relationship between porosity and depth. As recognised by Marie (1975), the cause is probably inversion (uplift following burial of an area).

The aquifer properties of the Sherwood Sandstone were examined in some detail at an outcrop site near Mansfield, Nottinghamshire (Williams and others, 1972). The geometric mean horizontal and vertical permeability of core samples from two boreholes were 3.93 and 2.72 m/d (5.7 and 3.9 D). In finer-grained sandstones the horizontal to vertical permeability ratio increased from 1.33 to 2, because of the increasing importance of clay particles orientated parallel to the bedding. The transmissivity of the aquifer at the site, as derived from pumping tests was 1500 m^2/d (2200 D m), but this was influenced by fissures and the intergranular transmissivity was only about 300 m^2/d (430 D m).

In general, these results are typical of the sandstone at and near the outcrop in Nottinghamshire, where the permeability is high. The most common rock type is a medium to coarse-grained, clean, well-graded sand with a permeability of 5000 to 10 000 mD. In this rock the porosity exceeds 30%. Values are lower in finer-grained lithologies — sometimes as low as 1000 mD — and they may also be reduced by cementation to some 500 to 2000 mD (Lovelock, 1977).

In Yorkshire, north of York, and in Durham there is a tendency for the sandstone to become finer-grained and this is also observed in this region with increasing depth. Generally the fine-grained sandstones have permeability values of 200 to 300 mD, medium-grained sandstones range from 2000 to 3000 mD and coarse-grained sandstones up to 10 000 or 15 000 mD, the grain size being the dominating factor controlling permeability (Lovelock, 1977).

The discussion above refers to the sandstone at or near the outcrop and there is only limited information for areas where the sandstones may contain water with a geothermal potential. The recent geothermal exploration well at Cleethorpes proved 400 m of Sherwood Sandstone with an average porosity of about 20% and a transmissivity of at least 60 D m and possibly as high as 200 D m. The Sherwood Sandstone is developed for gas in the Hewett Field, which is some 25 km off the north Norfolk coast and 90 km off the Lincolnshire coast. In this field the porosity of the pay section averages 25% and the average permeability is 474 mD, although individual layers have much higher values (Cumming and Wyndham, 1975).

Even allowing for the fact that the permeability of the sandstone probably declines towards the east below overlying sediments, the average permeability is still likely to exceed 200 mD and will probably be much higher in particular horizons. Fractures provide the main channels for groundwater flow in and near the outcrop and greatly enhance the permeability. The importance of fractures is likely to decline towards the east, intergranular permeability becoming more important. If the rock is cemented, the yield of water to wells would depend upon small fractures connected to less cemented zones of higher permeability and probably limited vertical distribution in the sequence.

Potable waters are found in the Sherwood Sandstone for some 15 to 25 km down-gradient from the outcrop, but further east the quality deteriorates, and eventually becomes saline.

The estimated temperatures at the mid-point of the sandstone calculated from the mean temperature gradients are shown in Figure 4.6. Few measurements of temperature were made in the formation itself so extrapolation from measurements in deeper formations was necessary. Only the most reliable classes of temperature measurement

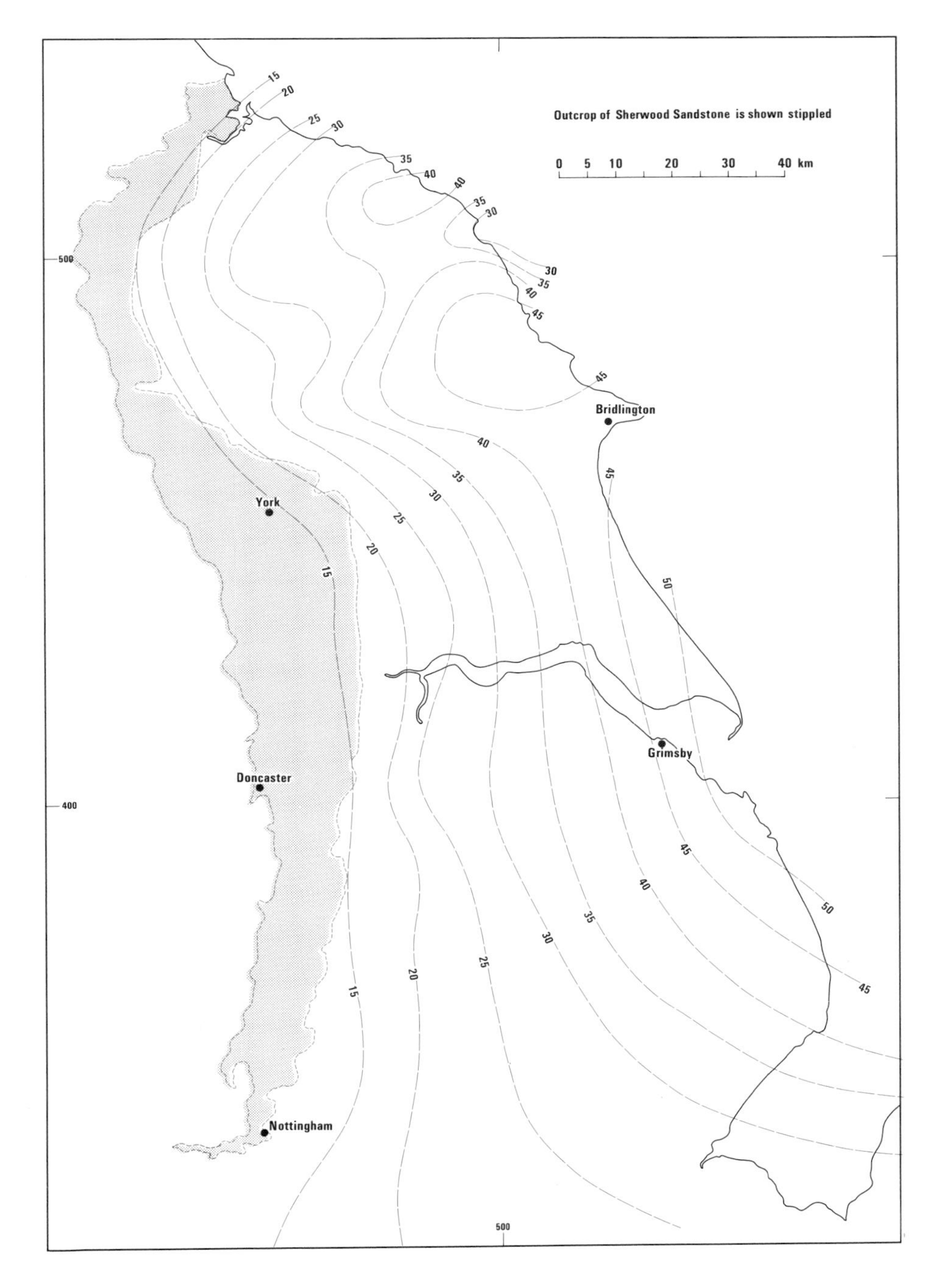

4.6 Estimated temperature at the mid-point of the Sherwood Sandstone Group in the East Yorkshire and Lincolnshire Basin (°C).

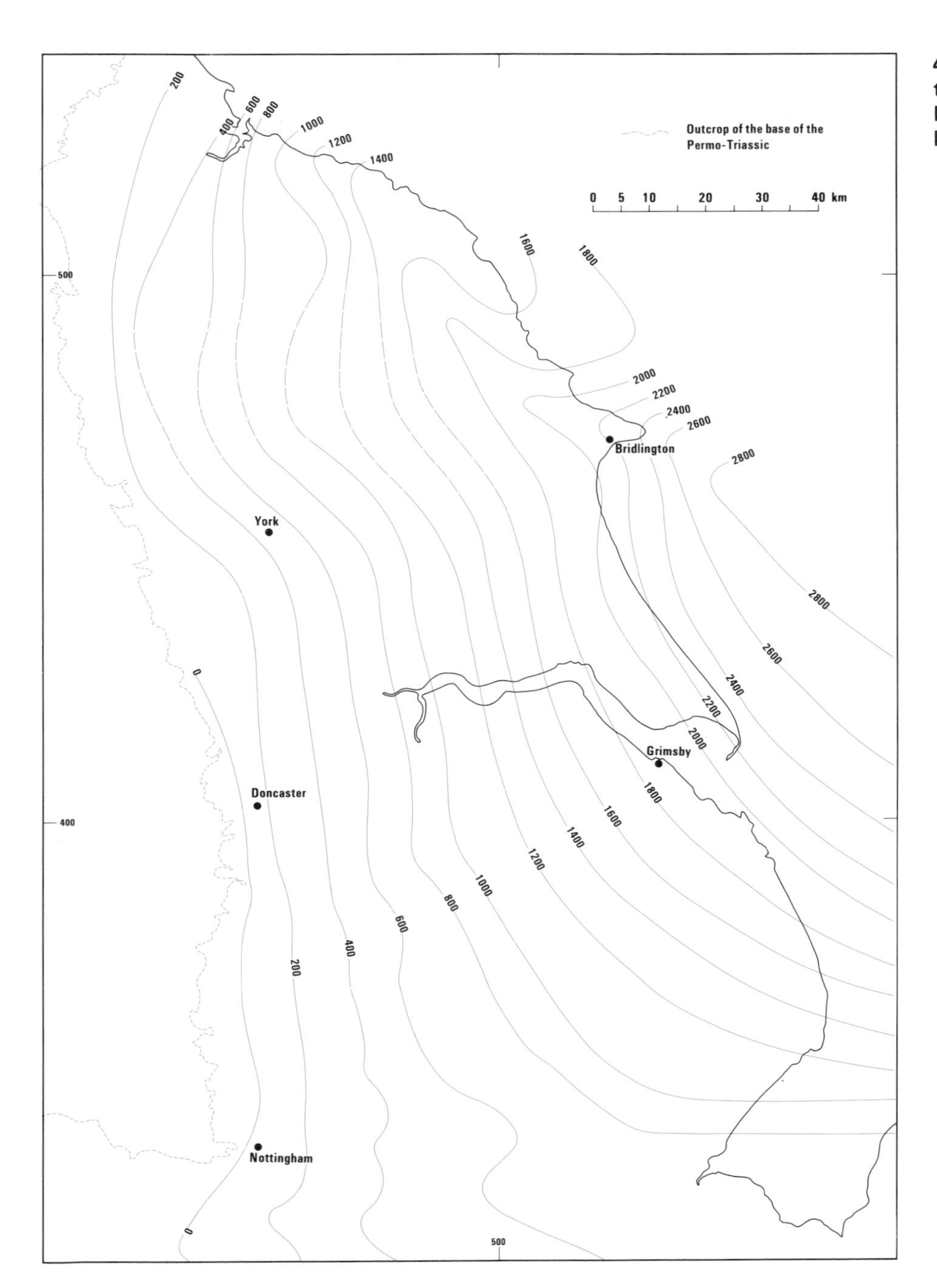

4.7 Structure contours on the base of the Permian in the East Yorkshire and Lincolnshire Basin (metres below sea level).

(Burley and others, 1984) were used in order to obtain the best estimate of the geothermal gradient. This method of calculation does not allow for local variability in thermal conductivity and heat flow and, therefore, provides only a mean estimate of temperature. The estimated mean aquifer temperature exceeds 40°C along the east coast from Scarborough to the Wash and probably exceeds 50°C around the mouth of the Humber.

Lower Permian

The Basal Permian Sands and Breccia rests unconformably on Carboniferous rocks at depths shown on Figure 4.7. It varies considerably in thickness but in general thickens towards the east and is better developed and more persistent under south-east Yorkshire, Humberside and Lincolnshire (Figure 4.8). However, it is only in east Lincolnshire that the thickness consistently exceeds 30 m. The Lower Permian sediments are believed to be mainly aeolian dune sands although some are probably of fluvial origin. They tend to be poorly cemented and this characteristic caused problems in the Durham Coalfield during shaft sinking, and may cause completion problems in boreholes.

The porosity values, deduced mainly from geophysical logs, lie between 20 and 25% in Lincolnshire, where the deposit is thickest, but maximum values of between 25 and 30% occur in Yorkshire, although there the thickness is only 10 to 15 m. Values from boreholes in the Cleveland Basin are lower than for equivalent depths elsewhere, a feature which Marie (1975) considered was due to inversion.

As already mentioned, much of the Lower Permian sandstone is of aeolian origin and is, therefore, likely to be 'clean'. Such deposits form good reservoirs and the high porosities support this view. When they were initially deposited, porosities were possibly of the order of 40% and permeabilities several tens of darcies (Beard and Weyle, 1973). During the course of diagenetic development such values can be modified considerably (Glennie and others,

1978; Kessler, 1978). However, in the absence of significant diagenesis, sandstones of aeolian origin are likely to have porosities of 20 to 30%, and this would imply permeabilities of at least 100 mD.

A few drill-stem tests (DSTs) have been carried out in the Lower Permian but with the exception of one borehole these are all to the north of the Humber where the aquifer is quite thin. Values derived from these tests, and from two carried out in the North Sea, range from 19 to 230 mD. There is no obvious correlation between the permeability and the porosity determined from sonic logs.

In east Lincolnshire, where the sandstones attain their greatest thickness, there is evidence for a more variable succession. Core samples show that the upper part of the sandstones are coarse-grained, with a porosity of 20%, and a permeability ranging up to 650 mD, but averaging 300 mD. Conglomerates may be more common in the lower part, with lower porosities (about 10%) and permeabilities (averaging 10 mD). There are interbedded thin marls and silts, and cementation is variable, ranging from indurated sandstone to unconsolidated running sand. In one borehole in this area, the pay thickness was estimated to be 38 m with an average permeability of 140 mD, giving an estimated transmissivity of 5.3 Dm.

In general, the nature and origin of the Lower Permian sandstones together with their known porosity values imply good permeabilities of between 100 and 200 mD with much higher values in some zones. The thickness of the aquifer is limited and the thickness of the effective aquifer may be reduced further by cementation. The greatest aquifer potential, therefore, lies in east Lincolnshire where the aquifer attains its greatest thickness, and the transmissivity may exceed 5 Dm. A geothermal exploration well at Cleethorpes proved a somewhat thinner sequence than had been anticipated comprising 26 m of fluviatile sandstones and conglomerates; only 8.5 m represented suitable reservoir rock

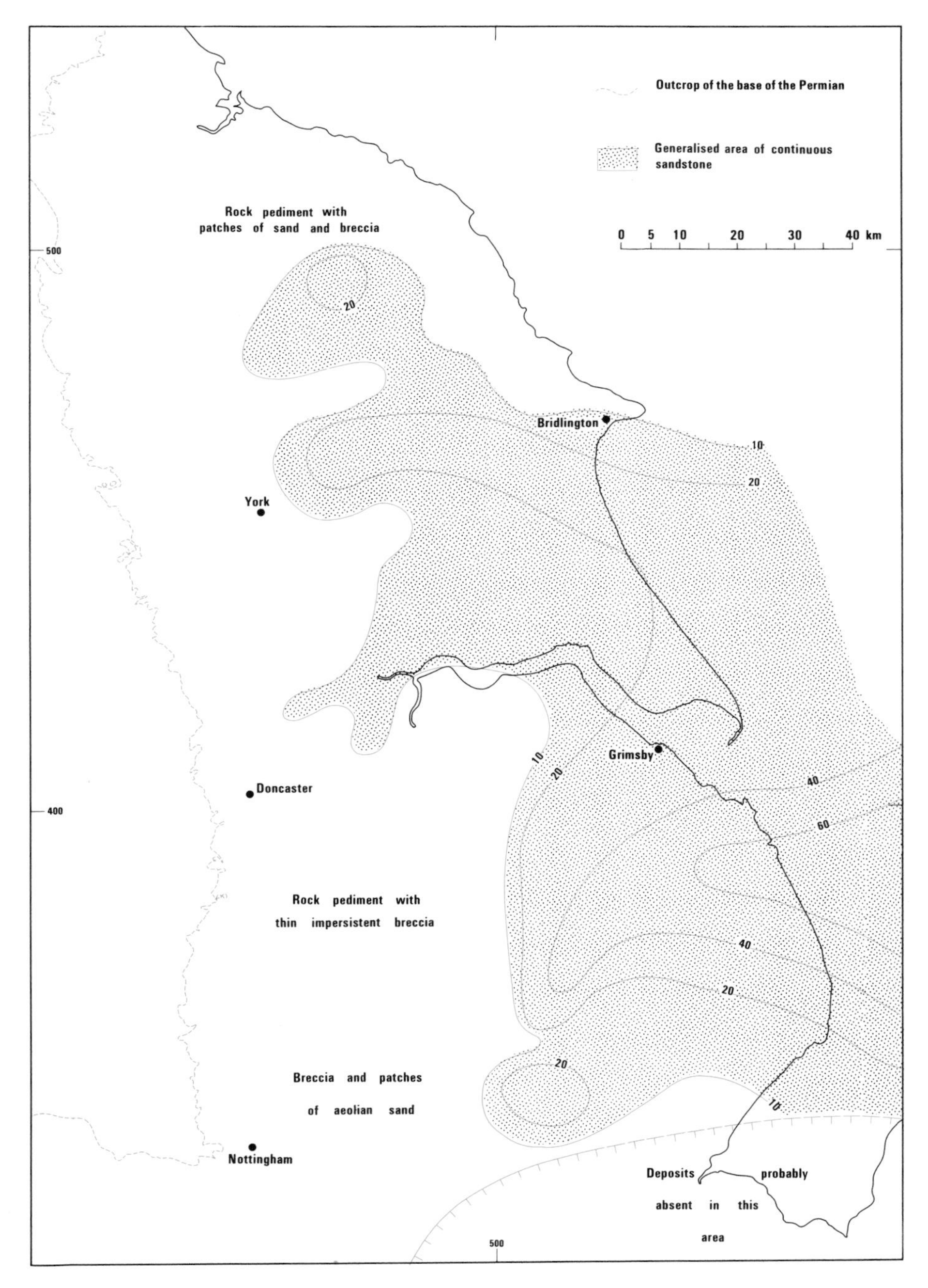

4.8 Isopachytes and lithofacies of the Basal Permian Sands and Breccia in the East Yorkshire and Lincolnshire Basin (metres).

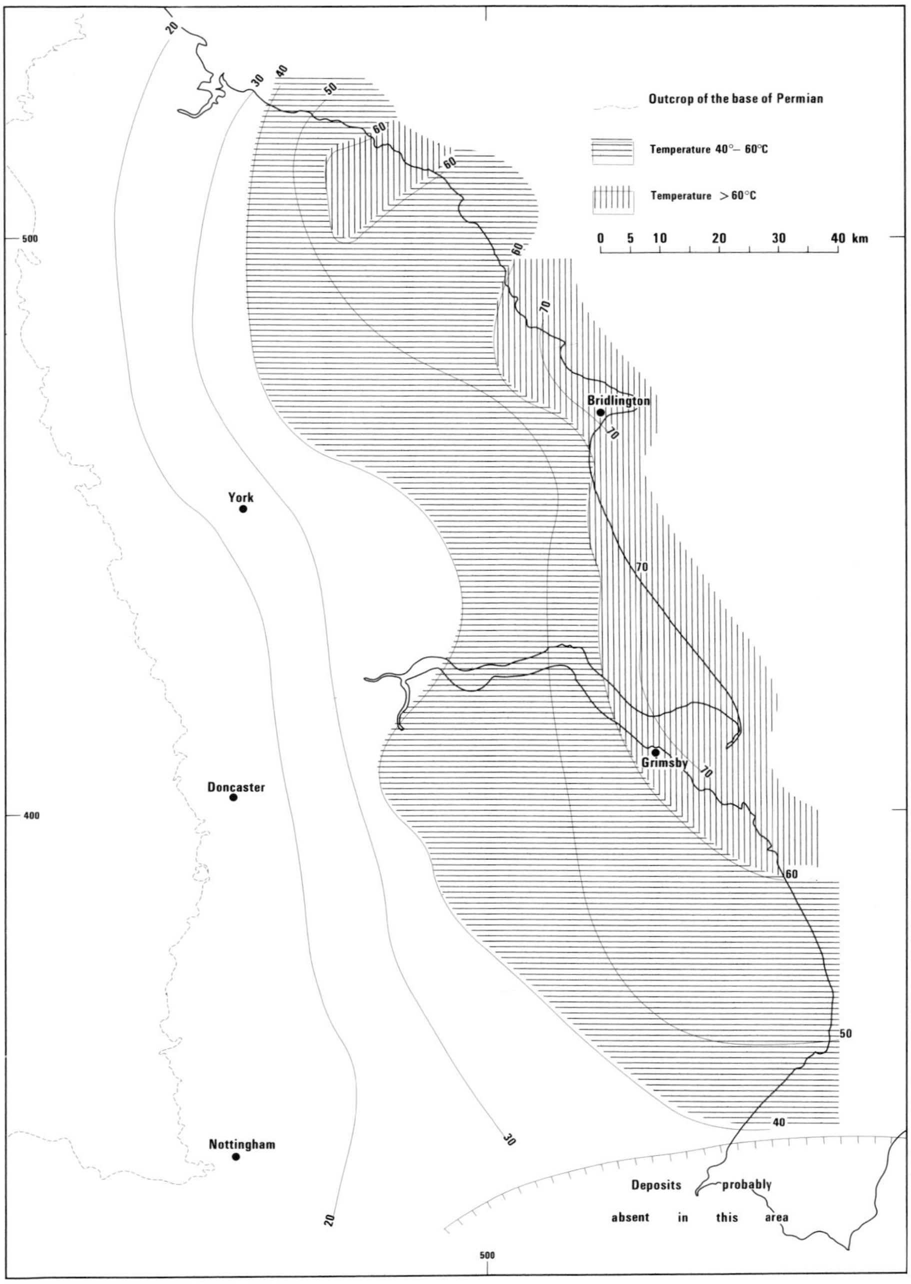

4.9 **Estimated temperature at the base of the Permian in the East Yorkshire and Lincolnshire Basin (°C).**

and the total transmissivity was less than 2 D m.

The estimated temperature at the base of the Permian is shown in Figure 4.9. As would be expected, there is a steady increase towards the east coast, with the coastal strip exceeding 50°C from the Tees to the Wash, and values in excess of 70°C around Flamborough Head and Holderness. However, as already pointed out, the aquifer only exceeds 30 m in east Lincolnshire and this is the area where most potential exists. Certain anomalous temperatures have not been included in Figure 4.9 as it is intended to show regional variations. At Cleethorpes the temperature, as expected, was 64°C.

WESSEX BASIN

General geology

The Wessex Basin is a structural and sedimentary basin containing rocks ranging in age from Permian to Tertiary (Table 4.2). The sediments were deposited on a complex basement, ranging in age from at least Ordovician to possibly Upper Carboniferous, which was very deformed during the Variscan Orogeny. The basin flanks the southern side of the London Platform and extends from Devon eastwards under the Weald. Although the structure also extends under the English Channel, only the land area is discussed here.

Most of the data available for interpretation have been provided by commercial and government organisations exploring for oil and from two geothermal wells in the Southampton area. Few boreholes in the basin exceed 1 km in depth and those that do are mostly clustered in relatively small areas, as, for example, to the west of Bournemouth. The basin contains several aquifers, but at the depths where temperatures would be appropriate for potential geothermal development the permeabilities of most of these are too low for economic exploitation. The only exceptions are the Triassic Sherwood Sandstone and the Lower Cretaceous sandstones, the latter on the south coast between Ports-

Table 4.2 Geological succession in the Wessex Basin

Stratigraphy	Lithology	Hydrogeology
TERTIARY	Unconsolidated clays of varying sand content. Clayey sands and sands.	Localised intergranular permeability.
CRETACEOUS:		
Upper Chalk	White, rather soft biomicritic limestone with common bands of flint modules.	Possible secondary permeability near top.
Middle Chalk	White, biomicritic limestone.	
Lower Chalk	White, rather argillaceous biomicritic limestone.	
Upper Greensand	Fine-grained, rather glauconitic sandstone with variable clay content.	Possible intergranular permeability.
Gault	Blue clay, sometimes sandy.	
Lower Greensand	Sandstone and sands, some very coarse-grained, some glauconitic. Occasional clay beds. Nodular chert horizons.	Some intergranular permeability.
Wealden Beds (absent over most of western part of Wessex Basin)	Mainly clays of varying sand content with some well-developed sands and pebble beds, particularly in the lower part.	Possible intergranular permeability.
JURASSIC:		
Purbeck Beds	Fine-grained limestones with interbedded lime mudstones and some evaporites.	
Portland Beds	Massive oolitic and bioclastic limestones with sandstone and mudstone in the lower part.	Where present the sands may have intergranular permeability. The massive limestones may have developed secondary permeability.
Kimmeridge Clay	Dark bluish grey clays and shale with some concretionary limestones.	
Corallian	A rather variable succession of oolitic and bioclastic limestones with sandstones, some of which are coarse, and mudstones.	
Oxford Clay and Kellaways Beds	Grey clays with large concretions often slightly sandy at base.	
Cornbrash and Forest Marble	Oolitic and bioclastic limestones and clays.	
Great Oolite	Massive oolitic limestones with clays. In the western part of the area the succession is mainly clays.	
Fuller's Earth	Clay	
Inferior Oolite	Oolitic and bioclastic limestone. Rather coarse-grained.	
Lias (with Bridport Sands at top in western area)	Fine grained silty sands (Bridport Sands developed in Dorset). Otherwise blue-grey clays, shales and argillaceous limestones. Some clayey sands in upper part.	Some intergranular permeability in Bridport Sands (dependent on the absence of total cementation).
TRIASSIC:		
Penarth Group	Very thin white limestones, black shales and silty sands.	
Mercia Mudstone Group	Red and greenish-grey sandy mudstones with some evaporites (anhydrite and halite). Possibly some fine-grained sandstones and siltstones.	
Sherwood Sandstone Group	Red to reddish-brown medium-grained sandstones. Some pebble beds at base.	Intergranular permeability dependent on absence of total cementation.
PERMIAN:		
Aylesbeare Group	Red to purple mudstones, argillaceous mudstones, and siltstones. Some breccias near base.	Secondary permeability probably negligible in deeper parts of the basin.
DEVONIAN AND CARBONIFEROUS:		
Variscan Basement	Shales, mudstones, sandstones and limestones affected particularly in the south by folding and cleavage.	

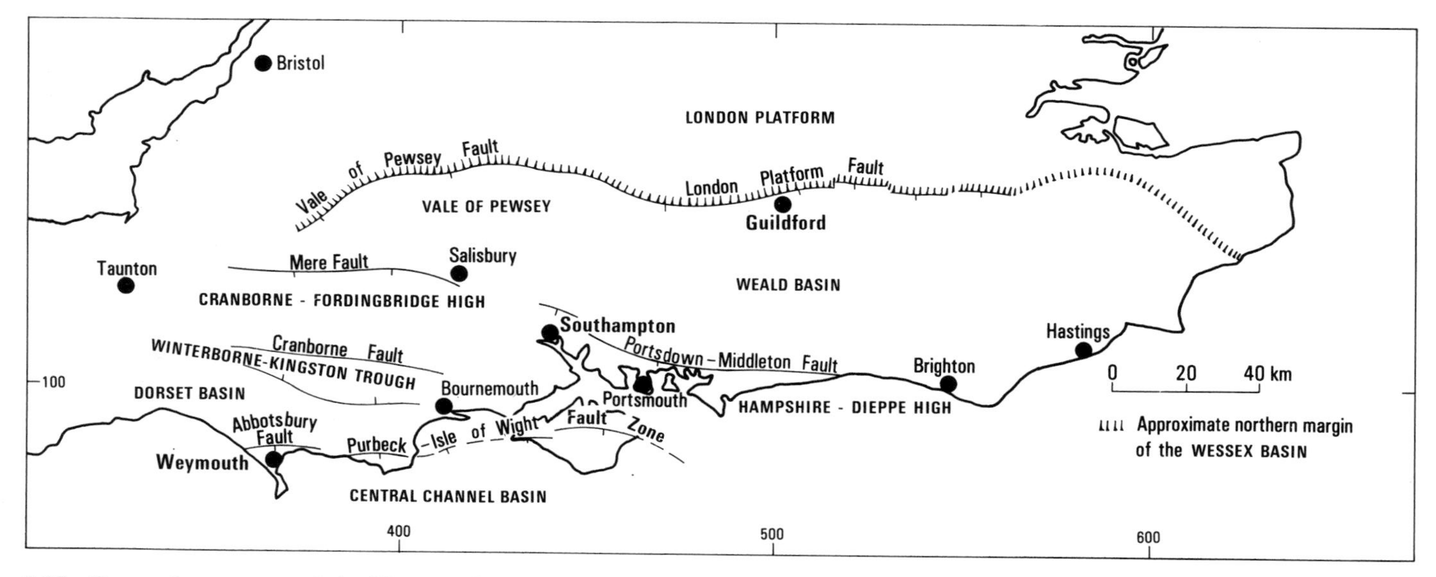

4.10 General structure of the Wessex Basin.

mouth and Worthing. As a consequence attention is focused on these two aquifers, particularly the Sherwood Sandstone which represents the main geothermal potential in the Wessex Basin.

The development of the Wessex Basin was initiated after the Variscan Orogeny, when an extended period of crustal subsidence began in southern England. This subsidence was not uniform across the whole area but was concentrated along major growth-faults, almost all of which had a southward downthrow when active. These growth faults were active at various times from the Permian to early Cretaceous and exercised considerable tectonic control over patterns of subsidence and sedimentation (Allen and Holloway, 1984).

The northern margin of the Wessex Basin is defined by a zone of growth-faults along the southern margin of the London Platform and the northern margin of the Vale of Pewsey (Figure 4.10). The southern margin of the basin lies to the south of southern England and it appears to form part of a major sedimentary basin extending into the Western Approaches and Northern France.

Further growth-faults (Figure 4.10) split the basin into structural provinces. The Portsdown-Middleton and Mere faults define the northern edge of a half-graben including the Hampshire-Dieppe and Cranborne-Fordingbridge structural highs. Between the Portsdown-Middleton Fault and the northern margin of the Wessex Basin is the Weald Basin, a sub-basin which continues west into the Vale of Pewsey. In Dorset another sub-basin occurs to the south of the Cranborne growth-fault and is here described as the Dorset Basin. The eastern end of this basin lies in the vicinity of the Isle of Wight. The deepest part, around the Winterborne Kingston Borehole, is described as the Winterborne Kingston Trough.

A further sub-basin is considered to lie south of the major growth-fault zone that runs beneath the Purbeck and Isle of Wight monoclines. This growth-fault zone is continued along the Abbotsbury-Ridgeway faults to the west and along the St Valery-Bembridge structural line offshore. As it is largely located offshore, this structure is known as the Central Channel Basin.

In addition, large numbers of predominantly east-west trending normal faults have cut the succession at different times. During the Alpine compressional tectonic episode, the sense of movement of some of these faults was reversed, and numerous anticlines and periclines were formed, typically with steeper northern limbs.

Geology of the Permo-Triassic

The Permo-Triassic sediments of the Wessex Basin can be divided into five broad lithological groups. The oldest of these is the breccias and sands of possible Lower Permian age, seen at outcrop in the Exeter, Tiverton and Crediton areas. They probably represent locally derived detritus laid down in interdigitating breccia fans (Smith and others, 1974 p.27), although some sands are aeolian in origin. They were deposited on the margins of the basin under a hot semi-arid climate (Henson, 1970) and are present in boreholes in the west of the Wessex Basin and in the Wytch Farm area, but are absent in the Marchwood and Cranborne boreholes.

The second group of Permian age is the Aylesbeare Group. This comprises principally mudstones with subordinate sandstones. It is of great thickness in Dorset (for example, greater than 575 m in the Winterborne Kingston Borehole). Depositionally the Aylesbeare Group may represent floodplain or floodbasin deposits, the coarser sands and gravels representing channel deposits (Henson, 1970).

The third group, the Sherwood Sandstone Group, is largely of Triassic age in the Wessex Basin, though the exact position of the Permian-Triassic boundary is unknown in the area. The Sherwood Sandstone Group broadly encompasses formations previously included in the Bunter and arenaceous lower part of the Keuper (Warrington and others, 1980, p.13), and consists of a series of breccias and conglomerates, overlain by a series of cyclically deposited sandstones.

The fourth group, the Mercia Mudstone Group, overlies the Sherwood Sandstone and, as the name implies, it is argillaceous in nature, being composed of red, green and grey mudstones and siltstones often with anhydrite.

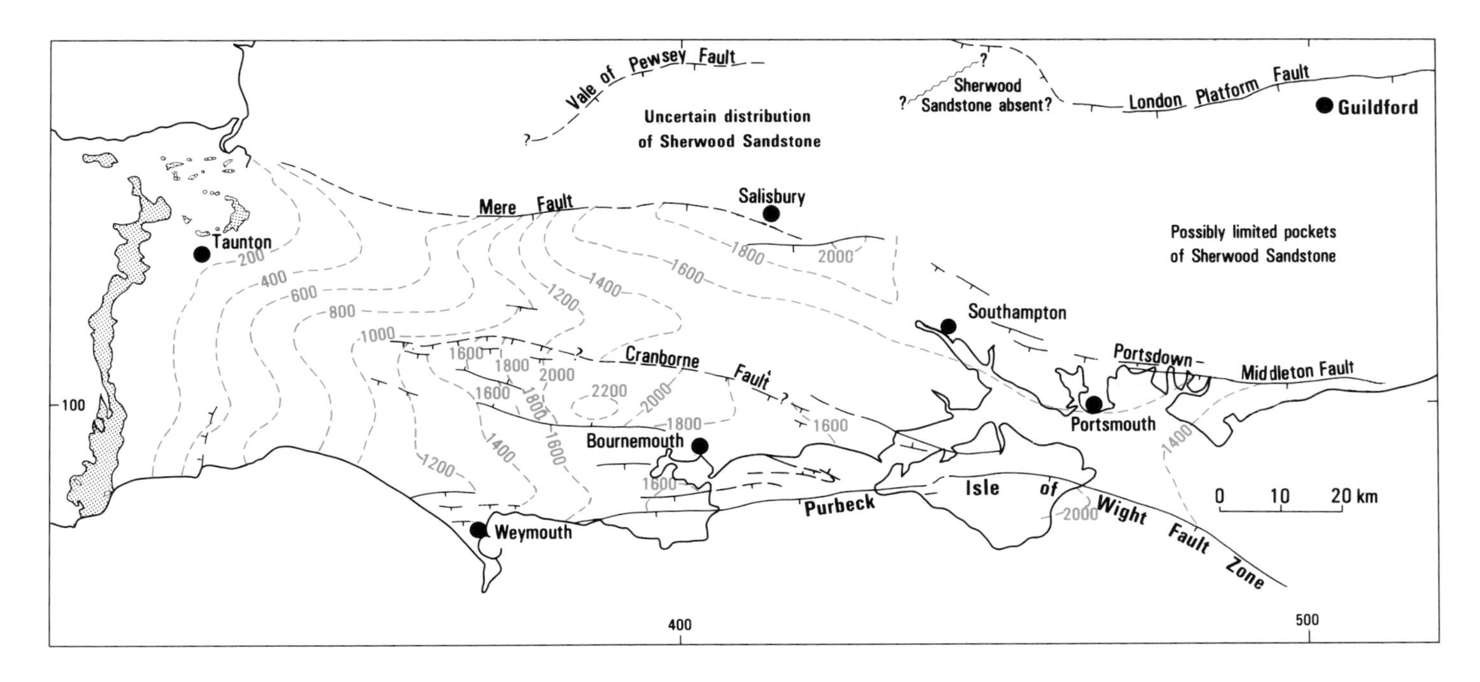

4.11 Structure contours on top of the Sherwood Sandstone Group in the Wessex Basin (metres below sea level).

Deposition was mainly subaqueous, probably partly in playas.

The fifth group, the Penarth Group, comprises sediments formerly described as 'Rhaetic', which are generally fine-grained and relatively thin.

There are no regional seismic reflections below the Jurassic and the top of the Sherwood Sandstone is not resolved by seismic data throughout most of the area. In the absence of a good seismic event, contours on the top of the Sherwood Sandstone have been estimated (Figure 4.11) by interpolation between boreholes using the broad form of the closest identifiable seismic reflector, which is at the top of the White Lias.

An isopachyte map, based on borehole thicknesses and inferences of thickening or thinning across the major structural elements of the basin, and a facies map, based on borehole information and comparisons with the outcrop of the Sherwood Sandstone Group are shown in Figures 4.12 and 4.13 respectively.

The major control on the deposition and facies of the Sherwood Sandstone was exercised by differential subsidence related to movement on the major growth-faults. Thus the various sub-basins or structural provinces have different depositional histories and are discussed separately below.

4.12 Isopachytes of the Sherwood Sandstone Group in the Wessex Basin (metres).

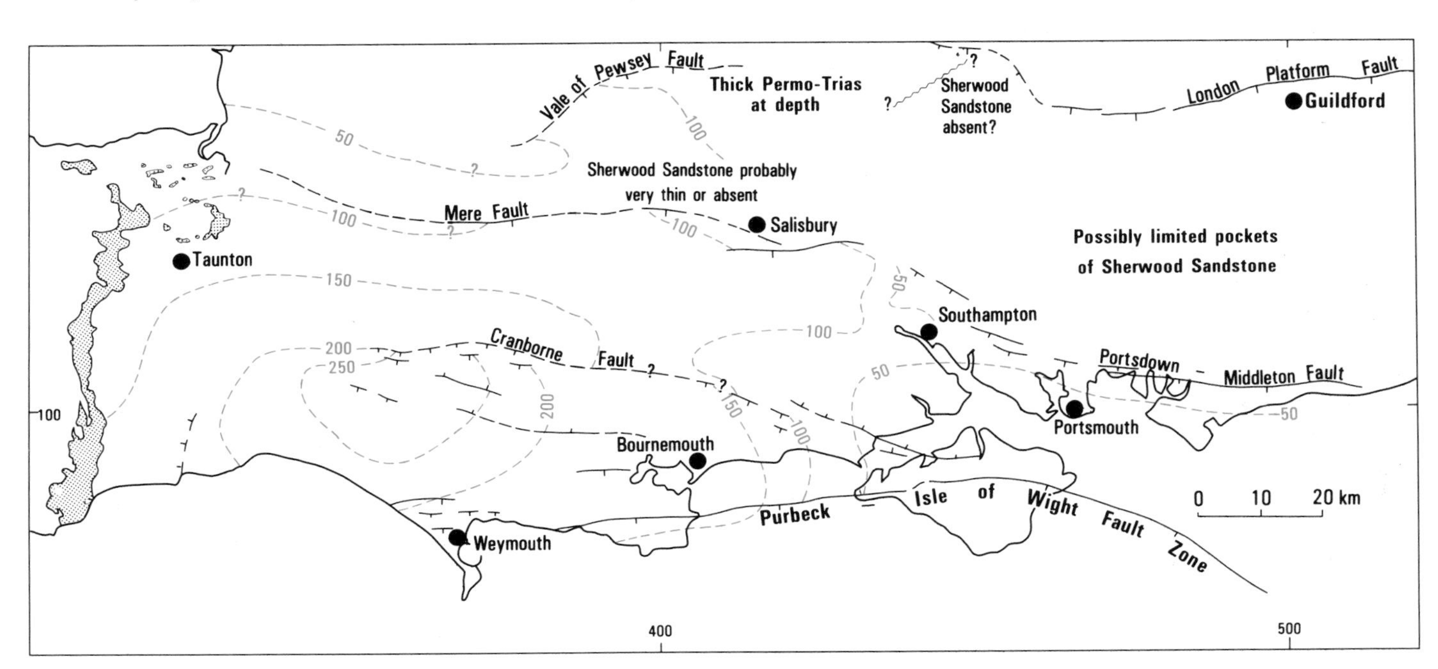

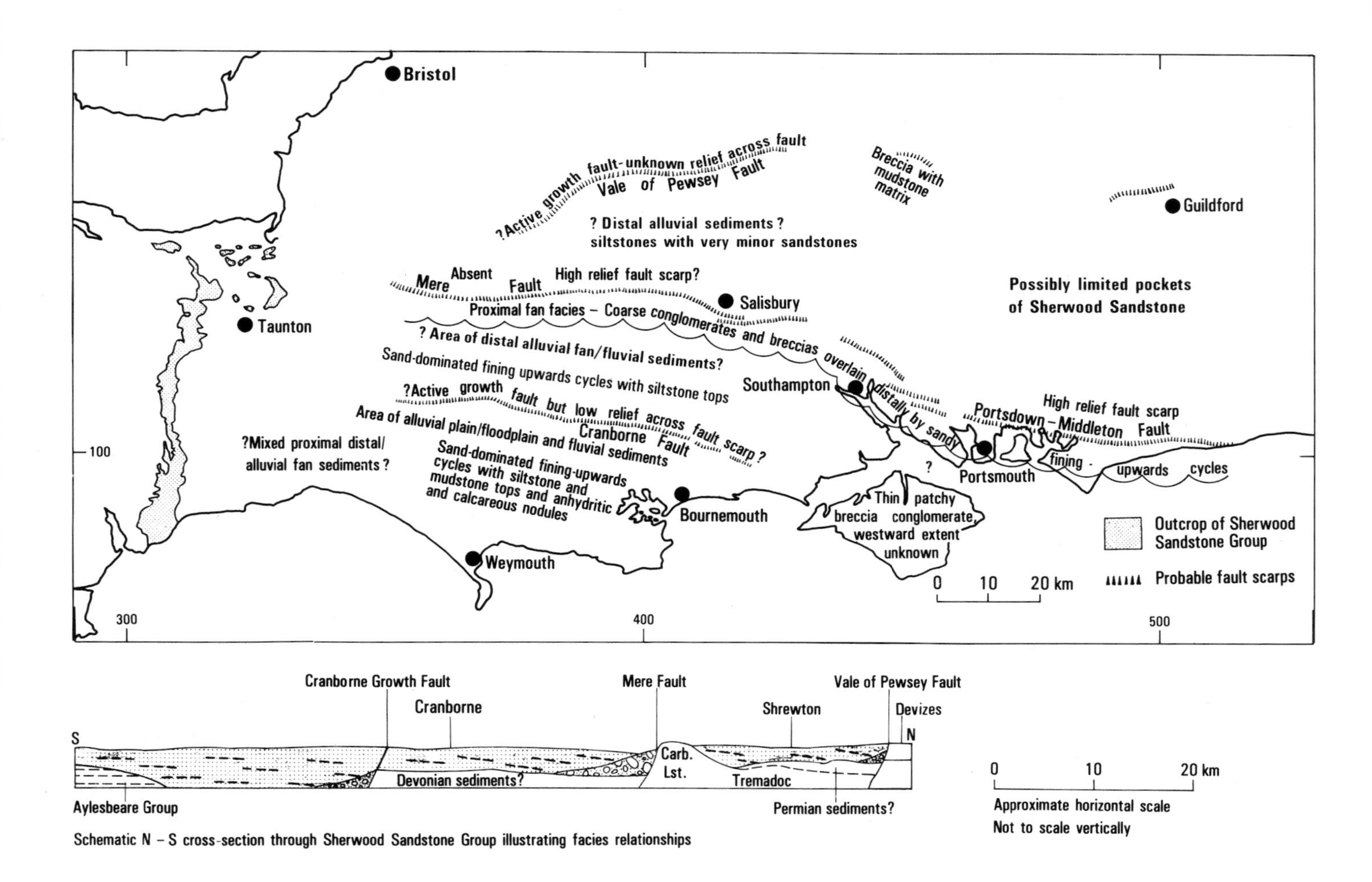

4.13 Facies variations in the Sherwood Sandstone Group in the Wessex Basin.

Vale of Pewsey-Weald Basin

On the northern margin of the Weald the growth-faults defining the edge of the London Platform do not appear to have been active during the Permian and early Triassic times. In general, Permian sediments and the Sherwood Sandstone Group are absent in the Weald Basin, although red sandstones and mudstones, possibly assignable to the Sherwood Sandstone, and breccias of uncertain age are recorded. These deposits may be present elsewhere on the down-throw sides of growth-faults, although they are likely to be of limited extent. In the Vale of Pewsey the Sherwood Sandstone Group is relatively thin (Figure 4.12), although a thickness of more than one kilometre of Permo-Triassic sediments may occur immediately to the south of the Vale of Pewsey Fault. During the deposition of the Sherwood Sandstone Group it appears probable that most of the Weald area was a topographic high, as was a narrow ridge running parallel to, and immediately north of, the Mere Fault. Sedimentologically the Sherwood Sandstone Group is represented by siltstones with only very fine-grained thin sandstones and breccias with a mudstone matrix.

Cranborne-Fordingbridge High and Hampshire-Dieppe High

In this structural province the Sherwood Sandstone Group has been proved in boreholes to range in thickness between about 30 and 110 m. Further evidence of a topographic high along the northern side of the Mere and Portsdown-Middleton faults is seen in the extensive development of coarse-grained sediments immediately to the south of the Portsdown-Middleton fault zone, representing proximal alluvial fan deposits. Thus breccias and conglomerates are found at the base of the successions in the Marchwood and Southampton wells. The conglomerates in Marchwood are thicker than those at Southampton, which is further from the growth-faults. Conglomerates and breccias are also found in a borehole on the Isle of Wight, and further east the group consists of pebbly sandstone.

Overlying the basal conglomerates and breccias in the Marchwood and Southampton wells is a series of small scale (approximately 2 m thick) sand-dominated, fining-upwards cycles, which are interpreted as distal alluvial fan deposits. Similar fine-grained sediments are found in the

Cranborne Borehole, further from the fault scarp. Distribution of rock fragments, reflecting the source of the sediments, may have an important bearing on porosity (Knox and others, 1984). It is the coarse-grained units in these cycles which form the main producing horizons in the Southampton and Marchwood geothermal wells and which are considered to represent the most promising targets within the Sherwood Sandstone Group.

Dorset Basin

In this basin the Sherwood Sandstone Group ranges from about 160 to 280 m in drilled successions. There are no boreholes immediately to the south of the Cranborne growth-fault, the nearest being Winterborne Kingston, some 10 km to the south. This borehole contains no basal conglomerates or breccias within the Sherwood Sandstone Group though the thickness difference in the group between Cranborne (114 m) and Winterborne Kingston (170 m) indicates considerable movement on the growth-fault. The absence of very coarse sediments in the boreholes of south Dorset suggests that either they were too far from the fault scarp to receive proximal fan sediments, or that regular small movements across the fault took place and thus a substantial fault scarp never developed.

The presence of thick successions of favourable facies within the Dorset Basin suggests that it has the best geothermal potential. The thickest successions would be expected in the Winterborne Kingston Trough, though these may not have the most favourable porosity and permeability characteristics. The following summary is likely to be typical of the Sherwood Sandstone in the eastern part of the Dorset Basin. It is based on an analysis of geophysical logs and core from the Winterborne Kingston Borehole (Lott and Strong, 1981). Regular fining upwards cycles up to 3 m in thickness are seen in cored intervals and together they represent a major fining upwards megacycle. They are probably fluvial in origin, with evidence of a hot climate. The cycles are composed of intraformational conglomerates, cross-bedded sandstones and laminated siltstones and mudstones. Concretionary carbonate patches are present, which probably represent caliche horizons. The sandstones dominate the cycles. They are red-brown to pink, ferruginous and feldspathic. They fine upwards, are poorly to moderately sorted and often have good intergranular porosity. They show low-angle cross-bedding, fine lamination and climbing ripples. Both quartz and feldspar grains have authigenic overgrowths. The conglomerates consist of mudstones and micritic intraclasts with sporadic quartz pebbles and quartz and dolomitic granules in a sandy matrix. They are usually poorly sorted but are often graded. The bases of conglomerates, being at the base of cycles, are usually erosive. Locally the sandstones are heavily cemented, which is possibly related to the proximity of major growth-faults.

Hydrogeology of the Sherwood Sandstone Group

Vale of Pewsey-Weald Basin

At present the hydrogeological information from the Vale of Pewsey-Weald Basin is very sparse and not encouraging for geothermal development, mainly because of the high proportion of siltstones and scarcity of uncemented sandstone in the basin. The potential for thick sediments to the east of the Devizes Borehole provides more ground for optimism, although present information suggests that only thin sandstones may be present.

Cranborne-Fordingbridge High and Hampshire-Dieppe High

The dominant control on the hydrogeological characteristics of the Sherwood Sandstone in this area of the Wessex Basin is the degree of cementation which significantly affects porosity and intergranular permeability (there is no evidence at present of fissure flow in the sandstone at depth). The diagenesis of the formation has an important bearing on the physical properties and hence on the water-bearing capacity of the formations (Burley, 1984; Knox and others, 1984). The degree of cementation is so variable that its effects on porosity and permeability are much more significant than those caused by variations in grain-size or sorting. Most of the presently available information concerning cementation at depth in the Sherwood Sandstone Group is derived from the Marchwood and Southampton wells which are to the east of the Cranborne-Fordingbridge High.

The Marchwood Well was the first exploratory well to be drilled near Southampton and was located in the expectation of a thick Permo-Triassic sequence, principally on the basis of gravity interpretation (Smith and others, 1979). The sequence in the well was found to consist of rather thin Triassic strata overlying the Devonian. The Sherwood Sandstone Group was 59 m thick but only in the upper 25 m were there bands of high permeability. Three such layers occur, each approximately 2 m thick, with porosities of about 25% and permeabilities of up to 5 darcies. Cementation had eliminated any aquifer potential in the basal conglomerate and the immediately overlying sandstone (Burgess and others, 1981). The Southampton Well proved 67 m of the Sherwood Sandstone Group (Downing and others, 1982a) and the upper 26 m included three zones, approximately 5, 7.5 and 4 m thick with porosities ranging between 20% and 26%; two of the zones contribute 90% of the total transmissivity of the Sherwood Sandstone.

The transmissivity of this thin, layered aquifer has been established as about 6 m^2/day (3.5 D m) by pumping tests at Marchwood (Price and Allen, 1982) and Southampton (Allen and others, 1983); a storage coefficient of 4×10^{-4} has been determined. The fluid in the aquifer is a brine with a salinity of over 100 g/l (105 g/l at Marchwood and 125 g/l at Southampton) and a temperature of about 75° C (73.6°C at Marchwood and 76°C at Southampton). The rest (or static) piezometric level is approximately at ground level. There is no evidence of fissure flow in the aquifer except perhaps immediately adjacent to the wells, but there is leakage from the low-permeability sandstones into the high-permeability layers — although this would not be expected to affect the rate of drawdown for more than a few weeks after the start of production (Allen and others, 1983).

The thin water-bearing layers found at Southampton are probably typical of the Sherwood Sandstone in Wessex. They may not persist for great distances because of lateral sedimentological or diagenetic change, or because of faulting. This appears to be the case near Southampton where evidence from an analysis of downhole pressure data collected during and after production tests on the Marchwood and Southampton wells led to the conclusion that the wells are either in a closed block of some 200 km^2 or in a narrow wedge (Allen and others, 1983).

Elsewhere in the Cranborne-Fordingbridge High and Hampshire-Dieppe High hydrogeological data are sparse. It does appear, however, that near to the Portsdown-Middleton-Mere Fault line, the more distal, finer sediments of the Cranborne-Fordingbridge High may have improved geothermal potential. Towards the south east of Southampton the sandstone thins and is reduced to zero within some 30 km.

Dorset Basin

At outcrop the Triassic sediments of east Devon and west Somerset cover an area of approximately 200 km^2 and are an important source of groundwater. The hydrogeological characteristics of the aquifer are variable, with high values of transmissivity and porosity in the south, decreasing northwards (Table 4.3). This change is primarily due to the degree of calcareous cementation apparent both in the pebble beds and the overlying sandstone. The change in intergranular permeability is reflected by a change in transmissivity (Table 4.3), although not of the same magnitude, because some of the transmissivity is provided by fissure flow. Two factors lead to a drastic reduction of transmissivity at depth. One is the loss of fissure permeability because the fissures are closed by overburden pressure. The other is the reduction in porosity with depth leading to a marked reduction in intergranular permeability, because the extent of solution of carbonate cement by acid meteoric water decreases away from outcrop.

Eastwards from the outcrop the Sherwood Sandstone Group thickens and deepens into the Dorset Basin, probably reaching a maximum thickness north west of Dorchester (Figure 4.12). Hydrogeological data are sparse at depth in that area, being limited to values for porosities derived from the sonic logs of three boreholes. The average porosities for these boreholes are 23%, 17% and 19% and the zones of high porosity are distributed throughout the sequence.

Further east the Winterborne Kingston Trough lies in the deepest part of the Dorset Basin, with the top of the Sherwood Sandstone Group at a depth of 2232 m below sea level. Samples of core from the Sherwood Sandstone from the Winterborne Kingston Borehole gave values of porosity ranging from 7% to 20% (the lower values corresponding to intercalated mudstones). Permeabilities ranged from less than 0.1 to 2330 mD with the higher values corresponding to the discrete, poorly cemented sandstones typical of the Wessex Basin and described above for the Southampton area. For example a 3.2 m section at a depth of 2424 m had an average permeability of 1695 mD, giving a value for the transmissivity of this short section alone of 5.4 D m (Burgess, 1982).

Intrinsic permeability data derived from pressure records of drill-stem tests have been extrapolated to the full sandstone sequence at Winterborne Kingston by inspection of geophysical logs. The weighted average of the bulk intrinsic permeability is 70 mD over some 150 m, corresponding to a transmissivity of 11 D m. However, the data were affected by the proximity of a hydraulic barrier and a more realistic indication of the transmissivity is 15 D m (corresponding to an average permeability of 100 mD over 150 m). As the lithology of the sandstone is similar to that close to outcrop, the much lower transmissivity of these estimates compared to those given in Table 4.3 is considered to be caused mainly by the presence of intergranular cement and the lack of fissures.

Between the Winterborne Kingston Trough and the Purbeck-Isle of Wight Monocline, the Sherwood Sandstone has been proved by several boreholes which indicate favourable hydrogeological characteristics with porosities exceeding 18% over total thicknesses commonly exceeding 50 m, and with transmissivities in places of 20 D m or more.

Near the Purbeck-Isle of Wight Monocline and the Cranborne Fault these favourable values of porosity and transmissivity are drastically reduced by increased cementation. These structures are therefore assumed to act as a hydraulic boundary to the Dorset Basin.

Table 4.3 Permo-Triassic sandstones: Hydraulic characteristics at outcrop in South-West England

Data		Field data	Laboratory Data
Porosity	(%)	—	25–35 (5–20)
Hydraulic conductivity	(m/d)	1.25	1.5–5.5 (0.1)
Intrinsic permeability (mD)		1950	1000–9000 (200)
Transmissivity		200 m^2/day 310 D m (30 D m)	—
Thickness	(m)	100–200	

Values refer to the loosely cemented sandstones at outcrop in east Devon, those in brackets refer to the more heavily cemented outcrop in west Somerset.

Piezometric surface

Piezometric data relating to significant depths in the Wessex Basin are sparse and confined to two areas, the Dorset Basin and the Southampton area. To the west of Bournemouth evidence from drill-stem tests indicates that the piezometric surface is of the order of 200 m below ground level. If it is assumed that the brine in this area (with salinities of the order of 200 to 300 g/l) is in hydraulic continuity with the fresh water at outcrop in Devon and that the salinity decreases linearly to the outcrop, then this is approximately the level of the piezometric surface that would be expected. In the Southampton area the static piezometric surface proved to be only a few metres below ground level in an area where the Sherwood Sandstone Group is 1700 m deep and contains a brine with a salinity of about 100 g/l.

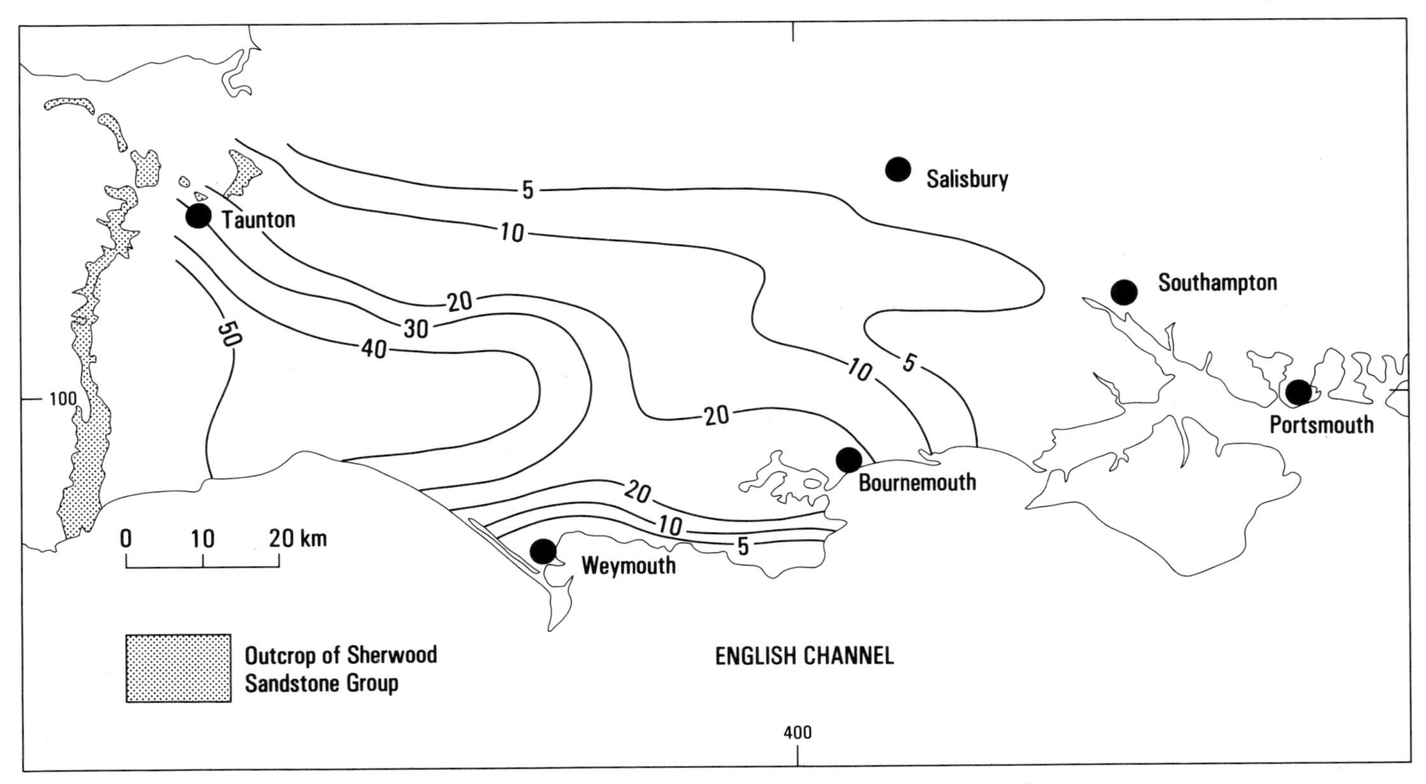

4.14 Estimated transmissivity of the Sherwood Sandstone Group in the Wessex Basin (darcy metres).

Transmissivity

There are very few direct measurements of the transmissivity of the Sherwood Sandstone at depth in the Wessex Basin, and those that have been made are mainly confined to the Southampton area, and to an area to the west of Bournemouth. In constructing a transmissivity map (Figure 4.14) much reliance has been placed on porosity data as a guide to transmissivity.

The transmissivity near Southampton is about 5 D m. It is believed to increase towards the west on the basis that, as the Sherwood Sandstone Group thickens in that direction,

4.15 Salinity of groundwater in the Sherwood Sandstone Group in the Wessex Basin (grams/litre).

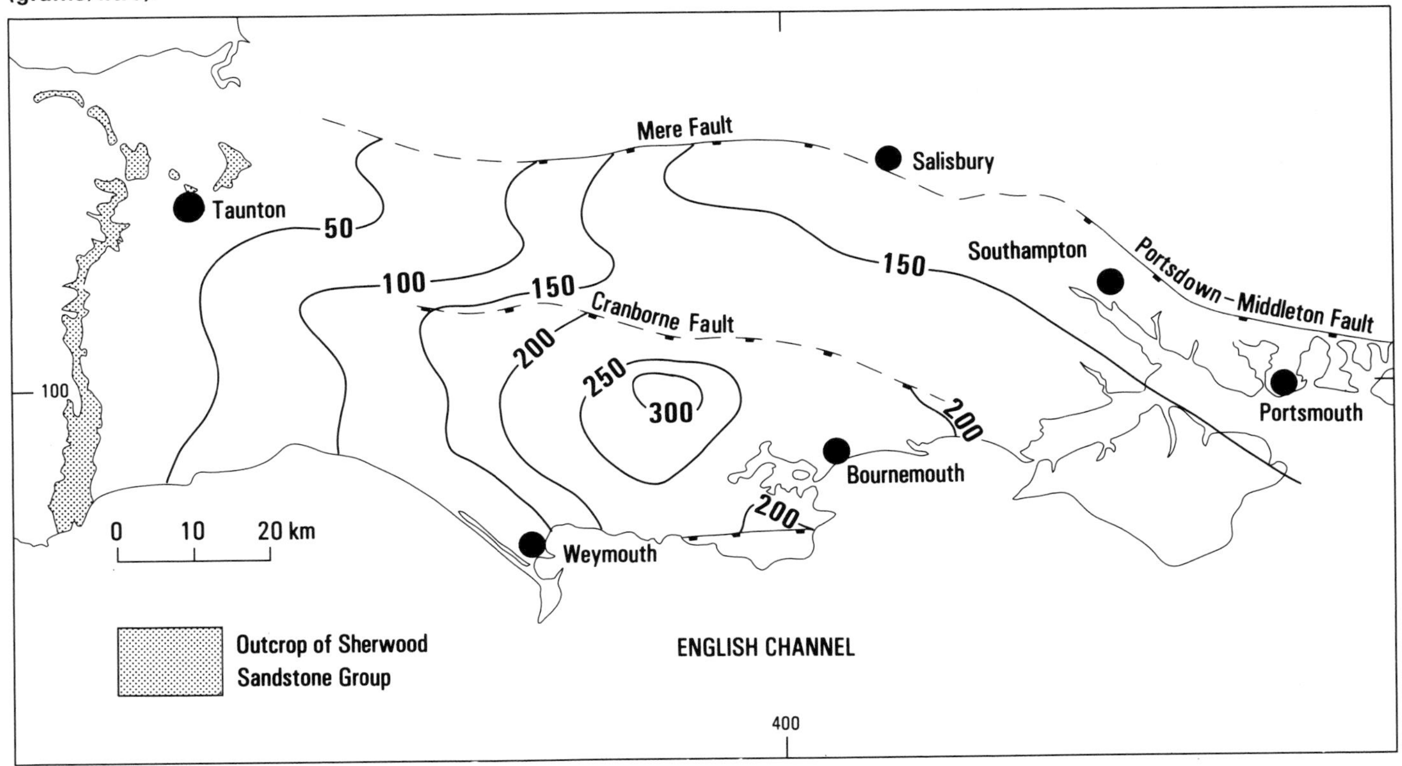

the transmissivity will increase proportionally. The underlying assumption is that the average permeability of the water-bearing horizon remains constant at about 300 mD. Values increase towards Bournemouth until some 20 D m are attained in an area to the west of Bournemouth in the Dorset Basin. There may be a significant reduction in permeability immediately to the south of the Cranborne growth-fault, and the Mere Fault system because of a higher degree of cementation.

South-west of Bournemouth the permeability of the Sherwood Sandstone rapidly decreases towards the Purbeck-Isle of Wight faults and it is considered that Weymouth lies in an area where values for transmissivity are less than 5 D m. To the west of Bournemouth, however, the thickness of the Sherwood Sandstone increases and with it the thickness of water-bearing material with porosities greater than 18%. Boreholes 50 km west of Bournemouth indicate twice the thickness of aquifer material that exists near Bournemouth, and it is assumed that the transmissivity is doubled accordingly — to more than 40 D m. Still further west, towards the southern outcrop of the Sherwood Sandstone Group, the transmissivity increases to values of the order of 300 D m (Table 4.3), but most of the increase is thought to occur relatively near to the outcrop.

Hydrochemistry

The variation in the salinity of groundwater in the Sherwood Sandstone is shown in Figure 4.15. Apart from the outcrop area and the zone immediately to the east, the waters are brines with total ionic concentrations exceeding 100 g/l over most of the basin. Maximum values of over 300 g/l are attained to the west of Bournemouth. In general values increase with depth and although the figure indicates the general regional variation, marked differences can occur within very limited areas. For example, in the Marchwood Well the salinity was 103 g/l, whereas 1.9 km away in the Southampton Well 125 g/l was measured. A similar difference was recorded between Winterborne Kingston and an oil well to the south east, although in this case the two wells are separated by a fault. The geochemistry of these brines is considered in more detail in Chapter 6.

Sub-surface temperatures in the Sherwood Sandstone Group

The temperature at the centre of the Sherwood Sandstone Group has been estimated from bottom-hole temperature measurements (BHT) made in boreholes (Figure 4.16). Representative thermal conductivities have been calculated for the major lithological groups and used to interpolate appropriate temperature gradients for each lithological group between BHTs. Because a value of heat flow is derived from such calculations, it has been possible to extrapolate beneath shallower boreholes to the required depth, again using assumed thermal conductivities for the expected strata. Thermal conductivities have been estimated from values available to BGS from earlier work in the area, which are listed in Burley and others (1984). Table 4.4 shows the estimated values for important rock groups in the Wessex Basin.

In addition to the thermal conductivity, at each borehole site the thickness was obtained from the borehole record,

4.16 Estimated temperature at the centre of the Sherwood Sandstone Group in the Wessex Basin (°C).

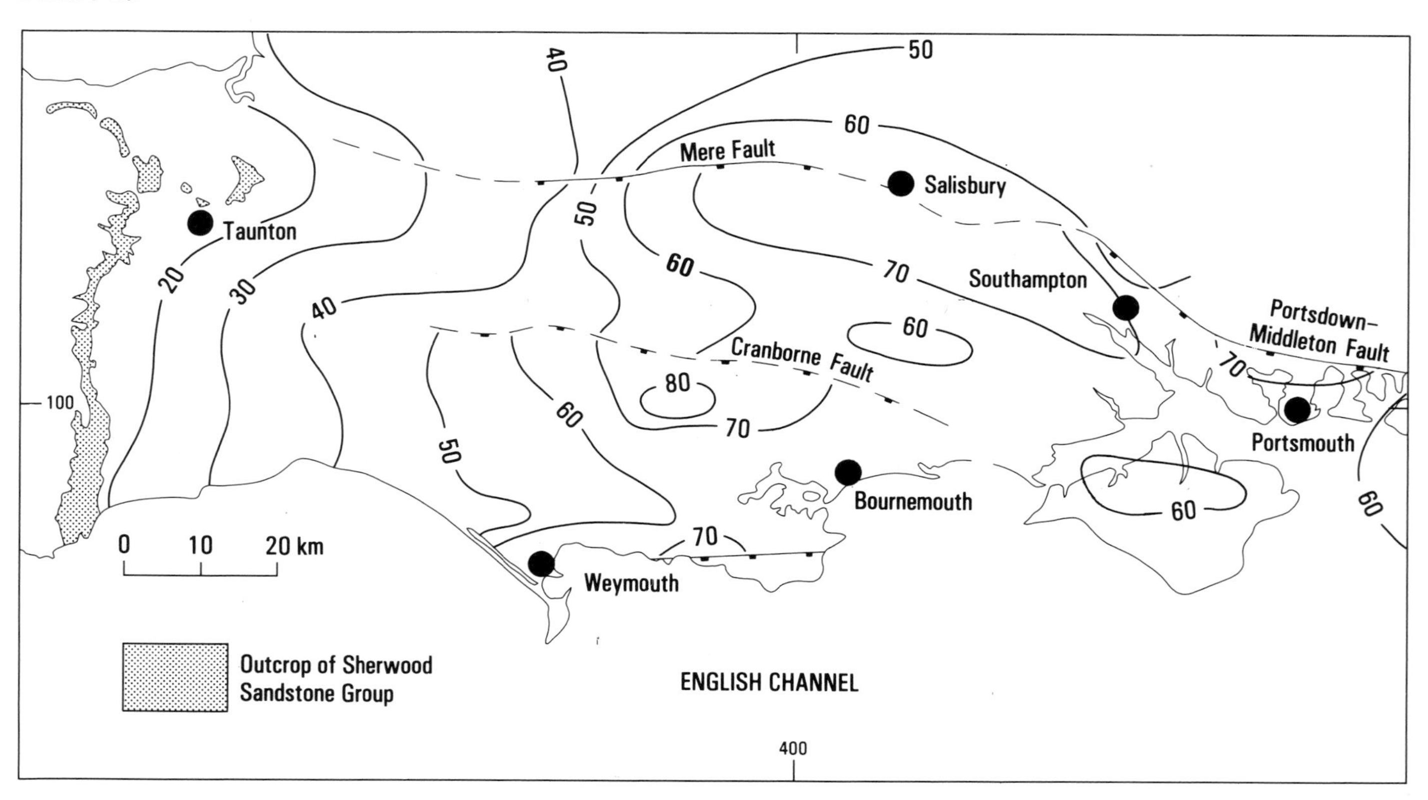

Lithological group	Thermal conductivity (W/m K)
Tertiary	1.8
Cretaceous	2.0
Jurassic	1.8
Mercia Mudstone	2.0
Sherwood Sandstone	3.1
Permian	2.1
Carboniferous	2.9
Devonian	2.6

Table 4.4 Values of thermal conductivity adopted for major lithological groups in the Wessex Basin

from seismic reflection interpretations or from geological extrapolation. The BHTs and ground level temperatures were taken from Burley and others (1984). These data have been used to estimate the temperature at the top of each lithological group, to satisfy the BHTs closest to the Sherwood Sandstone. It is considered that the deepest BHT is the least likely to be affected by fluid movement through permeable formations (although it is not suggested that such flows have not affected some of the values). However, the best temperature estimate for a particular level will be given by the closest measurement. This compromise is satisfactory in practice because in many cases the two considerations overlap, since the Sherwood Sandstone has been the target of many boreholes, and the completed depth reflects the depth of the sandstones. Where heat flow determinations are available, the local values of conductivity have been used in the calculations, and the heat flow values have been used for extrapolation to the appropriate depth.

The temperature distribution at the centre of the Sherwood Sandstone (Figure 4.16) reflects to a large degree the depth of the reservoir. Highest temperatures are attained in the Dorset Basin with more than 70°C, and a maximum of 80°C, occurring to the south of Blandford Forum. Over 70°C is attained to the south of the Mere Fault and the Portsdown-Middleton Fault line, including the Southampton area where temperatures of water derived from the sandstones during pumping tests were about 75°C. However, near the fault line itself reservoir properties are not likely to be suitable for development. In the Central Channel Basin, to the east of Kimmeridge Bay and including the southern part of the Isle of Wight, values also exceed 70°C. More than 60°C is attained over very extensive areas to the south of the Mere-Portsdown-Middleton Fault line.

Geology and hydrogeology of the Lower Cretaceous sandstones

Lower Cretaceous sandstones, referred to as the Lower Greensand, occur in the Wessex Basin to the south-west of the Weald, at depths where temperatures exceed 20°C. They are glauconitic and calcareous sandstones with minor siltstone and mudstone horizons.

In this area the depth to the Lower Greensand gradually increases from the outcrop to over 1000 m in a shallow syncline on the Isle of Wight, which lies to the north of the major east-west structure that intersects the island (Figure 4.17). Unfortunately, as the depth increases the thickness reduces from 150 m at and near the outcrop to no more than 10 to 20 m on the Isle of Wight and along a line extending north-west from Portsmouth (Figure 4.17).

4.17 Depth to the top and thickness of the Lower Greensand between Bournemouth and Worthing.

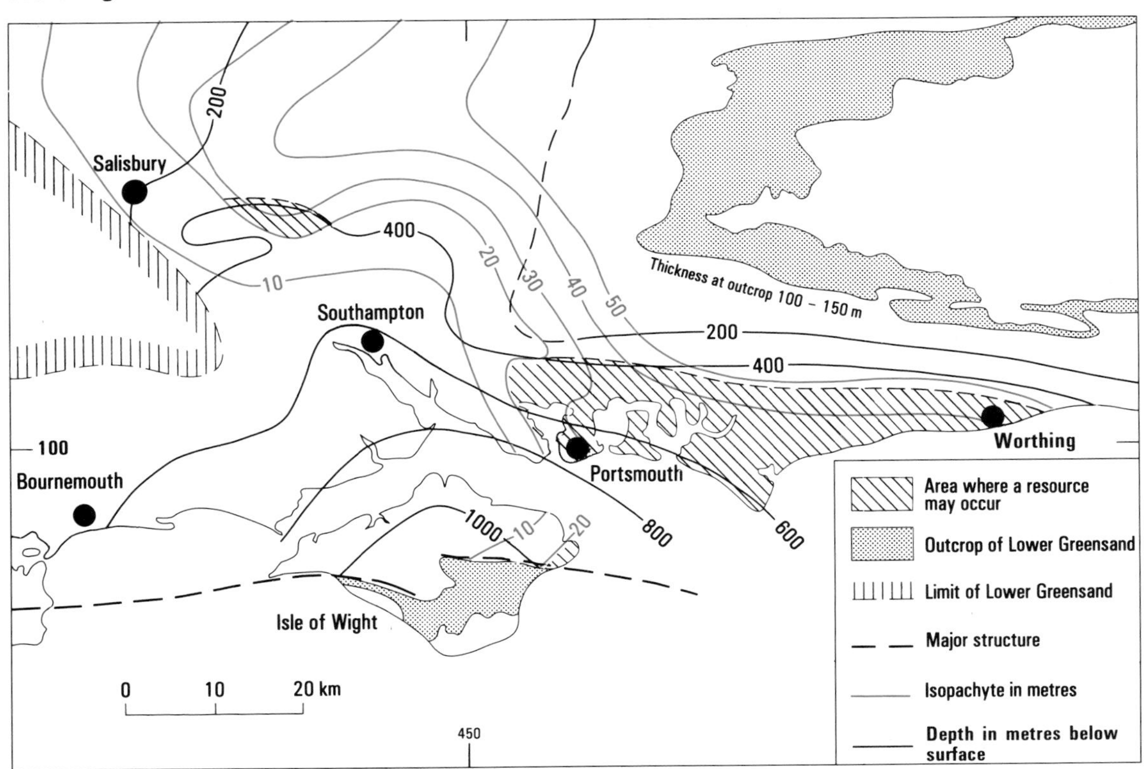

Information about the distribution and thickness of the Lower Greensand away from the outcrop is based on very limited data, but the evidence currently available implies that there are small areas where the formation is between 20 and 50 m thick at depths of more than 400 m (Figure 4.17). The principal area where these conditions obtain is along the south coast between Worthing and Portsmouth, although a small area also exists near Salisbury. In the north-east of the Isle of Wight the thickness may amount to 20 m at a depth of over 900 m. This area is adjacent to the major fault-structure in the Isle of Wight which is likely to act as a flow barrier and thereby reduce the rate of replenishment and increase the drawdown for a given yield from a well. However, the aquifer extends below the sea and this would increase the size of the resource available.

A borehole at Sompting, near Worthing (Young and Monkhouse, 1980) proved almost 50 m of Lower Cretaceous sandstone at a depth of 404 m. The upper 35 m were medium- to coarse-grained, uncemented or poorly cemented sandstones. The underlying 14m were very fine-grained clayey, glauconitic sandstone. The sequence yielded 1000 m^3 of water per day at a temperature of 21°C, indicating an average thermal gradient of about 25°C/km, assuming an average surface temperature of 11°C. This gradient would give a temperature of 35°C at a depth of 1 km. Test-pumping of the borehole at Sompting gave a transmissivity of 70 m^2/d (Young and Monkhouse, 1980) which implies an average permeability of some 2 darcys. This obtains at a depth of 400 m and is likely to decrease with increasing depth. The only other value for permeability that is available in the area is based on core analyses for the Marchwood Well, where the thickness is only 8 m; values ranged up to 1 darcy, but the average is probably of the order of 10 mD (Allen and Holloway, 1984).

The temperature of water in the Lower Greensand could be 20 to 30°C in the coastal zone between Portsmouth and Worthing. In view of the relatively high transmissivity at Sompting, permeability in this coastal zone may be about 250 mD, in which case a transmissivity of 5 Dm would be attained with a thickness of 20 m. On this basis a geothermal resource, requiring development with heat pumps, would exist in the Portsmouth-Worthing area with a mean temperature of approximately 25°C over an area of about 550 km^2. Clearly, exploration would be necessary to obtain more data to test the assumptions just given.

The water from the Lower Greensand at Sompting is fresh with a salinity of only 100 mg/l (Young and Monkhouse, 1980). As fresh water commonly extends some distance down-gradient from the outcrop in this aquifer, it is possible that water in the aquifer along the south coast is also fresh and, if developed, could be used for water supply after the heat had been extracted.

WORCESTER BASIN

This basin extends from near Droitwich in the north to close to Cirencester in the south, and lies to the east of the Malvern Hills. The form of the basin is not evident from the geological map (Figure 4.18) because of the overlapping Lower Jurassic succession, but a general indication of the form is given by the −20 mGal contour line (Figures 4.18 and 4.20) which encloses the area where development of the Permo-Triassic sandstones is greatest and where the deepest aquifers may be expected (Smith and Burgess, 1984). There are four main centres of population in the Worcester Basin: Worcester, Evesham, Cheltenham and Cirencester. Evesham, Cheltenham and Cirencester are all situated near the edge of the basin, in less favourable positions than Worcester from the point of view of developing and using geothermal energy. Most of the area consists of rolling agricultural countryside, with the escarpment of the Cotswold Hills forming a limit in the east.

General geology

A map of the solid geology of the Worcester Basin is presented in Figure 4.18 and Figure 4.19 shows the stratigraphical correlation of the Permian and Triassic rocks between key boreholes. About two-thirds of the area is covered by Lower Jurassic rocks, which are mainly argillaceous but locally include limestone beds. They contain no significant aquifers and do not constitute a geothermal resource. Beneath the Jurassic is the Penarth Group, of Rhaetic age,

4.18 Geological map of the Worcester Basin.

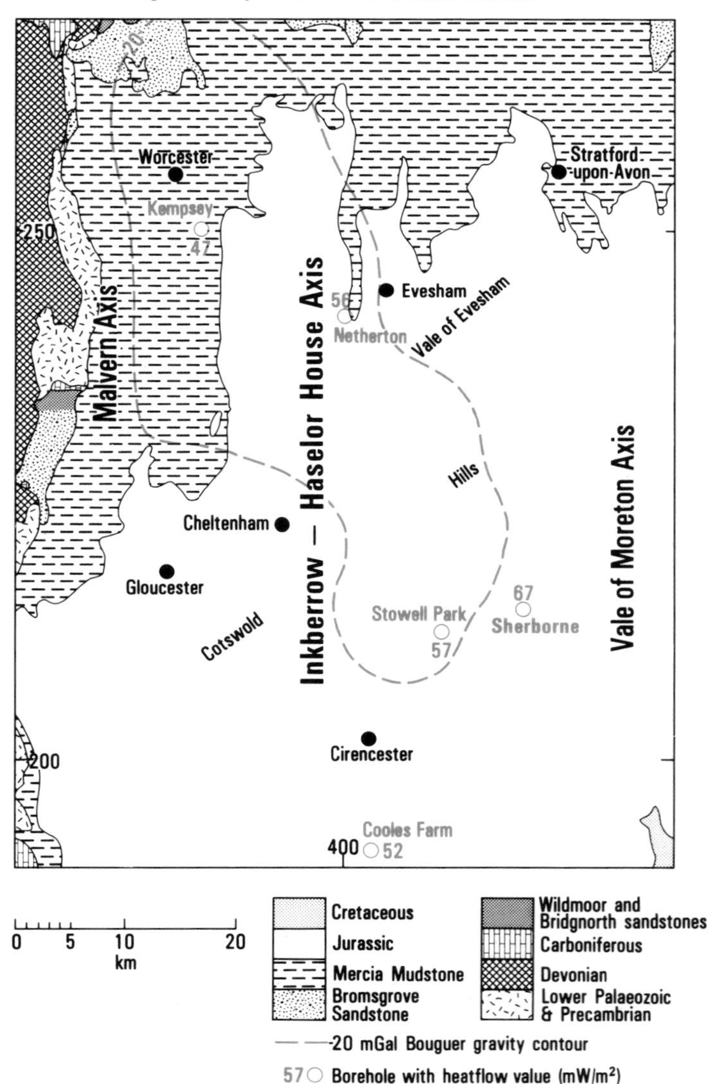

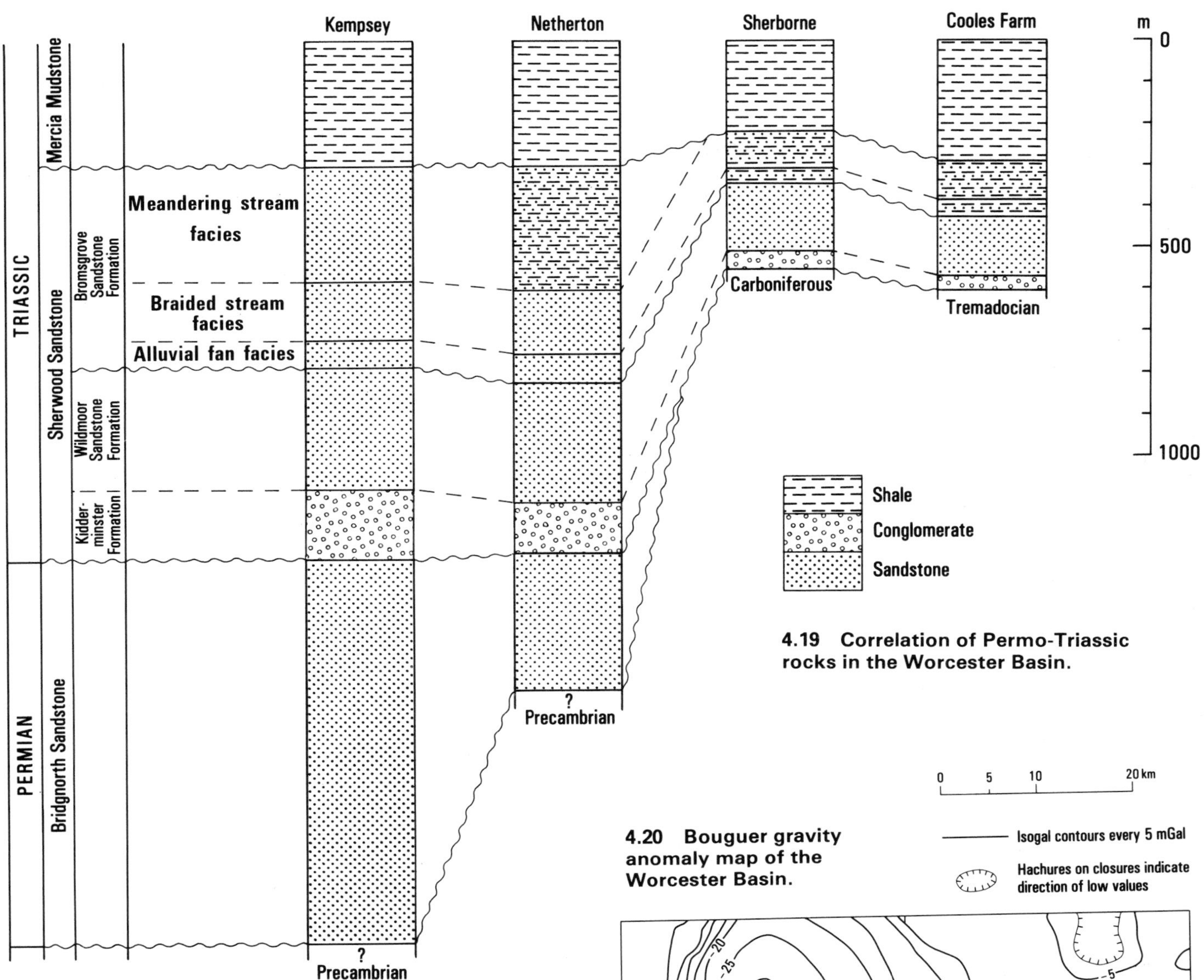

4.19 Correlation of Permo-Triassic rocks in the Worcester Basin.

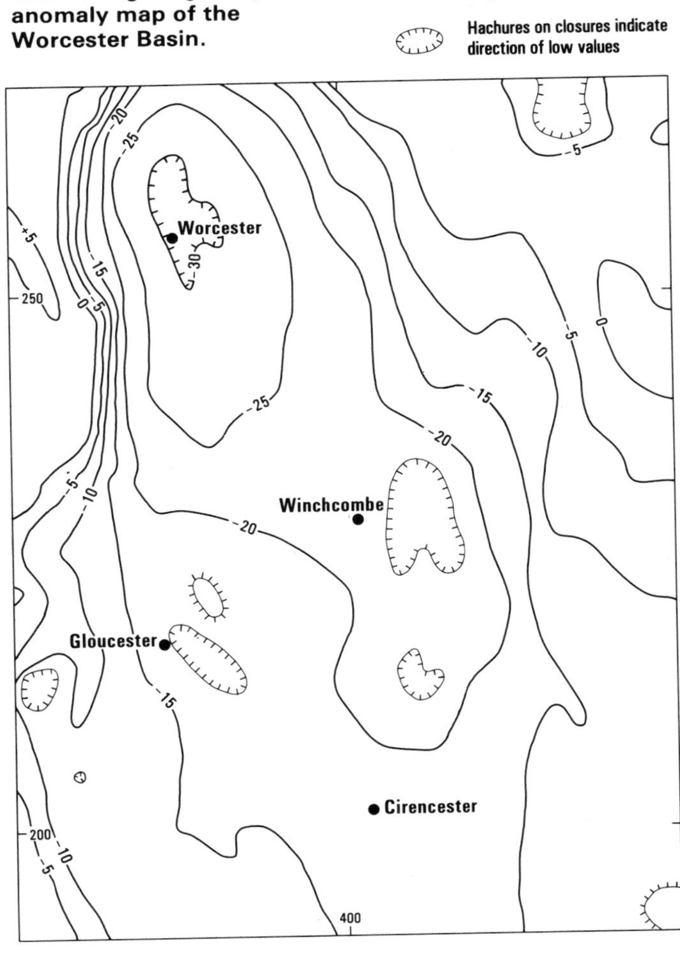

4.20 Bouguer gravity anomaly map of the Worcester Basin.

which probably nowhere exceeds a thickness of 10 m. This in turn overlies the largely argillaceous Upper Triassic Mercia Mudstone Group which has a uniform thickness over much of the Worcester Basin, averaging about 300 m.

The Permo-Triassic sandstones lie beneath the Mercia Mudstone Group. The sandstones have restricted outcrops along the western and northern boundary of the basin, adjacent to the Newent Coalfield, and the Forest of Wyre and South Staffordshire coalfields respectively. However, away from the margins, boreholes have proved great thicknesses of sandstone with considerable lateral variations. The Permo-Triassic sandstones overlie beds ranging in age from Precambrian to Carboniferous. The four major arenaceous formations in the Permo-Triassic are briefly described below, from the oldest upwards.

The Bridgnorth Sandstone, of assumed Permian age, is similar to, and may correlate with, the Collyhurst Sandstone of the Cheshire and West Lancashire basins and the Penrith Sandstone of the Carlisle Basin. It consists of bright red sandstones, with large scale cross-bedding and many rounded, wind-polished grains indicating a continental, aeolian deposit. The grains are lightly cemented by a continuous coating of iron oxides which does not greatly reduce the porosity or permeability. Local thin marl bands occur.

The formation has a restricted outcrop around the basin edge, but it has been intersected by several deep boreholes. It infills the irregular pre-Permian topography and is consequently variable in thickness, reaching a recorded maximum of 938 m in the Kempsey Borehole (Figures 4.18 and 4.19). Locally basal breccias, several tens of metres thick, underlie the sandstone: these may mark the Carboniferous-Permian boundary in the English Midlands.

Unconformably overlying and overstepping this unit is the Kidderminster Formation, which is the lowest Triassic division. Locally it infills steep-sided channels cut into the underlying sandstones. It consists of coarse, cross-bedded sandstones, which are locally heavily cemented. It characteristically contains large rounded pebbles, the number of which vary considerably from place to place. The formation has been proved by boreholes throughout the basin and is up to 175 m thick. Both the top and the bottom are believed to be diachronous.

The Wildmoor Sandstone Formation oversteps the Kidderminster Formation, although commonly the base is gradational. In appearance and lithology it resembles the Bridgnorth Sandstone, but is considered to be fluviatile in origin. It attains 430 m in one borehole, but is thinly developed on structural highs.

The Bromsgrove Sandstone Formation rests unconformably upon the Wildmoor Sandstone Formation, from which it is lithologically distinct. There is evidence of rejuvenation of the river system that existed at the time of its formation, as a consequence of the relative uplift of the source area, resulting in an unconformable contact between the two. The formation consists of fluvial, fining-upwards cyclic deposits, which include conglomerates, sandstones, siltstones and mudstones. This cyclicity is superimposed on a gradual upwards fining, interpreted as a result of evolution from an alluvial fan, through a braided river environment, to flood plains with meandering rivers. Its thickness, which is notably uniform except where it laps onto older rocks at the edges of the basin, averages 500 m and attains a maximum of 531 m in the Netherton Borehole. In the distal part of the formation, in the south of the area, the beds consist of mudstone, with little resource potential. Overlying passage beds are included with the Mercia Mudstone.

The basin is bounded by three major axes or belts of normal faulting (Figure 4.18), trending generally north-south, which have controlled the development of the deposits. The fault belts appear to have variable displacements along their lengths and, indeed, the sense of throw changes along the Inkberrow-Haselor Hill Fault Belt

4.21 Structure contours (metres below sea level) and temperature at the base of the Mercia Mudstone Group in the Worcester Basin.

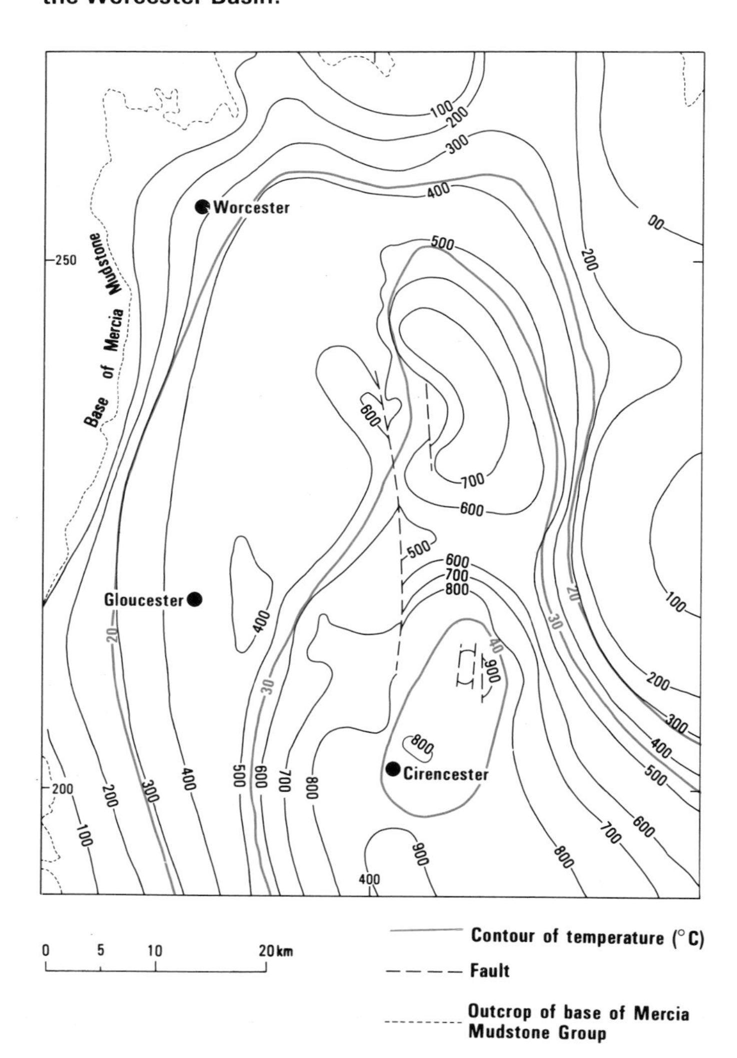

4.22 Structure contours (metres below sea level) and temperature at the base of the Permo-Triassic in the Worcester Basin.

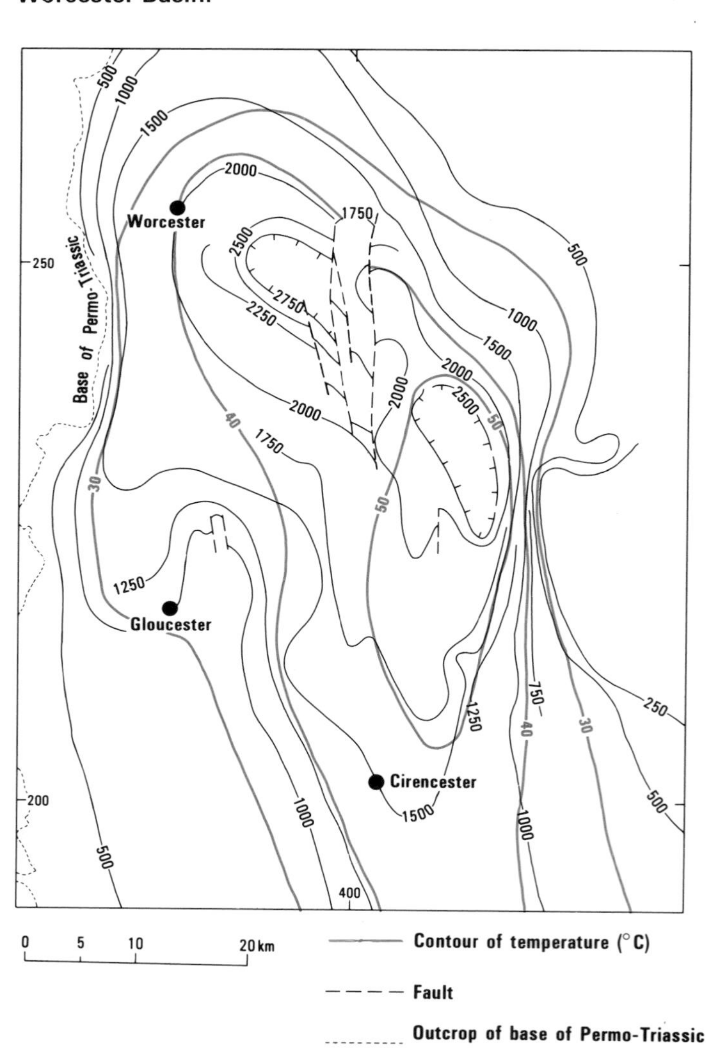

from a westerly downthrow in the north to an easterly downthrow in the south. This results in two *en échelon* basins (one beneath Worcester and the other beneath Winchcombe) which are indicated by the shape of the Bouguer gravity anomaly contours (Figure 4.20).

The depth to the base of the Mercia Mudstone Group, including local mudstone facies of the Bromsgrove Sandstone in the south of the basin, is shown in Figure 4.21. The depth to the horizon increases from both east and west into a trough which runs between the Inkberrow/Haselor Hill and Moreton axes. The greatest depth is in the south, reaching over 900 m around the Stowell Park Borehole.

The depth to the base of the Permo-Triassic is shown in Figure 4.22. Two sub-basins are revealed on either side of the Inkberrow-Haselor Hill Axis. The deeper, which may extend to 2 750 m, lies to the west of the faults, some 8 km to the east of the Kempsey Borehole, which proved a depth of 2285 m. The subsidiary basin to the east reaches about 2250 m. The sub-basins are separated by a horst-like structure at a depth of about 1700 m, as proved by the Netherton Borehole.

The isopachyte map of the Permo-Triassic sandstones (Figure 4.23) shows that the basins were actively subsiding along the major faults while the sandstones were being deposited. Considerable variations in the thickness occur along the faults but the greatest thicknesses are found in the centre of the basins. The maximum thickness of sandstone appears to occur between the Kempsey Borehole and the Inkberrow-Haselor Axis, and may attain more than 2250 m.

4.23 Isopachytes of the Permo-Triassic sandstones in the Worcester Basin.

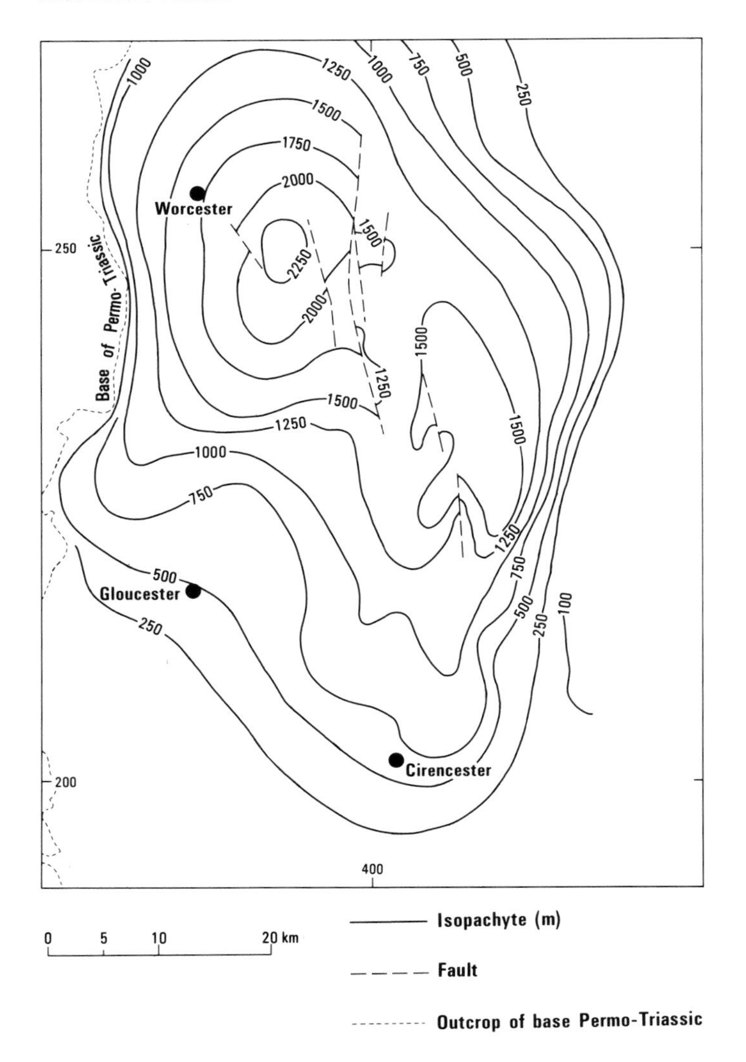

Sub-surface temperatures

Temperatures of rocks within the Worcester Basin have been determined using three different types of data: heat flow determinations, bottom-hole temperatures and temperature logs. These demonstrate in different ways how the temperature varies within the basin.

Bottom-hole temperatures have been measured in twelve boreholes in the basin (Burley and others, 1984) and values selected as being suitable are presented in Table 4.5. These data provide the best coverage of the basin and therefore the most effective way of calculating the temperature distribution in aquifers. To achieve this, thermal conductivities for the major lithological groups have been estimated from various sources including Poole (1977 and 1978), Richardson and Oxburgh (1978) and Bloomer (1981). A value of 1.97 ± 0.15 W/m K has been assigned to mudstones and 'marls' and 3.30 ± 0.2 W/m K to sandstones. These values have been used to interpolate an appropriate gradient between measured temperatures and thus provide values at the top and bottom of the sandstone. Temperature contours based on these calculations are shown in Figures 4.21 and 4.22. Those temperature gradients which have been judged to be of local significance only have not been given their full weight. The contours represent, therefore, the predicted regional temperature variations. It should be noted that the temperature gradients in the Mercia Mudstone Group are nearly double those in the Sherwood Sandstone Group, because the thermal conductivity of the former is approximately half that of the latter. Temperatures increase relatively slowly with depth within the sandstone formation.

Heat flow determinations have been made in the basin at the Malvern Link Borehole, which has a value of 34 mW/m^2, and in the Worcester Borehole, with a value of 41 mW/m^2. The temperature calculations also provide approximate values of heat flow. They indicate a variation between 45 and 70 mW/m^2. There are some substantially higher values as in the upper section of the Cooles Farm Borehole and some low values, as at Kempsey. These variations probably result from groundwater movement and should be regarded as local effects. Values determined in shallow boreholes such as Eldersfield, Malvern Link and Twyning must be expected to be only approximate, because of simplifications in the calculation, as may be judged by the difference between the experimentally determined value of 34 mW/m^2 and the estimated value of 45 mW/m^2 for the Malvern Link Borehole. If the disturbed values are omitted the regional heat flow is probably between 50 and over 60 mW/m^2 and is similar to the national average.

Hydrogeology

Geological and geophysical analyses indicate that the Worcester Basin contains a thick Permo-Triassic sedimentary

Borehole	Data type	Ground level (m above OD)	Corrected temp (°C)	Depth (m)	Average gradient (°C/km)	Mercia Mudstone Depth to base (m)	Mercia Mudstone Temperature (°C)	Permo-Triassic sandstones Depth to base (m)	Permo-Triassic sandstones Temperature (°C)	Heatflow mW/m²
Cooles Farm	BHT	90	65.9	1298	42.3		56.2		60.9	90
			80.5	2740	25.4	1022	37.0	1198	39.7	52
			90.7	3431	23.2					
Eldersfield	Log/BHT	42	19.4	250	34.8					
			24.5	398	34.7	368	24.2	(1500)	(49.2)	74
Highworth	BHT	104	39.3	1053	28.3		37.4		39.5	58
			41.0	1160	27.2	932	36.0	1061	38.2	55
Kempsey	BHT	20	37.2	1565	16.8					
			63.1	3003	17.4	421	20.0	2301	46.8	47
Malvern Link	HF	50	15.0	245	17.6	215	14.3	(1300)	(29.4)	45; *34
Netherton	BHT	51	61.8	2324	22.0	474	23.5	1759	54.3	56
Ombersley	EQM	41	12.6	175	10.3	—	—	(1600)	(33.5)	49
Sherborne	BHT	191	43.1	1055	31.6					
			55.2	1939	23.4	767	36.1	974	40.3	67
Stowell Park	BHT	171	42.8	1169	27.9	1086	41.4	(1750)	(52.9)	57
Twyning	Log	34	18.1	257	28.4	(420)	(23.2)	(1800)	(49.1)	62

* Indicates heat flow determination
Values in brackets are estimates of thickness and temperature.
Depths are given below ground level. In some boreholes two values of estimated temperatures are given, resulting from different evaluations of the BHTs.

HF is heat flow
BHT is bottom hole temperature
Log refers to temperature log
EQM is equilibrium temperature

Table 4.5 Temperature measurements and calculated heat flow values for boreholes in the Worcester Basin

sequence, which includes aquifers with potential to supply thermal water in sufficient quantities and at temperatures suitable for the use of their thermal energy; these aquifers are the various Permo-Triassic sandstones discussed above.

The geological sequence in the Worcester Basin is set in a hydrogeological context in Table 4.6, and the characteristics of the major arenaceous and rudaceous divisions are described below. Information on the hydraulic properties of Permo-Triassic sandstones from outcrop areas is available in Skinner (1977), Lovelock (1977) and Ramingwong (1974). Information about the sandstones in the deeper parts of the basin is available as a result of seven deep boreholes and their associated geophysical logs. Data from laboratory measurements on core material, including that from the Kempsey Borehole and from drill-stem tests at Kempsey, are also available. Black and Barker (1983) have described an

Table 4.6 Hydrogeological units of the Permo-Triassic in the Worcester Basin

	Formation	Thickness (m)	Hydrogeological notes
TRIASSIC			
Mercia Mudstone Group	Mercia Mudstone Passage beds (equivalent to "Keuper Waterstones")	300, uniform	Aquiclude but includes the Arden Sandstone
Sherwood Sandstone Group	Bromsgrove Sandstone Fm.	500, uniform 531, maximum	Multiple aquifer; <50% contributory sandstone; fluvial sandstone with transmissivity up to 80 D m
	Wildmoor Sandstone Fm.	430, maximum	
	Kidderminster Fm.	170, maximum	Aquitard; low porosity, well-cemented, conglomeratic sandstone
PERMIAN	Bridgnorth Sandstone	up to 940 but variable	Aquifer: aeolian sandstone, transmissivity up to 115 D m

analytical model which highlights the factors controlling the vertical movement of groundwater in the Mercia Mudstone of the Worcester Basin.

Bridgnorth Sandstone

The porosity of the Bridgnorth Sandstone near the outcrop is between 25 and 30%, and the limited information available from the deeper parts of the Worcester Basin indicate these values are apparently maintained. Groups of abstraction boreholes at outcrop can yield 10Ml/d where a contribution to the total transmissivity is made by secondary permeability of fractures. The only in situ measurement of permeability of the Bridgnorth Sandstone in the deeper parts of the Worcester Basin is a single drill-stem test in the Kempsey Borehole at a depth of 1370 m where the average over a 6.1 m interval, selected as typical of the formation, was 150 mD (corresponding to 0.17 m/d under the prevailing conditions). The piezometric surface was calculated to be about 20 m above ground level. Laboratory tests on core material over an interval of 8 m with rather lower porosity than the majority of the formation, showed the intergranular permeability varies markedly; while the average porosity was 20%, the horizontal permeability ranged from 3 to 3000 mD. Of the total thickness of 938 m at Kempsey, the geophysical logs suggest that 925 m would be likely to contribute water. In the light of the drill-stem test just referred to, the total transmissivity may be about 115 D m, equivalent to 130 m^2/d.

Kidderminster Formation

The thickness of the Kidderminster Formation is extremely variable, but it has a maximum proven thickness of 175 m. A combination of very poor sorting and a ubiquitous carbonate cement reduces the porosity and permeability of the conglomerates considerably. Surface outcrops of the formation are often decalcified and may give a misleading impression of conditions at depth. The geophysical logs from boreholes in deeper parts of the basin commonly indicate a porosity of less than 15%, which is significantly less than might be expected. It is likely, therefore, in the absence of fractures, that the Kidderminster Formation will act as an aquitard between the more porous and permeable arenaceous units above and below.

Wildmoor and Bromsgrove Sandstone formations

In the absence of sufficient evidence to the contrary, the Wildmoor and Bromsgrove Sandstone formations are considered as a single hydrogeological unit, with a maximum proven combined thickness of 894 m. Both the Wildmoor and the Bromsgrove sandstones constitute a series of minor aquifers, in the former case on account of discrete but numerous argillaceous interbeds and in the latter case on account of the more laterally extensive argillaceous units of the fluvial cyclothems. The sandstones in both cases are generally lightly cemented with a red iron oxide cement, but are finer grained than the aeolian sandstones of the Bridgnorth Sandstone. Their porosity is high, between 20 and 30%, and close to outcrop their intergranular permeability ranges from 0.5 to 5 m/d (Skinner, 1977). Under normal groundwater conditions this implies an intrinsic permeability in the range 780 to 7800 mD. Individual abstraction boreholes at outcrop may produce between 4 and 10 Ml/d, the variation depending upon the local cementation and jointing characteristics of the formation. In the Droitwich area the permeability of the Bromsgrove Sandstone is remarkably consistent, and the average of 28 in situ determinations is 650 mD (Ramingwong, 1974). Deeper in the basin the porosity of the Wildmoor and Bromsgrove sandstones does not appear to be greatly reduced. Their permeability, although rather less than at outcrop, is still significant.

At Kempsey the average porosity of a representative 18 m interval was measured as 23%, and the respective average horizontal permeability was 690 mD. A drill-stem test over part of this cored interval (which had an average permeability as determined in the laboratory of 450 mD) indicated a permeability of 290 mD. The geophysical logs show that sandstones which are likely to contribute water make up less than half of the Wildmoor Sandstone Formation at Kempsey, on account of the regularly occurring argillaceous interbeds. From its 285 m total thickness, beds of clean sandstone make up only 120 m (42%), the overall transmissivity of which has been estimated at 34 D m. Under the prevailing conditions of temperature and groundwater density this represents a transmissivity of 39 m^2/d. The Bromsgrove Sandstone contains contributory sandstone in a similar proportion. Geophysical logs from other deep boreholes in the basin generally indicate a component of contributory sandstone of rather less than 50% for both the Wildmoor and the Bromsgrove Sandstone formations. Laboratory determinations of the hydraulic properties of other core material from the Bromsgrove Sandstone, from a borehole where its total thickness is 450 m, indicated an average porosity of 20% and an average horizontal permeability of 1440 mD. In the light of these determinations and an assessment of the geophysical logs, the transmissivity of the Bromsgrove Sandstone is estimated to be in the order of 80 Dm.

The Wildmoor and Bromsgrove Sandstone formations, therefore, constitute a series of thin sandstones which may locally be isolated but which combine to form a multi-aquifer of considerable transmissivity, likely to be several tens of darcy-metres even in the deeper parts of the basin.

Hydrochemistry

Data on the chemistry of formation waters from the Permo-Triassic sandstones in the deeper parts of the Worcester Basin are very limited. Formation waters from the Kempsey Borehole (Table 4.7) are not so highly mineralised as groundwaters from the Permo-Triassic sandstones in the Wessex Basin (Allen and Holloway, 1984) or the East Yorkshire and Lincolnshire Basin (Gale and others, 1983). The formation waters listed in Table 4.7 were obtained by centrifuge extraction of pore-water from core material, or collected in bulk from drill-pipe subsequent to drill-stem

	Triassic Wildmoor Sandstone		Permian Bridgnorth Sandstone	
	DST 1	Core 1	DST 2	Core 2
Depth (m)	936–942	936	1393–1399	1486
pH*	8.05	7.35	6.90	7.30
Total mineralisation (mg/l)	6010	5580	22260	28850
Ca	340	500	1500	2100
Mg	90	140	290	450
Na	1800	1320	6600	8000
K	80	20	195	160
SO_4	820	420	1560	1580
Cl	2840	3130	12000	16300
HCO_3†	N/A	2	N/A	96
Sr	21	28	41	59
Li	1.00	0.95	3.70	3.80
Rb	0.088	N/A	0.260	0.120
Br	14.8	20.1	66.5	104
I	0.014	0.120	0.225	N/A
F	N/A	0.10	N/A	0.45
SiO_2	51.4	21.4	45.0	21.0
Fe total	0.57	0.011	1.70	0.1
Mn	0.50	0.20	0.98	0.91
Cu	N/A	0.010	N/A	0.038
^{18}O ‰ SMOW	−7.9	−8.0	−7.4	−7.0
^{2}H ‰ SMOW	−53	−55	−50	−48
Ionic balance	+3.74	−1.38	+2.61	−0.02

* On-site measurements from DST samples; otherwise laboratory measurements
† Laboratory determinations by the back titration method
N/A not analysed

Table 4.7 Analyses of representative groundwaters from the Triassic and Permian sandstones in the Kempsey Borehole. Units mg/l unless specified

tests. No other information about the deep formation water of the Worcester Basin is available.

If the formation waters at Kempsey are representative, they indicate the aquifer system has been considerably flushed by meteoric water. The relatively low salinity of the water from the Wildmoor Sandstone may indicate that direct recharge by groundwater moving down through the overlying Mercia Mudstone, which includes halite, is not significant. The higher salinity of the Bridgnorth Sandstone formation water supports the suggestion from geophysical logs that the intervening Kidderminster Formation restricts flow between the Wildmoor and Bridgnorth sandstones.

Piezometric surface

The regional groundwater flow pattern in the Worcester Basin is controlled by piezometric levels in the high ground surrounding the basin to the west, north and east. Direct natural recharge of the Permo-Triassic sandstones occurs in the west and north of the basin where groundwater levels are 60 to 100 m above sea level and indirect recharge of groundwater by downward vertical flow occurs below the high ground of the Cotswolds where the water level is 160 to 200 m above sea level. Most of the water entering the Permo-Triassic sandstones in the outcrop areas is discharged to the local river systems, but a limited amount flows below the overlying argillaceous beds into the confined groundwater zone that occupies most of the central part of the basin and the area of geothermal interest.

This pattern of basinward groundwater flow from the recharge areas results in predominantly horizontal flow in the Permo-Triassic sandstones, and vertical flow in the Mercia Mudstone and argillaceous Jurassic strata. Water from the sandstone ultimately moves upwards through the overlying Mercia Mudstone and Jurassic clays in the central parts of the basin to outlets in the Severn flood plain.

The flow balance in the sandstones is controlled by the ratio of the horizontal permeability of the Permo-Triassic sandstones to the vertical permeability of the Mercia Mudstones. At present the only determinations of piezometric pressure in the centre of the basin are from drill-stem tests in the Kempsey Borehole, where the piezometric surface of the Wildmoor Sandstone appears almost to coincide with that of the deeper Bridgnorth Sandstone and both are approximately 20 m above ground level (40 m above sea level).

CHESHIRE AND WEST LANCASHIRE BASINS

The Cheshire and West Lancashire basins comprise Permo-Triassic rocks which underlie the relatively flat ground between the Carboniferous uplands of northern England on the east and the Irish Sea and Welsh Borderlands on the west. The Liverpool and Manchester conurbations are situated along the northern rim of the Cheshire Basin but, apart from these and towns like Crewe and Chester, the area is principally agricultural. The West Lancashire Basin is also largely agricultural but Blackpool and Southport, as well as Liverpool on the southern margin, lie in this area (Figure 4.24).

The structure and geology of the basins with particular reference to the distribution and properties of the sandstones in the lower part of the Permo-Triassic succession are described in the following section, which is based on a report by Gale and others (1984a). The structural interpretations are largely from geophysical data, there being relatively few boreholes which penetrate the full Permo-Triassic succession.

General geology

There are three major Permo-Triassic basins in north-western England: the Solway-Carlisle Basin, the Northern Irish Sea Basin (Colter and Barr, 1975) and the Cheshire Basin. The West Lancashire Basin and the narrow coastal strip to the west of the Lake District Massif are the onshore extensions of the Northern Irish Sea Basin, which is connected *en échelon* to the Cheshire Basin. In general each basin is fault-bounded on its eastern margin, typically by NNE to SSW faults. The Triassic sediments of the narrow coastal strip west of the Lake District dip to the west, away from the Palaeozoic uplands and attain thicknesses of over 1 km (Black and others, 1981), although in a zone with a maximum width of only 9 km at Barrow-in-Furness.

The geology of the Cheshire and West Lancashire basins is shown in Figure 4.24. The former is bounded by Carboniferous rocks of the Pennines to the east and of the Wrexham Coalfield to the west. The eastern margin of the West Lancashire Basin is formed by the Lancashire Coalfield, which lies on the flank of the Pennines. The Permo-Triassic sequence rests unconformably on folded Carboniferous rocks. These, in turn, rest unconformably on Lower Palaeozoic shales, seen in the Prees Borehole, and in the extreme south-west of the Cheshire Basin, where the Carboniferous rocks are absent. The Permo-Triassic deposits infill the pre-Permian topography, which results in considerable lateral thickness variations (Poole and Whiteman, 1955). The entire succession was penetrated in the Prees Borehole in the south of the Cheshire Basin. The correlation between it, the Knutsford Borehole in the north of the basin, and the Formby No. 1 Borehole in the West Lancashire Basin is shown in Figure 4.25.

The Collyhurst/Kinnerton sandstones unit consists of aeolian sands, with dune bedding and 'millet seed' grains and, as a consequence, might be expected to have favourable hydrogeological characteristics. The Manchester Marl represents a marine incursion from the north-west and

4.24 Geology of the Cheshire and West Lancashire basins.

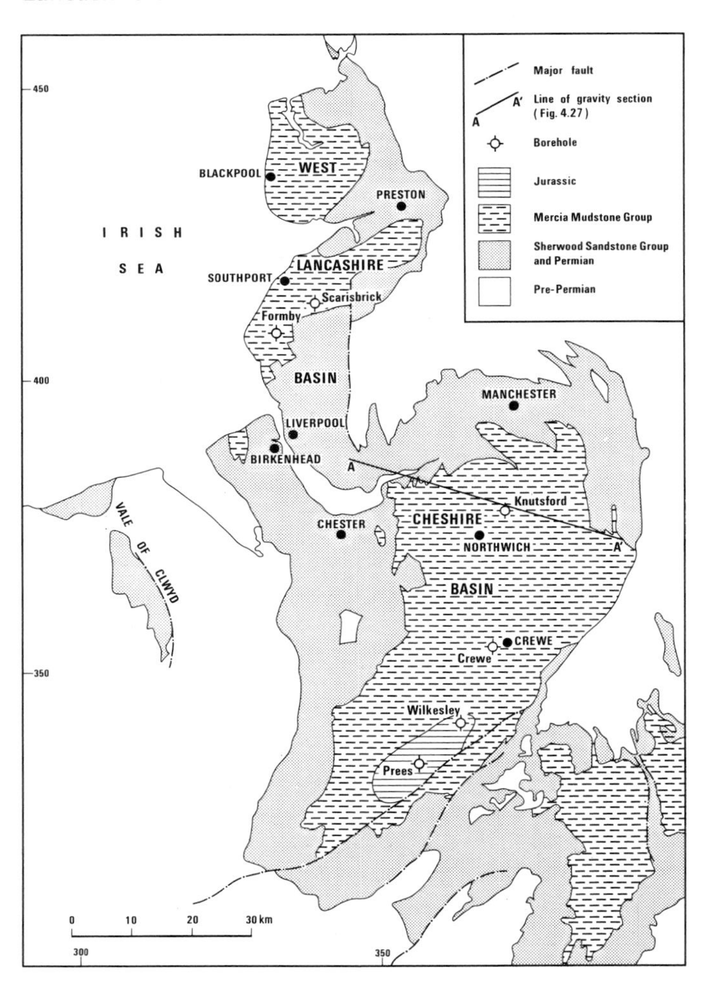

4.25 General correlation of the Permo-Triassic between the Cheshire and West Lancashire basins.

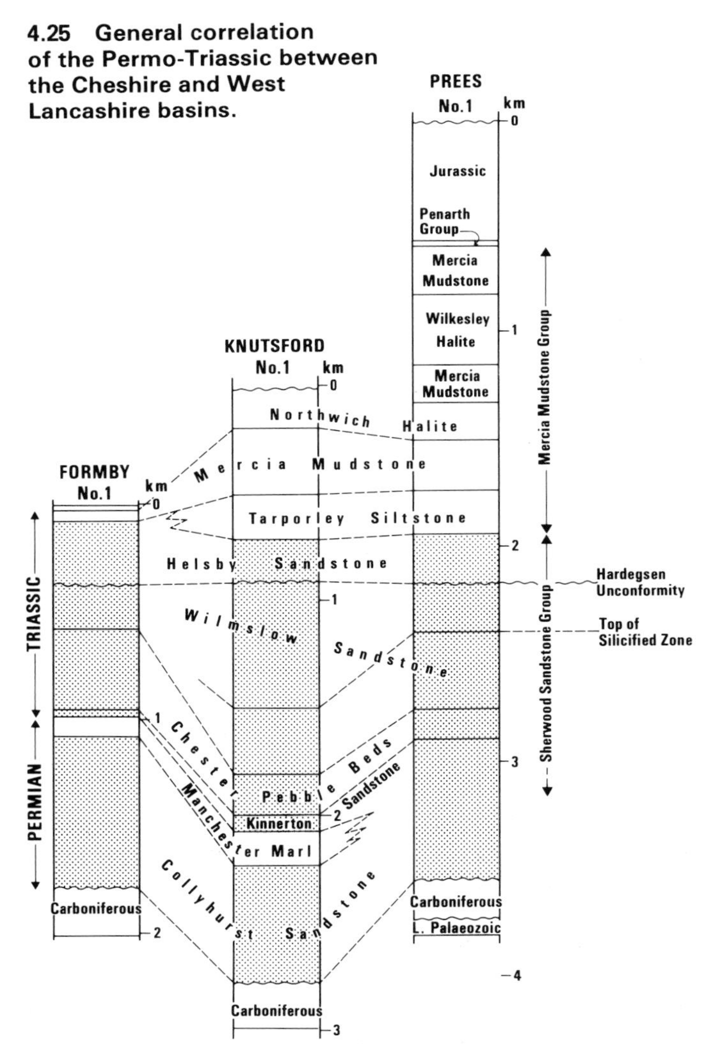

consists of argillaceous and calcareous rocks which become more sandy to the south-east, as may be recognised in the Formby and Knutsford boreholes, so that in the Prees Borehole it is not discernible (Colter and Barr, 1975). It may act as an aquiclude or partial aquiclude separating the aquifer unit into the Collyhurst Sandstone below and the Sherwood Sandstone above.

The Chester Pebble Beds Formation consists of sandstones and conglomerates but its top and bottom are not well defined. It passes upwards into the Wilmslow Sandstone Formation which was described by Audley-Charles (1970) as a fluviatile and lacustrine deposit with some material of aeolian origin (Colter and Barr, 1975). Colter and Barr also described a thick, silicified zone which affects both this formation and the Chester Pebble Beds and which would be expected to modify the aquifer properties greatly where it exists.

The Hardegsen Unconformity occurs at the base of the Helsby Sandstone Formation. Although regionally extensive, it is unlikely to prevent hydraulic communication between the Wilmslow Sandstone and the Helsby Sandstone above. Three main depositional units within the Helsby Sandstone Formation, described by Thompson (1970) in the Cheshire Basin, have been correlated by Colter and Ebburn (1978) with units in the Northern Irish Sea Basin using geophysical borehole logs. Although there is likely to be hydraulic communication between the units, their physical properties are markedly different, being related to primary depositional controls and to secondary diagenetic processes. They are likely to be more arenaceous in the west of the basins (G. Warrington, personal communication).

At the top of the Sherwood Sandstone Group the Helsby Sandstone passes gradationally into the Mercia Mudstone Group. The Tarporley Siltstone occurs at the base of the group and consists of fluviatile sandstones, siltstones and mudstones. In general it is not water-bearing but Colter and Ebburn (1978) have shown that in the Northern Irish Sea Basin it is porous locally and it is therefore included in the discussion of the aquifer units.

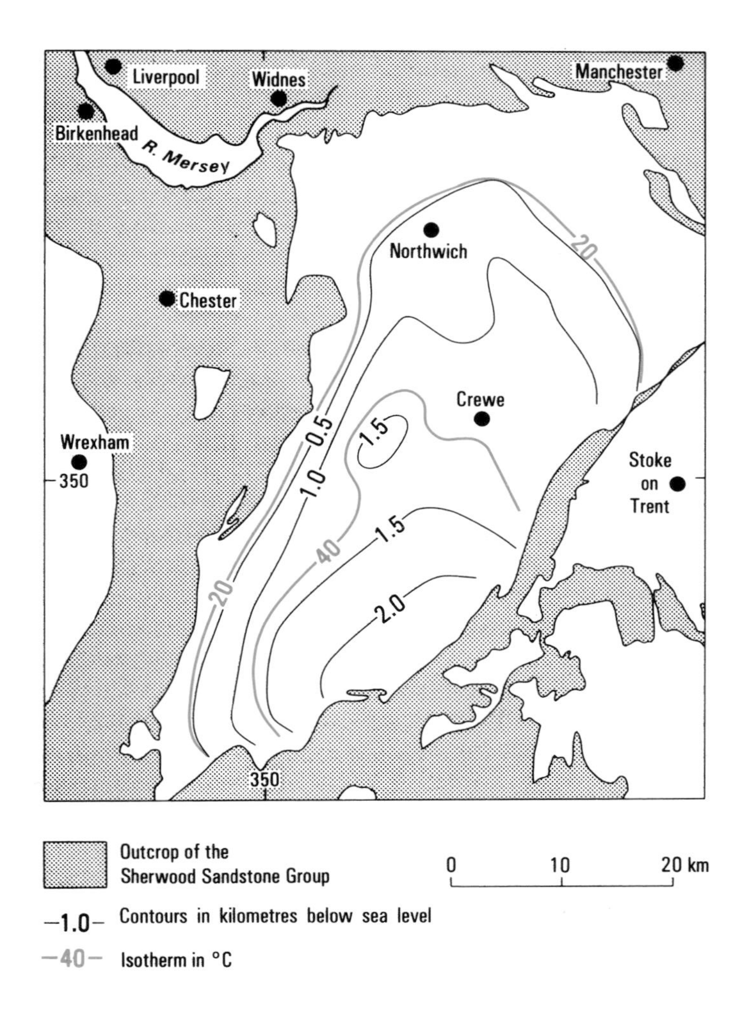

4.26 Structure contours on and estimated temperatures at the top of the Tarporley Siltstone in the Cheshire Basin.

The Mercia Mudstone Group confines the Helsby Sandstone in the deeper parts of the basins. It is largely argillaceous, with two thick evaporite sequences, mainly of halite, identified in Cheshire. These are not seen at outcrop

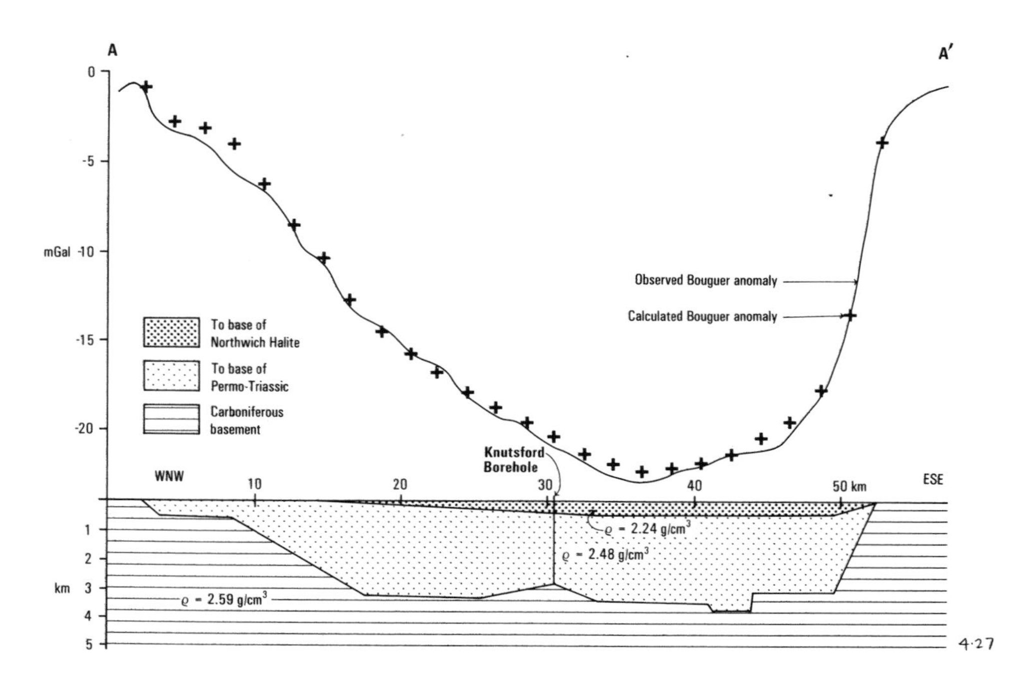

4.27 Gravity interpretation of the structure of the Cheshire Basin.

Formation	Cheshire Basin				West Lancashire Basin			
	Prees No 1		Knutsford No 1		Formby No 1		Scarisbrick	
	Depth (m)*	Thickness (m)	Depth (m)*	Thickness (m)	Depth (m)*	Thickness (m)	Depth (m)*	Thickness (m)
Mercia Mudstone	1737	1141	493	533	73	49	254	258
Helsby Sandstone	2169	432	912	419	369	296	452	198
Wilmslow Sandstone	2591	422	1507	595	572	203	—	}
Chester Pebble Beds	—	}	1997	490	948	376	—	} 621
Kinnerton Sandstone	—	} 949	2072	75	985	37	1073	}
Manchester Marl	—	}	2220	148	1073	88	1298	225
Collyhurst Sandstone	3540	}	2775	555	1788	715	1653	355

* Depths are below sea level

Table 4.8 Depth to the base and thickness of Permo-Triassic strata as recorded in boreholes in the Cheshire and West Lancashire basins

but their positions are recognised as subsidence features and they have been proved in boreholes. The Penarth Group and Jurassic rocks are preserved in an outlier at Prees. Over both basins much of the solid geology is masked by a thick layer of Pleistocene glacial deposits.

4.28 Structure contours on and estimated temperatures at the base of the Manchester Marl in the Cheshire Basin.

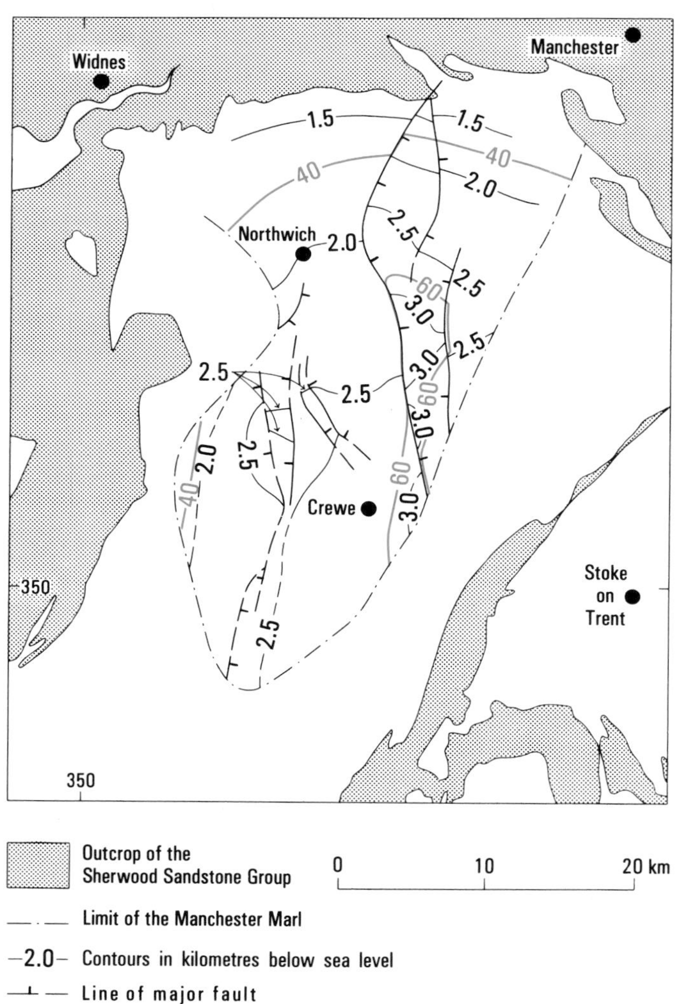

Faulting is the major structural control in the basins, although Kent (1975) suggested that contemporaneous subsidence significantly affected the thickness of the strata. The nature of the faulting has been debated; White (1949) suggested, from the evidence of a gravity survey, that the Cheshire Basin is a graben, bounded on the east and west margins by faults; Colter and Barr (1975) considered it to be a one-sided graben with the Red Rock Fault on the eastern margin controlling subsidence and the western side being a steep unconformity covered by sedimentary onlap of Permo-Triassic rocks. Evans and others (1968) proposed that the Red Rock Fault may be a complex fault belt with a throw estimated at 'as little as a few hundred feet [say 100 m] or as great as 7000 feet [approximately 2100 m].' Challinor (1978) doubted that the structure is a fault at all and suggested that it may be a steep unconformity.

The regional structure presented here has been based largely on geophysical interpretation. In the Cheshire Basin, two deep boreholes have been drilled through the Permo-Triassic rocks to reach the underlying Carboniferous strata, and geophysical logs have been recorded. Available seismic reflection data have been used to compile maps showing depths to the top and bottom of the Sherwood Sandstone aquifer (Figures 4.26 and 4.28); Table 4.8 gives the depth and thickness of the formations observed in boreholes.

The gravity data corroborates, to a great extent, the model produced by the seismic interpretation (Figure 4.27). The greatest depth of the base of the Mercia Mudstone (Figure 4.26) is about 2 km in the south-east and suggests a regional dip in that direction. The Manchester Marl (Figure 4.28) can only be traced over a limited area but a maximum depth of over 3 km is attained in the east and centre of the basin. The base of the Permo-Triassic rocks (Figure 4.29) reaches a depth of over 4 km to the north-east of Crewe and it is over 3 km deep in an area of some 350 km^2 centred on Crewe.

In the West Lancashire Basin, estimates of depth have been based on gravity data. The results of this interpretation suggest that the maximum depth of the Permo-Triassic below the Ormskirk-Southport Plain is 2 km and in the Fylde, 3 km. Other data indicate that depths may in fact be

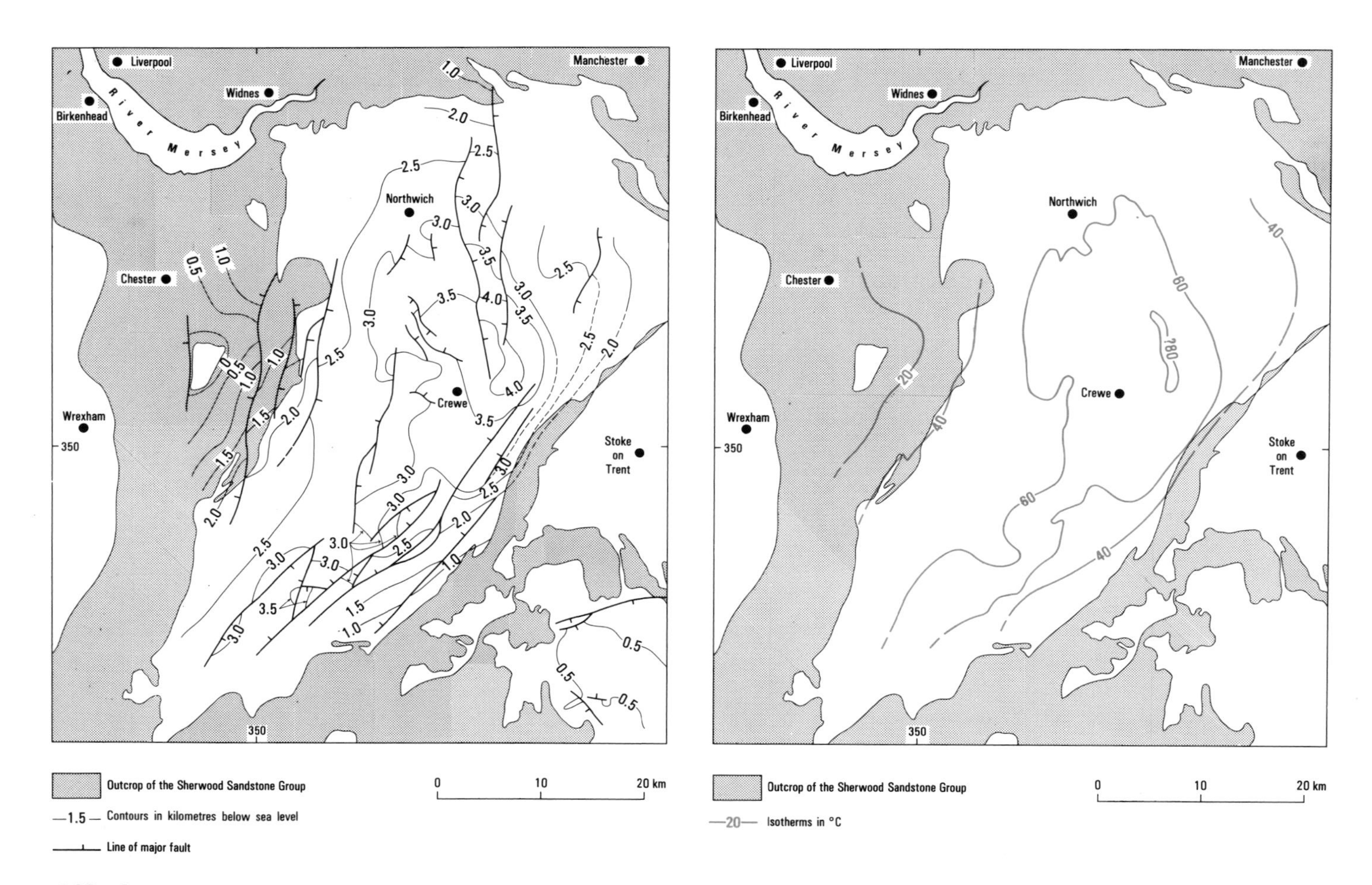

4.29 Structure contours on and estimated temperatures at the base of the Permo-Triassic in the Cheshire Basin.

less than this in the Fylde, possibly not reaching 2 km. The depths and thicknesses of formations in the Permo-Triassic, as recorded in boreholes, are given in Table 4.8.

Heat flow and sub-surface temperatures

In the Cheshire and West Lancashire basins, heat flow measurements, calculated from equilibrium temperatures in a borehole with thermal conductivity measurements from the same borehole, have been made at only four locations: at Holford and Crewe in the Cheshire Basin, and at Thornton Cleveleys and Weeton Camp in the West Lancashire Basin; the values are 34, 56, 50 and 50mW/m^2 respectively.

Oxburgh and Richardson (quoted in Burley and others, 1984) recorded heat flow values in the range 25 to 58 mW/m^2 in four boreholes in the Cheshire Basin and 71 mW/m^2 for a borehole in the West Lancashire Basin using assumed thermal conductivities. Temperature gradients given in Burley and others (1984) have been combined with geological information to obtain heat flow estimates by assuming thermal conductivity values are uniform for stratigraphical units (see Chapter 2). Nearly all the estimates for the Cheshire Basin are in the range of 40 to 60 mW/m^2.

From the evidence available at present, the Cheshire and West Lancashire basins can be considered a region of average heat flow, similar to the currently accepted national average for the UK of about 54 mW/m^2. Variations in heat flow values from around 50 mW/m^2 in the West Lancashire Basin to around 35 mW/m^2 in the Cheshire Basin, might be due to groundwater movement in aquifers below the depth of investigation of the heat flow boreholes, rather than any crustal heat flow variations.

Mean temperature gradients for the Cheshire and West Lancashire basins taken from Burley and others (1984) show variations between 9°C/km and 23°C/km, with the exception of Kirkham, which as new temperature information has become available appears to be atypical of the region.

The following uniform temperature gradients were chosen for the various rock types in order to predict temperatures at specific stratigraphical horizons:

Mercia Mudstone Group	23°C/km
Sherwood Sandstone Group	15°C/km
Thick Permo-Triassic including both the above groups	17°C/km

It is important to emphasise that these gradients have been calculated on very limited borehole information and the temperature contours drawn from them are at best only approximate estimates. The temperatures at the base of the Mercia Mudstone, the base of the Manchester Marl and the base of the Permo-Triassic have been calculated assuming the ground surface is close to sea level with a mean surface temperature of 10°C.

The temperatures at the base of the Mercia Mudstone are given in Figure 4.26. Temperatures attain a maximum of slightly over 40°C on this surface over a large area of the southern Cheshire Basin between Crewe and Prees. The temperatures at the base of the Manchester Marl, in the restricted area where it exists, are given in Figure 4.28. Temperatures on this surface are generally at least 40°C with 60°C attained in a narrow zone only a few kilometres wide extending 20 km northwards from a position just east of Crewe. The depth to the base of the Permo-Triassic and the temperatures on this surface are given in Figure 4.29. The temperatures at the base of the Permo-Triassic exceed 60°C over about 500 km^2 and they possibly attain 80°C over a few square kilometres about 5 km north-east of Crewe.

In the West Lancashire Basin the temperatures at the base of the Permo-Triassic are estimated to be approximately 36°C and, from the limited seismic evidence considered, it is unlikely that 40°C is attained anywhere in the basin. In the north-central part of the Fylde a narrow basin of Mercia Mudstone is believed to attain a maximum thickness of 950 m (A. Wilson, personal communication). Temperatures estimated on the basis of the gradient in the Mercia Mudstones in the nearby Thornton Cleveleys Borehole suggest a maximum of 30°C over a few square kilometres at the base of the mudstones to the east of Fleetwood. The base of the Permo-Triassic in this area is estimated to be at a depth of just over 1 km and temperatures on this horizon are likely to be around 35°C, similar to the maximum temperatures on this surface elsewhere in the basin.

Hydrogeology of the Permo-Triassic sandstones

The majority of data available on the aquifer properties of the Permo-Triassic sandstones is from outcrop or shallow boreholes. However, information at depths more relevant to geothermal investigation is available from three deep boreholes and from work carried out offshore in Morecambe Bay. Detailed descriptions of the aquifer properties are given below by formations for easy comparison, the data being mainly derived from core analyses and interpretation of downhole geophysical logs.

Tarporley Siltstone

At outcrop the fine-grained sandstones and mudstones which comprise the Tarporley Siltstone are not considered to be an aquifer and in boreholes, geophysical logs indicate low porosity implying low permeability. On this basis alone, therefore, the Tarporley Siltstone would not be considered to be a potential aquifer in the Cheshire and West Lancashire basins. However, from laboratory tests on core material from the Northern Irish Sea Basin, Colter and Ebburn (1978) have shown that fine- to coarse-grained sandstones, correlated with this formation, retain porosities of up to 21% and may still have an intergranular permeability of up to 400 mD (Figure 4.30). Calcite cementation was observed to reduce the permeability, in many of the samples examined, to less than 10 mD.

Helsby Sandstone

The Helsby Sandstone of the Cheshire Basin has been divided into three (Thompson, 1970): the Thurstaston, Delamere and Frodsham members. Colter and Barr (1975) stated that the more porous sandstones are restricted to the lower members, a fact attributable presumably to their partly aeolian origin. Lovelock (1977) suggested that in mid-Cheshire, where the Delamere Member is absent, both porosity and permeability are greatest in the middle of the Helsby Sandstone sequence. Porosity may reach 30% and permeability may be as high as 4000 mD. More data are available in north Cheshire, where the permeability at outcrop and in shallow boreholes ranges up to 15 000 mD, although the porosity is reduced by suture contact pressure solution to 25 to 30%. Geophysical logs indicate that, at depth, primary porosity is in the range of 14 to 21% averaging 17% in the upper and 15% in the lower parts.

Cores from the Helsby Sandstone of the Northern Irish Sea Basin have been tested in the laboratory and examined petrographically to investigate the relationship between their porosity, permeability and lithology (Colter and Ebburn, 1978). The porosity/permeability plots are reproduced in Figure 4.30, from which it can be seen that only in the highest Frodsham Member could permeabilities of more than 100 mD be expected in sandstones with greater than 15% porosity.

Wilmslow Sandstone

The Wilmslow Sandstone is predominantly a fine-grained red sandstone with argillaceous laminae and varying degrees of cementation, resulting in a wide range of physical properties, between more than 10 000 mD with 30% porosity and less than 10 mD with 13% porosity (Lovelock, 1977). Laboratory tests on core samples taken from the 'St Bees' Sandstone of the northern Irish Sea (where the Kinnerton and Wilmslow sandstones cannot be distinguished because of the absence of the intervening Chester Pebble Beds) show that a combination of quartz overgrowth, calcite cementation and neoformation of platy clay minerals reduces the intergranular permeability to less than 10 mD, even when the porosity is still at 20% (Colter and Ebburn, 1978). Geophysical logs in deep boreholes indicate porosities ranging from 8 to 17% with an average of about 13%. The more porous sections are in the upper part of the sequence.

Chester Pebble Beds

The Chester Pebble Beds are predominantly coarse-grained, cross-bedded, reddish sandstones with frequent interspersed quartzitic pebbles. They represent fluvially reworked sediments of a large alluvial fan derived from the south. The unit thins to the north-west, with the pebbles becoming correspondingly smaller and less frequent as it does so and it is entirely absent in the Northern Irish Sea Basin (Colter and Barr, 1975).

At outcrop, the porosity of the Chester Pebble Beds is in the range of 20 to 30% and bulk field permeability ranges from 800 to 8000 mD (Skinner, 1977). In Cheshire and South Lancashire, 50% of samples from outcrop and shallow borehole cores have intergranular permeability greater than 670 mD as measured in the laboratory; those

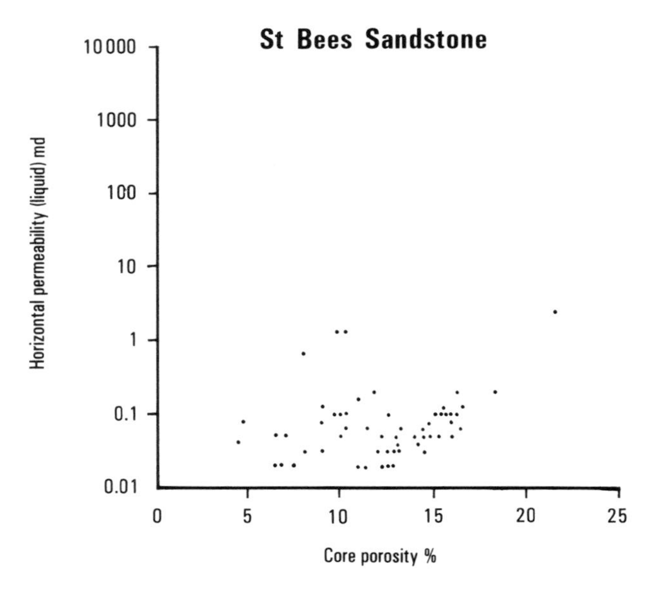

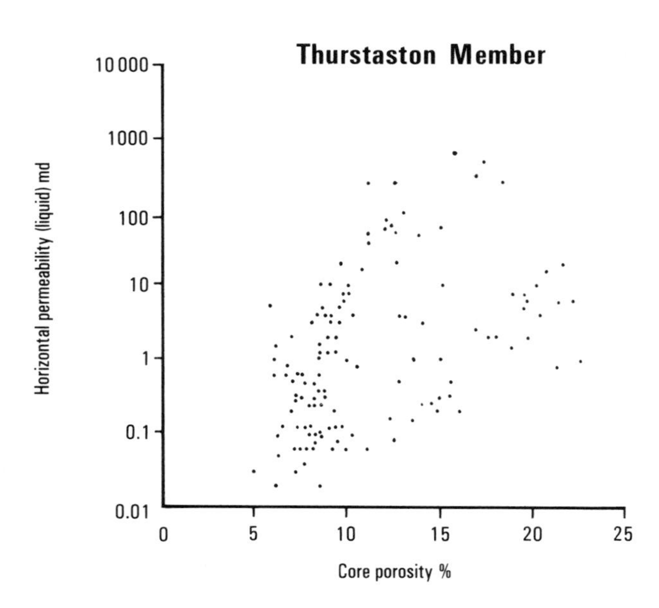

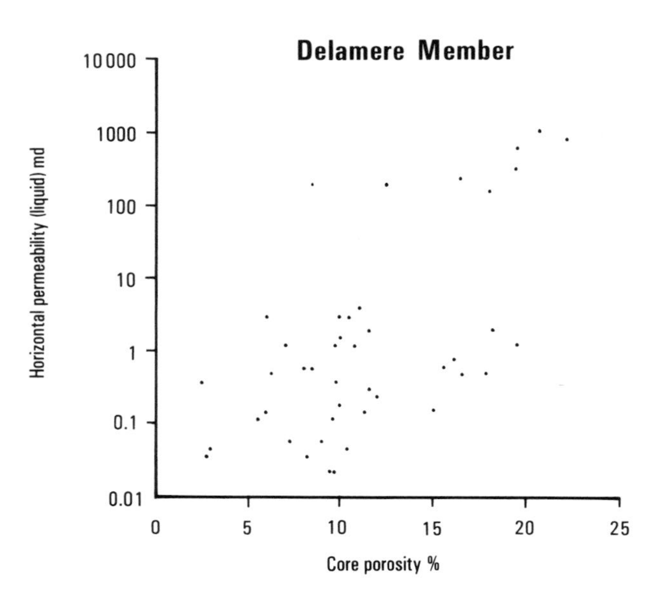

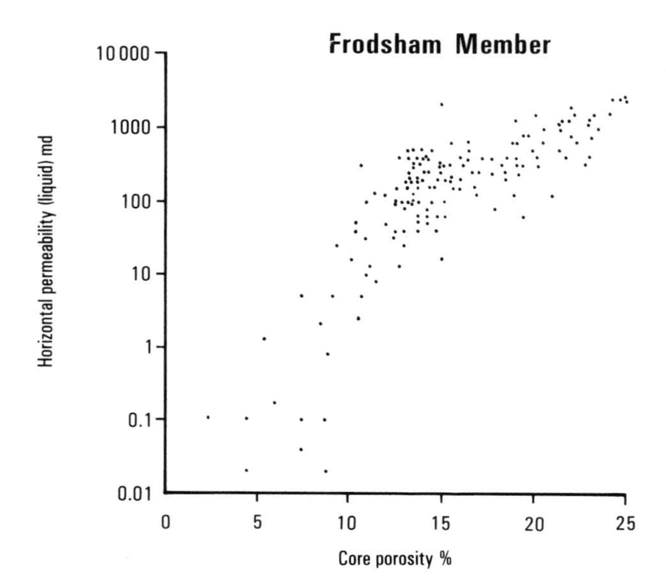

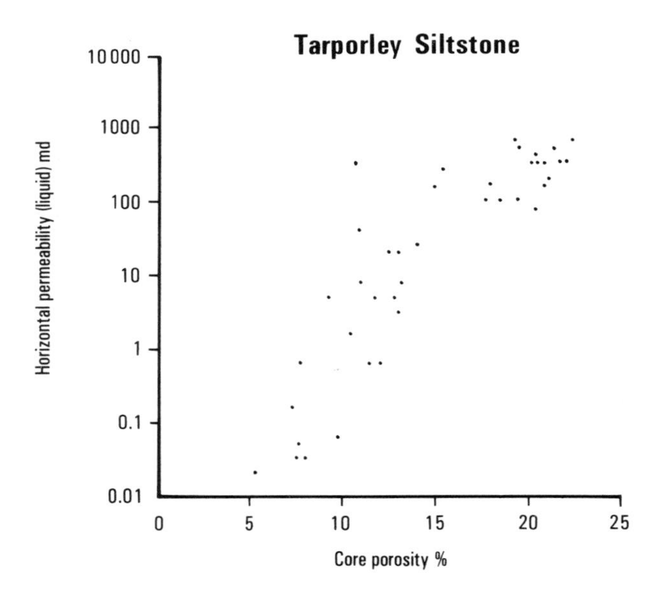

4.30 Porosity-permeability relationships for the Triassic sandstones in the Northern Irish Sea Basin (after Colter and Ebburn, 1978).

from south Cheshire are considerably more permeable (Lovelock, 1977, and Figure 4.31).

At depth geophysical logs indicate generally low porosities ranging from 9 to 14%, so from this evidence the Chester Pebble Beds cannot be considered to be potential geothermal aquifers.

Collyhurst and Kinnerton Sandstones

In southern Cheshire, the Collyhurst and Kinnerton sandstones merge into one unit with the disappearance of the Manchester Marl which separates them further north. As a consequence, they are treated here as a single unit.

Information about the hydraulic properties of these sandstones is limited. Laboratory studies on core material from shallow boreholes in south Cheshire and Shropshire have indicated promising intergranular permeabilities (Lovelock, 1977), with half the specimens having permeabilities in the horizontal direction of over 2000 mD (Figure 4.31). At depth, evidence from geophysical logs indicates porosities from 9 to 17%, the higher porosities being found

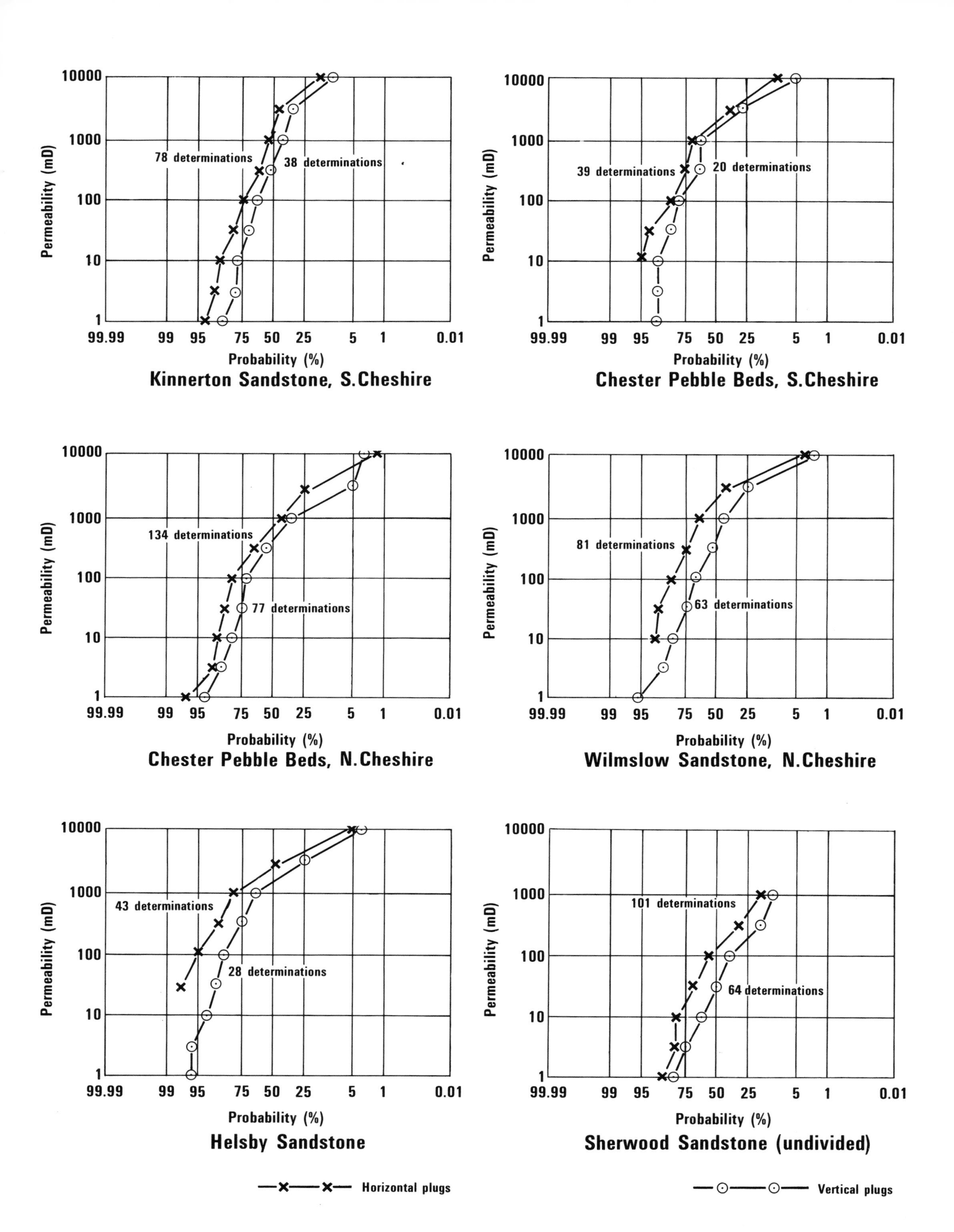

4.31 Probability distributions of permeability in the Triassic sandstones in Cheshire and Shropshire (after Lovelock, 1977; the horizontal axis shows the percentage probability that permeability will equal or exceed the value indicated).

in the upper part of the section beneath the Chester Pebble Beds. This implies a reduction of permeability at depth which is probably caused by cementation and the formation of secondary clay minerals. It may not be ubiquitous throughout the strata at depth, because, at higher levels in the Northern Irish Sea Basin, Colter and Ebburn (1978) have observed erratic and patchy diagenetic changes and the consequent irregular preservation of favourable hydraulic properties in strata between those rendered less porous and permeable by secondary mineral formation.

Transmissivity

Although there is a lack of data in the deeper parts of the Cheshire and West Lancashire basins relating to hydraulic properties, evidence from the adjacent Northern Irish Sea Basin and from outcrop indicates that this very thick sequence of sandstones, siltstones and mudstones does contain substantial thicknesses of permeable sandstone. Estimates of transmissivity of the various sandstone units have not been made as there are insufficient data, but it is apparent that the Helsby Sandstone and the upper Wilmslow Sandstone are permeable layers in the Triassic, and that the upper Kinnerton/Collyhurst sandstones form a Permo-Triassic aquifer. It is estimated from the limited data available that, because of the great thickness of the sandstone sequence and its nature and origin, the minimum transmissivity is likely to be at least 10 D m, and possibly more than this for both aquifers.

Piezometric surface

In the West Lancashire Basin gradients on the piezometric surface indicate a gross movement of groundwater from the east, which diverges to the north-west and south-west towards outlets in Morecambe Bay and the Ribble estuary respectively. High groundwater salinities indicate restricted groundwater flow westwards between these flow directions under the Mercia Mudstone (Sage and Lloyd, 1978). An apparent groundwater gradient of 1:3000 in the Triassic

4.32 Geological map of the Carlisle Basin and adjacent areas.

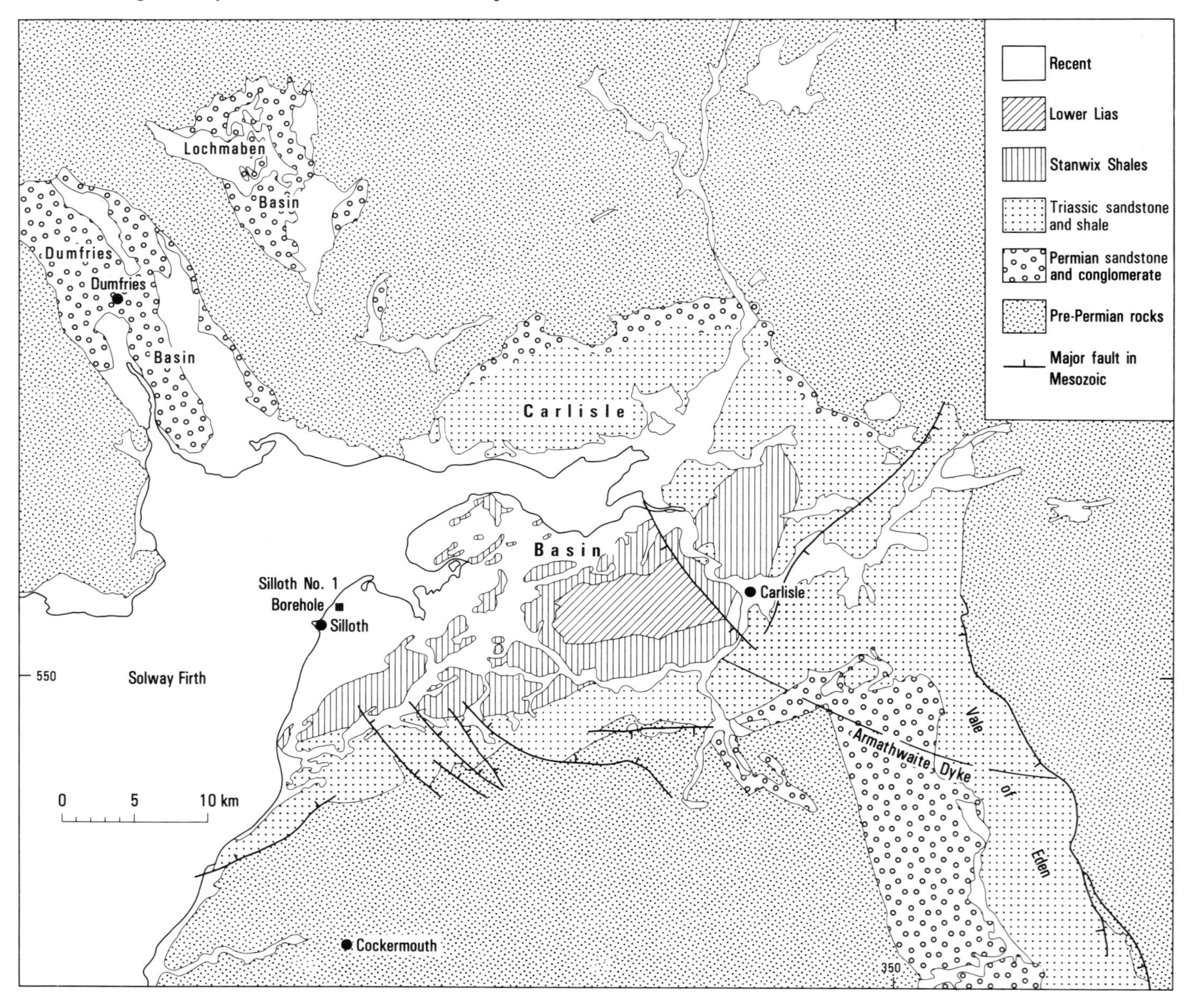

sandstones into Morecambe Bay is not a true piezometric gradient, for it is actually zero when account is taken of the groundwater salinities in the aquifer in this area. The distribution of saline groundwater bodies in the north of the Cheshire Basin is not fully documented, but is known to be complex. In the Liverpool area saline intrusion has occurred along the tidal reaches of the Mersey. In the Manchester area saline groundwater has been encountered which is probably connate, and therefore is likely to persist at depth. It is probable that the salinity of groundwater in the deeper part of the West Lancashire and Cheshire basins will be very high, possibly several tens of thousands of milligrams per litre. In view of this, although the water table reaches 25 m above sea level in the West Lancashire Basin (Sage and Lloyd, 1978) and 100 m above sea level in north Cheshire (Crook and Howell, 1970), at depth in the central areas of the basins the piezometric surface may be several tens of metres below ground level.

CARLISLE BASIN

The low-lying, drift-covered region of the Solway Plain, which is the surface expression of the Carlisle Basin, is bordered by the Lower Palaeozoic rocks of the Southern Uplands to the north-west and of the Lake District to the south, with Carboniferous outcrops predominant in the Pennines to the east. Interest in the geothermal potential of the area arose from the existence of permeable and porous Permo-Triassic sandstones in the Carlisle Basin.

The available geological and geophysical data for the region have been studied with the objective of recognising the main structural features and making a preliminary assessment of the geothermal potential (Gale and others, 1984b). In addition to examining data for the Carlisle Basin, several smaller basins in the surrounding area have also been studied. These are the Vale of Eden, and the Dumfries and Lochmaben basins. Because of the very limited geological information relating to the deeper parts of the various basins, the structural interpretation is mainly based on geophysical data, but also on limited borehole information which mainly defines the top of the Sherwood Sandstone. Only one deep borehole penetrating the Permo-Triassic succession has been drilled near the centre of the Solway Plain and the structural history of the area is a matter of conjecture, based on the evidence from outcrops around the margin of the basin. In particular the nature and extent of Carboniferous strata beneath the Permo-Triassic is unknown. Geophysical information for the area is derived from regional aeromagnetic coverage and gravity surveys compiled by the British Geological Survey, together with an additional detailed gravity survey and seismic reflection data.

General geology

The Carlisle Basin is a broad structural depression at the eastern end of the Solway Firth, containing a mainly Permo-Triassic sequence. It is the on-shore expression of the larger Solway-Carlisle Basin. The Vale of Eden structural depression extends in a south-south-easterly direction from the eastern end of the Carlisle Basin. North of the Solway Firth two small isolated Permo-Triassic basins occur in a region of Lower Palaeozoic rocks; these are the Dumfries and Lochmaben basins. The general geology is shown in Figure 4.32.

The post Carboniferous sequence in the Carlisle Basin is:

Jurassic	Lower Lias	
Triassic	Stanwix Shales (Mercia Mudstone Group)	
	Kirklinton Sandstone	(Sherwood Sandstone Group)
	St Bees Sandstone	
Permian	St Bees Shales	
	Penrith Sandstone and Brockrams	

The Lower Lias outlier at Great Orton forms the higher ground to the west of Carlisle and represents the only known occurrence of post-Triassic sedimentary rocks in the region. The Permo-Triassic sequence in the various basins referred to above exhibits large rapid variations in thickness and lithology, particularly in the lower part of the sequence. Breccias were deposited in the valleys and on the flanks of the high ground throughout the Permian and into the Triassic. Deposition of dune sands occurred throughout the Lower Permian in some areas and may have persisted into Upper Permian and Triassic times. Contacts between the sands and breccias are markedly diachronous.

The principal aquifers in the Carlisle Basin are the Kirklinton, St Bees and Penrith sandstones, the last of these being the most promising aquifer from the geothermal point of view because of its position at the base of the sequence, and hence its tendency to occur at greatest depths. Its origin, in part as a dune sandstone, has resulted in favourable hydraulic characteristics including a high porosity, although locally it is silicified. In the Carlisle-Brampton area there is less silicification and the principal rock type is a soft, medium- to coarse-grained, red, millet-seed sandstone. The St Bees Sandstone is primarily a flat-bedded deposit of finer and more uniform grain-size than the Penrith Sandstone. It is commonly micaceous, moderately hard, but does not contain secondary silicification; shale partings are common. The Kirklinton Sandstone is a soft, fine-grained, bright red sandstone.

The Permo-Triassic sequence in the Carlisle Basin is known from outcrops and was proved by the Silloth No. 1 Borehole to a depth of 1312 m, overlying what are believed to be Lower Carboniferous rocks. A summary geological log and calculated porosity of the sandstone bodies are given in Figure 4.33.

It was previously thought that both the Kirklinton and Penrith sandstones thinned and were absent to the west, but the occurrence of a significant thickness of Permian sandstone in the Silloth No. 1 Borehole confirms the view that conditions of deposition varied significantly within the area, and increases the possibility that aquifers might occur at depth elsewhere. The absence of the Upper Carboniferous is consistent with the interpretation of the gravity and seismic data. The borehole was located on a small local uplift, identified from the gravity data and seismic surveys, with the primary intention of testing the stratigraphy of the Permo-Triassic. Borehole geophysical logs suggest that all the sandstones are water-bearing and there were no indications of the presence of hydrocarbons. The effective

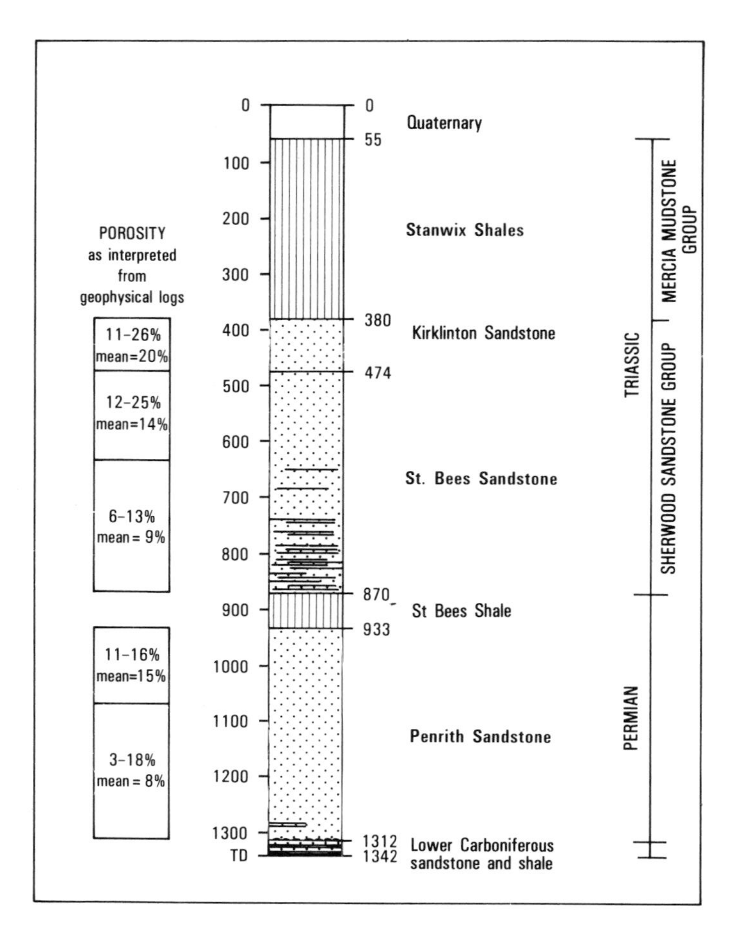

4.33 Geological log of the Silloth No. 1 Borehole (depths in metres).

porosity is up to 26% (Figure 4.33), the highest values being in the Kirklinton Sandstone (mean value of 20%) and the upper Penrith Sandstone (mean value of 15%), whilst the upper St Bees Sandstone also has a relatively high porosity (mean value of 14%).

The borehole is sited 7 km west of the centre of a Bouguer anomaly low (Figure 4.34) from which the values have increased by nearly 5 mGal. For a density contrast of 0.25 to 0.35 Mg/m^3 between the Permo-Triassic and Lower Palaeozoic rocks this might represent an additional 325 to 450 m of Permo-Triassic at the centre of the basin, while the seismic evidence indicates a thickening of at least 200 m. A range of 1550 to 1900 m is therefore given for the depth to the base of the Permo-Triassic beneath the Bouguer anomaly minimum. Further to the north-east of the basin the Permian thins as the pre-Permian floor rises to within 650 m of the surface.

Although the thickest part of the Permo-Triassic basin might have been expected to coincide with the youngest Mesozoic outcrop in the Great Orton outlier, the sequence is almost certainly thicker near the Bouguer anomaly minimum some 10 km further west along the direction of elongation of the Lias outcrop. From the seismic evidence it appears that lateral variations are also occurring in this direction so that the thickness of Permo-Triassic beneath the outlier may be significantly reduced, with little or no Permian sandstone, while the Upper Carboniferous is more fully developed. White (1949) considered the possibility of a Carboniferous synclinal axis trending south-west to north-east through the gravity lows A and E (Figure 4.34). The formation of the basin attributed to the gravity low may be explained by the intersection of north-north-westerly trending Hercynian fault lines with a broad Caledonian synclinal structure.

In order to account for the thickness of Permo-Triassic rocks found in the Silloth No. 1 Borehole, on the edge of the gravity low A (Figure 4.34) without invoking a steep regional gradient, it is necessary to assume either variations within the Carboniferous subcrop, as mentioned above, or the mean density of the Permo-Triassic increases to the west, implying either a higher proportion of shale and breccia relative to porous sandstone, or that salt occurs beneath the centre of the low.

Of the five Bouguer anomaly lows (Figure 4.34) associated with steep gradients, two, labelled F and G, can be attributed to the Criffel granodiorite and Skiddaw granite intrusives, while the other three (D, B and C) have been related to Permo-Triassic basins, namely the Vale of Eden, the Dumfries and the Lochmaben basins respectively. The thicknesses of Upper Carboniferous underlying the Permo-Triassic are not known and their presence would lead to a reduction in the estimated thickness of the Permo-Triassic but a greater total thickness for the basins. The amplitudes of these Bouguer anomaly lows exceed the values of the Carlisle Basin and the associated gradients are relatively steep, reflecting a different tectonic setting. In the case of the Scottish basins it seems likely that deposition occurred over a relatively short time-scale, while the Solway Plain was subject to subsidence over a longer period and wider area. The lower anomaly values found in the Vale of Eden probably reflect a thicker development of Carboniferous, as well as density variations within the underlying rocks. The three basins are all elongated along a NNW to SSE axis and, by analogy, the Permo-Triassic within the Carlisle Basin may be affected by the same trend. The seismic data

4.34 Bouguer gravity anomaly map of the Carlisle Basin and adjacent areas (mGal). The letters A to G indicate gravity lows referred to in the text.

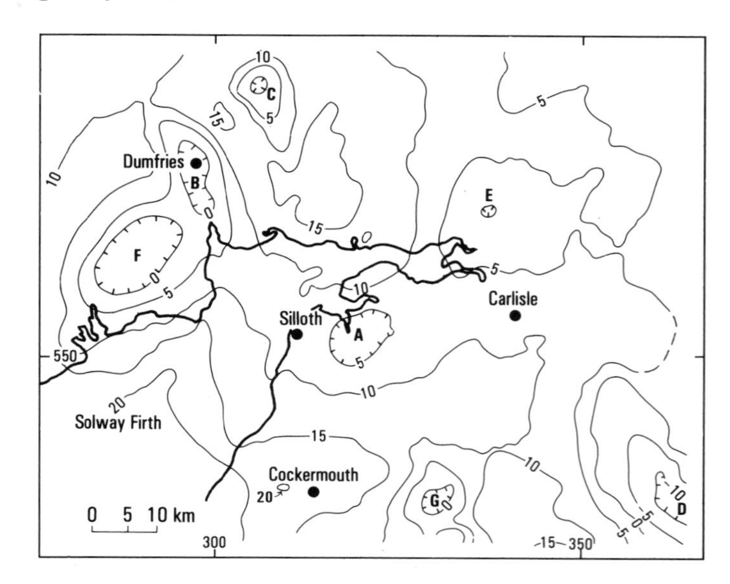

offer some evidence for this as do the gravity anomaly contours (although subordinate to the north-east trend).

The Vale of Eden is a faulted trench infilled with Permo-Triassic sediments, similar to those of the Carlisle Basin. The deposits are represented by the St Bees Sandstone and the underlying Eden Shales (equivalent to the St Bees Shales), Penrith Sandstone and Brockrams; the Penrith Sandstone passes laterally into the Brockrams which were deposited at the margins of the basin and are thickest on the south-west margin. The Penrith Sandstone is found in thicknesses up to 300 m, the Eden Shales between 45 and 180 m and the St Bees Sandstone is estimated to be between 500 and 600 m thick (Arthurton and Wadge, 1981). The maximum depth to the top of the Penrith Sandstone is approximately 700 m and a maximum Permo-Triassic thickness of 800 to 900 m has been estimated from gravity and seismic data (Collar in Arthurton and Wadge, 1981). Near Penrith the upper part of the Penrith Sandstone, in particular, tends to be silicified.

The isolated Dumfries and Lochmaben basins contain a Permian sequence comprising red dune sandstones with bands of coarse breccia. Thicknesses for both of about 1100 m have been derived by interpretation of the Bouguer anomalies (Bott and Masson Smith, 1960).

Sub-surface temperatures

In order to predict temperatures at specific stratigraphic horizons uniform temperature gradients for particular formations have been chosen. These gradients are as follows:

Geological formation	Estimated temperature gradient
Mercia Mudstone Group	31°C/km
Sherwood Sandstone Group and Permian sandstone	16°C/km

It is important to emphasise that these gradients have been calculated on very limited borehole information and are based largely on the equilibrium temperature measurements in the Silloth No. 2 Borehole for the Mercia Mudstone Group, and the equilibrium temperature measurement in the Becklees Borehole for the sandstones of the Permo-Triassic. If there were significant intercalations of shale and mudstone in these sandstones then the temperature gradients would be increased towards the value for the Mercia Mudstone. Temperature estimates based on these data are at best only approximate but are conservative. The temperature at the top of the Sherwood Sandstone beneath the insulating cover of the Stanwix Shales is estimated to be very slightly over 20°C in the Silloth area and in the sedimentary basin to the east. At the top of the Penrith Sandstone it may be about 30°C near Silloth, and might just attain 40°C in the deepest part of the Carlisle Basin east of the town, particularly if there proved to be an increase in the insulating cover of St Bees Shales over the Penrith Sandstone. Temperatures of more than 20°C at the top of the Penrith Sandstone are likely over a large area of the Carlisle Basin and may also occur in the eastern part of the Vale of Eden beneath the thickest sequence of the Eden Shales.

Hydrogeology

In general, the Penrith Sandstone has good hydraulic properties and is an excellent aquifer. It is, however, an heterogeneous deposit characterised by a very variable degree of cementation and a wide range of grain-sizes. Only limited data relating to aquifer properties are available (Lovelock, 1977) and these are for samples from surface outcrops. The permeability is believed to range from 100 to several thousand millidarcys while the porosity is between 18 and 26%. Between Penrith and Appleby, friable, well-sorted millet-seed sands are common. These deposits have a high intergranular permeability and where the sands are virtually uncemented the intergranular permeability may be as high as 10 000 mD and in places more than 20 000 mD (Lovelock, 1977).

In contrast to the Penrith Sandstone, the St Bees Sandstone is a very fine to fine-grained deposit with angular grains of uniform size and degree of consolidation. Typical intergranular permeability values for outcrop samples are 200 to 600 mD with porosities in the range 24 to 27%. In better-sorted samples the permeability may attain maximum values of 2 000 mD and porosities of 30 to 35% (Lovelock, 1977). On the north side of the Solway Firth, near Annan, where the St Bees Sandstone is generally harder and more cemented, the aquifer properties are less favourable. On the basis of a limited number of outcrop samples, the permeability appears to be no more than 10 to 80 mD with porosities of between 17 and 23%.

Information about the hydraulic properties of the Kirklinton Sandstone is also very limited. In some areas, although it is a soft, friable sandstone with an appreciable porosity, the intergranular permeability is low. However, it also contains horizons of very permeable, friable, millet-seed sand but the relative importance of this rock type in the deposit as a whole is uncertain.

The studies made by Lovelock (1977) suggested that at shallow depths the mean porosities of the Penrith and St Bees sandstones are very similar with values between 25 and 30%. The average intergranular permeability of the Penrith Sandstone is about 1000 mD (which would give an intergranular transmissivity of about 100 m^2/d) while that for St Bees Sandstone is about 350 mD (giving a transmissivity of about 75 m^2/d). The large differences observed between the intergranular and field transmissivities in the Penrith Sandstone reflect the extent that fissures control the rate of flow through the aquifers (Lovelock and others, 1975).

The only data on aquifer properties at depth are from the geological and geophysical logs of the Silloth No. 1 Borehole. Gypsum is present in the upper St Bees Sandstone in addition to patchy silica, dolomite and calcite cements. The Kirklinton Sandstone has a small amount of anhydrite cement, and in the Penrith Sandstone the proportion of anhydrite cement increases with depth with a consequent reduction of porosity (Figure 4.33). Waugh (1970) estimated the permeability of the Kirklinton and Penrith sandstones to be of the order of 'a few hundred millidarcys' but the anhydrite cement of both of these units is likely to reduce the value. The upper St Bees Sandstone is reputed to have a mean porosity of 14% and the permeability will therefore be considerably less than at outcrop, probably of the order of 50 to 100 mD. The lower part of the St Bees Sandstone is cemented with silica and some anhydrite and is interbedded with shale, indicating a low permeability.

As stated earlier the degree of cementation will determine the permeability of the sandstones and this cannot be predicted with any certainty from the data available. However, if the mean permeabilities of the Kirklinton and Upper Penrith sandstones are estimated to be 100 mD and that of the Upper St Bees Sandstone 50 mD, then the calculated transmissivities of the sandstones are 9 Dm, 13 Dm and 8 Dm respectively.

Near the outcrop, the groundwater in the Permo-Triassic sandstones has a low total mineralisation of some 200 mg/l, particularly in the Penrith Sandstone. However, sulphates derived from evaporites in the Eden and St Bees shales can be quite high. The salinity will certainly increase with increasing depth below the Mercia Mudstone in the Carlisle Basin but the general form of the basin suggests it may never become very high and is probably less than 100 000 mg/l.

NORTHERN IRELAND

General geology

The volcanic activity which occurred in the north-western parts of the British Isles in Tertiary times was very extensive in Northern Ireland. Central intrusion complexes were emplaced in Lower Palaeozoic greywackes in counties Down and Armagh, while a thick pile of plateau basalt lavas was built up over most of County Antrim and parts of counties Londonderry, Tyrone and Armagh. Erosion, including glaciation, has reduced the basalt sheet to a total area of some 3980 km² and a maximum known thickness of almost 800 m, although this figure could well be exceeded under Lough Neagh. However, the basalts have protected the underlying Cretaceous, Jurassic, Triassic and Permian sediments from erosion.

Strata of these ages crop out around most of the periphery of the basalt area. The relatively small outcrop areas of the Permo-Triassic rocks surround the Antrim basalts as shown in Figure 4.35, but the basalts conceal deep sedimentary basins which are infilled mainly with Triassic sediments, extending to depths of about 3000 m below sea level and underlying a total area of about 4000 km² (Bennett, 1983). The main centre of deposition extends from beneath Lough Neagh north-eastwards into the North Channel (Figure 4.35), lying more or less astride the continuation of the Midland Valley of Scotland. A narrower trough with a similar trend occurs further north-west reaching the north coast around Port More and is probably connected with the main basin along a NNW to SSE trending axis, which breaches a NE to SW structural high connecting Dalradian basement rocks on either side of the basalt outcrop. Further north-west again the basin appears to be locally deepened along another NE to SW axis in Lough Foyle.

Along their western boundary and in the Belfast Harbour area, the Permo-Triassic rocks rest mainly on Carboniferous sediments, but basement rocks form the south-eastern and north-eastern margins, where Triassic sandstones rest directly on Lower Palaeozoic greywackes and Dalradian schists respectively. The only borehole which has definitely penetrated below the base of the Permo-Triassic strata in the central area, at Langford Lodge, did not encounter Carboniferous strata, although the latter probably occur beneath the basin further east and beneath the northern trough. They were encountered in the Magilligan Borehole (Figure 4.35) beneath 977 m of Jurassic and Triassic sediments.

The sediments in the basal layers of the basins are generally arenaceous: Permian sandstones, which are often coarse-grained and uncemented, and brockrams are overlain by Upper Permian Marls and then by the Triassic Sherwood Sandstone Group, which is composed mostly of medium-grained sandstones (often with a calcareous cement), with marl and mudstone intercalations. In the deeper parts of the basins the combined thickness of the Permo-Triassic sandstones may exceed 1000 m. At Larne the Permian sandstones include more than 600 m of interbedded volcanic tuffs and basalts, while a halite bed 113 m thick underlies the Permian Upper Marls: neither the volcanic rocks nor the halite have been encountered in the Permian elsewhere in Northern Ireland (Penn and others, 1983). The Sherwood Sandstone is overlain by the Mercia Mudstone, which consists predominantly of marls and mudstones with sandstone intercalations. It includes thick halite deposits in the south-east part of the main basin, where the total thickness is almost 1000 m. The Triassic succession is completed by sediments of the Penarth Group which are not

4.35 Permo-Triassic basins in Northern Ireland.

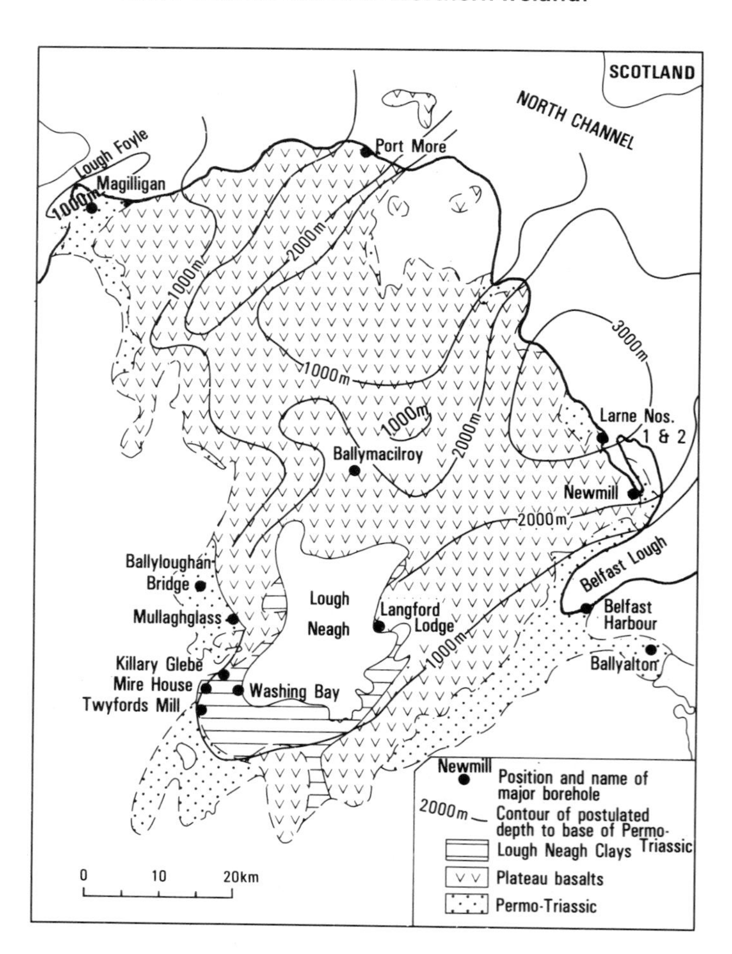

always present. They comprise both argillaceous and arenaceous materials but never exceed a few tens of metres in thickness. The Triassic strata are usually (but not always) overlain by Jurassic argillaceous sediments, the Lias, which can be up to about 100 m thick. In the basalt covered parts of the basins a layer of Chalk (the Ulster White Limestone) up 50 m thick usually occurs immediately beneath the basalt sheet. Around Lough Neagh, and presumably beneath much of the lough, basins of Tertiary sediments, the Lough Neagh Clays (Figure 4.35), and Quaternary sediments rest on the basalt sheet. They are predominantly argillaceous, attaining thicknesses of almost 400 m on land and are probably much thicker beneath the lough.

Sub-surface temperatures

Burley and others (1984) list temperature data for Northern Ireland. The general pattern of geothermal gradients is greatly influenced by the fact that several of the major formations which usually overlie the Permo-Triassic sandstones, namely the Lough Neagh Clays, Antrim Basalts, Lias, and Mercia Mudstone, have low thermal conductivities and so produce high temperature gradients of up to 58°C/km. This 'blanket' effect enhances the prospects of locating hot water at relatively shallow depths.

In the arenaceous formations which occur below the low conductivity strata, and which include the Triassic and Permian sandstone aquifers, thermal and hydraulic conductivities are both much higher, so that heat can not only be conducted more readily through the constituent minerals of the rocks, but can also be transported between the mineral grains by circulating fluids. Temperature gradients are therefore relatively low through these formations. For example, at Ballymacilroy it was 13°C/km; beneath them the gradient was greater.

Heat flow determinations have been made to date at only six sites in Northern Ireland, all by J. Wheildon (personal communication) in association with the Geological Survey of Northern Ireland and the British Geological Survey. With the exception of Port More, where the value is 80 mW/m^2, the Permo-Triassic basin values are closely in accord with the UK average of rather less than 60 mW/m^2. the Annalong Valley and Seefin Quarry boreholes (in the Mourne Mountains, to the south of the Permo-Triassic basins) were drilled in granite within a Tertiary central intrusion complex. The higher heat flow values at these sites of 87 and 84 mW/m^2 may be attributable in part to radioactive heat production from the granites themselves.

The temperature distributions observed in the Permo-Triassic strata at Ballymacilroy indicate that groundwater is circulating naturally in the Sherwood Sandstone and the Permian sandstone aquifers, bringing relatively cool water from shallower depths at the edge of the basin below impermeable cover. They also suggest a degree of hydraulic connection between the aquifers which would not be expected since they are separated vertically by apparently impermeable formations.

Figure 4.36 is a map showing the temperature distribution on the top of the Sherwood Sandstone. It has been compiled from the data discussed above. To a great extent it reflects

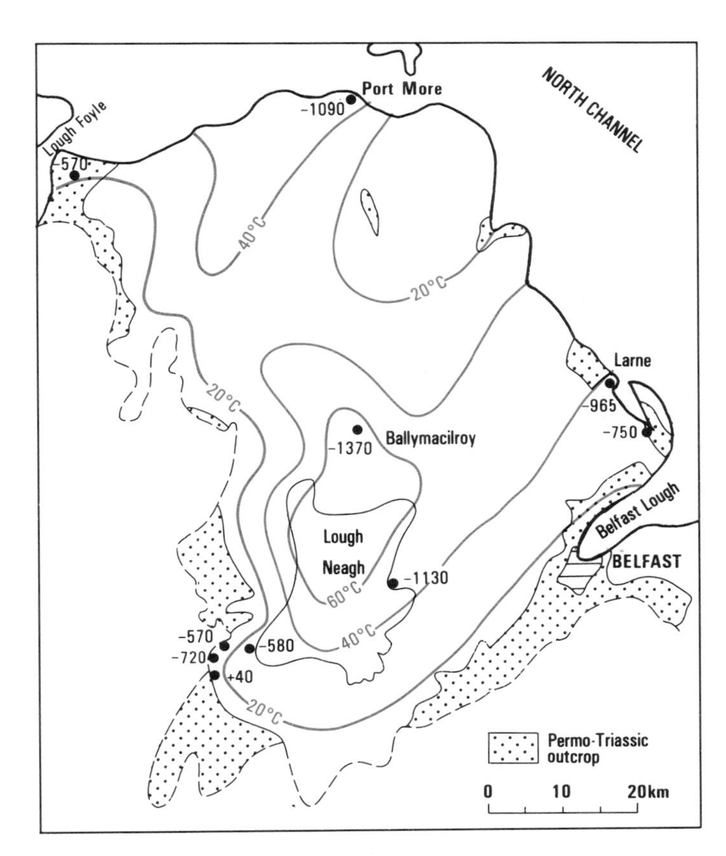

4.36 Estimated temperature at the top of the Sherwood Sandstone in Northern Ireland (figures give the top of the Sherwood Sandstone in metres relative to sea level).

the depth of burial of the sandstone, being greatest towards the centre of the two basins.

Geology and structure of the Permo-Triassic sandstones

The configuration of the buried Permo-Triassic basins and the Sherwood Sandstone and Permian sandstone aquifers within them has been investigated by only a limited number of deep boreholes. Seismic surveys have so far not been successful in unambiguously defining the base of the Permo-Triassic rocks in Northern Ireland. Aeromagnetic surveys have been made of the whole of Northern Ireland but they provide almost no information concerning the Permo-Triassic basins because the basalt cover is so intensely, and variably, magnetised that any anomalies produced by deep sources are totally obscured by the very strong effects from the basalt near the surface.

So, the only geophysical evidence which can supplement the borehole information over most of the area in question is provided by gravity surveys. A detailed interpretation of the gravity anomalies due to the Permo-Triassic basins has been carried out by Carruthers (1978). However, interpretation of gravity anomalies in Northern Ireland is complicated by several factors in addition to the usual uncertainties which normally contribute to the ambiguity of gravity interpretation. Probably the greatest difficulty arises from lack of information concerning the regional background field upon which the negative anomalies caused by the relatively low density Permo-Triassic sediments are superimposed.

If the high regional field, observed in the Midland Valley of Scotland, continues across the North Channel into Northern Ireland, then the apparent gravity anomalies due to the Permo-Triassic rocks would be significantly smaller than the observed anomalies. Also the densities ascribed to various major formations are very uncertain. In particular, the basalt sheet is problematical because of both lateral and vertical density variations, and irregular thickness changes from zero to at least 800 m. Carruthers based his interpretation on fresh basalt densities (between 2.80 and 2.85 Mg/m^3) and thicknesses provided by the Geological Survey of Northern Ireland. The effect of the basalt sheet was to increase the anomaly attributable to the Permo-Triassic rocks. Recent borehole density logs have suggested that the bulk density of the basalts may be close to the basement density of 2.70 Mg/m^3 so that the effect of the basalt on the interpreted thickness of Permo-Triassic strata may be almost neutral.

There is still uncertainty about the densities of the arenaceous Sherwood Sandstone and Permian formations, especially when depths of burial exceed 1000 m. But from the various formation density logs which are available, it would appear that in Northern Ireland only relatively small proportions of the arenaceous fill, in fact only the more open sandstones, have a density as low as 2.40 Mg/m^3 which is the value generally attributed to Permo-Triassic rocks in other parts of Britain. Density logs show that in the deeper parts of the basins considerable thicknesses of the Sherwood Sandstone, usually the basal part, and most of the Permian succession have densities in the range 2.55 to 2.60 Mg/m^3. Since the density contrast between such material and the basement is not as large as has been assumed in previous gravity interpretations the basins must be deeper than has been suggested in the past.

The approach used to postulate the contours of depth to the base of the Permo-Triassic strata in Figure 4.35 was therefore to modify earlier interpretations accordingly, and to calibrate the residual gravity anomalies in terms of the relevant deep boreholes. The end product is a crude representation of the 'regional' shape of the basins, which are almost certainly more complex in reality. In particular, subsidiary and more localised deepening and thinning is probably present along NNW to SSE trending structures, which are only vaguely apparent from the gravity anomalies. The findings of the Larne No. 2 geothermal exploration borehole, which was drilled after the contours of Figure 4.35 had been deduced, suggest that this gravity interpretation is reasonably accurate. Larne No. 2 was completed at a depth of 2880 m when still in Permian strata but may have been close to the base of the Permian succession (Downing and others, 1982b; Penn and others, 1983); Figure 4.35 shows a depth slightly in excess of 3000 m at Larne. However, neither the low density halite nor the high density lavas which were encountered in the Permian had been anticipated, and although the gravity effects of the two more or less cancel each other at this locality, their areal distributions should be independent of one another and the presence or absence of either must reduce the accuracy of the interpretation at other localities.

The gravity evidence cannot be used to provide any information about the actual proportion of aquifer sandstones occurring within the sedimentary fill of the basins. Deductions concerning the depth of burial and thickness of the arenaceous formations, which include the open sandstones, must therefore be made relative to the results of the few deep boreholes which have been drilled in conjunction with the anticipated total depth of the basin in any given area. The sandstones usually comprise at least 50% of the total thickness of fill of the basins, and locally may reach 75% in the deepest parts. The known depths to the top of the first prospective aquifer — the Sherwood Sandstone — are shown in Figure 4.36. Relatively open sandstones invariably occur within this formation, but the same cannot be said of sandstones within the underlying Permian succession. In fact, sandstone of this age which are suitable aquifers have only been encountered in the eastern part of the main basin, with none known south-west of the Ballymacilroy Borehole. In the Langford Lodge Borehole, which appears to be situated on a buried basement high (trending NNW–SSE), the Sherwood Sandstone was found to rest directly on Lower Palaeozoic greywackes with no Permian strata apparent. However, Permian sediments (Magnesian Limestone) are present to the west of Lough Neagh so it is possible that arenaceous rocks of this age, including aquifers, could occur under Lough Neagh.

Hydrogeology of the Permo-Triassic sandstones

Although porous, permeable, fluviatile sandstones of Triassic age appear to occur throughout all the basins this is certainly not true of the Permian succession. In any of the supposed Permian sequences the open sandstones, which are probably of aeolian origin, comprise only a small proportion of generally tight arenaceous material of mainly terrestrial origin, usually overlain by fluviatile marls. The open sandstones therefore tend to be impersistent laterally, so that the Triassic and Permian sandstones are grouped together and regarded as a single aquifer sequence in terms of their geothermal potential.

In situ determinations of transmissivity from pumping tests on shallow water wells and laboratory measurements of porosity and permeability on related core samples have been reviewed by Bennett (1976) and Lovelock (1977). The Permian sandstones only occur near the surface in a small area south-east of Belfast, where porosity is in the range 25 to 30% and transmissivity is usually about 120 m^2/d from pumping tests. This is also the magnitude which would be expected from laboratory measurements of the intergranular permeability, so fissure flow does not appear to be important in these relatively clean, non-indurated sandstones. The Sherwood Sandstone is fairly extensively used as a source of potable water along the western, southern and south-eastern margins of the basins. Porosity is usually between 15 and 25%, but because of poor sorting, fine grain-size, and widespread induration it has a low intergranular permeability. However, permeability is lowest in the area west of Lough Neagh and appears to increase gradually towards the east with transmissivity increasing from about 20 m^2/d to 150 m^2/d. The change is probably due to better sorting, slightly coarser grain-size, and less induration because there is no significant change in thickness. The transmissivity is locally greater than that of the Permian sandstone, which is an intrinsically more permeable material, because the Sherwood Sandstone shows a combination of greater thickness plus fissure flow. Field transmissivity values are usually about an order of magnitude higher than would be expected from laboratory measurements. Lovelock (1977, p.48) concluded that, although the intergranular permeability of the Sherwood Sandstone in Northern Ireland is lower than in other parts of the United Kingdom, 'porosity remains moderate, suggesting that ramifying systems of pore channels are present in the sandstones but that these have extremely small dimensions'.

Drill-stem tests have only been attempted in Northern Ireland at Ballymacilroy (Burgess, 1979) and at Larne No. 2 (Downing and others, 1982b). At Ballymacilroy they comprised two successful tests on a Triassic sandstone interval and on a relatively open sandstone which has been ascribed to the Permian but could almost equally be Sherwood Sandstone, and an unsuccessful test on indisputable Permian material below the Magnesian Limestone, which failed when the soft sandstone caved. The results of the Triassic sandstone test are assumed to be representative of the 215 m thick Sherwood Sandstone aquifer by analogy with the geophysical logs, suggesting a transmissivity of about 15 D m for this interval. Below this the geophysical logs indicate that the Sherwood Sandstone strata become much tighter, but an intermediate interval could contribute a transmissivity of several darcy-metres to a geothermal production well. The so-called Permian sandstone that was tested together with open sandstone of indisputed Permian age, has a total thickness of 45 m. The transmissivity of this combined aquifer should be about 3 D m.

In the Larne No 2 Well two successful tests were carried out on the Sherwood Sandstone. They indicated very low values of transmissivity which are in agreement with the low permeability measurements. Again it is apparent from the geophysical logs that the topmost Sherwood Sandstone is the most open aquifer material of that formation, contributing a transmissivity of about 7 D m to a total value of only 8 D m for the whole formation. The Lower Permian Sandstone at Larne also has a low transmissivity, with an estimated value of about 0.5 D m from geophysical logs and the few core data. The analyses of the results of the gas-lift pumping test from the Lower Permian Sandstone and the volcanic sequence below it gave estimates of transmissivity compatible with this value. The yield during pumping fell to 1.5 l/s, which is negligible as far as geothermal potential is concerned (Downing and others, 1982b).

Interpretation of the geophysical logs from the Newmill Borehole suggests that the Sherwood Sandstone succession there is very similar to that studied at Ballymacilroy. Again a tripartite subdivision of the Sherwood Sandstone is apparent on the basis of physical properties: the uppermost is relatively open sandstone, the next is of intermediate nature, and the lowest division is apparently tight material, resting on the 'Magnesian Limestone Equivalent' and open Permian sandstones. A similar picture also prevails at Killary Glebe where the respective thicknesses of the three units are, from top to bottom, 60, 90 and 86 m.

No reason has been found to explain why the permeability, and hence transmissivity, of both the Sherwood and Permian sandstone aquifers at Larne are so reduced relative to those at Ballymacilroy and Newmill. The porosities of the Sherwood Sandstone at Larne are comparable to those elsewhere so only minor diagenetic or sedimentological variations may account for the reduction in permeability. It may be relevant that the Larne Borehole is the first located in a central area of the main basin in which physical properties have been studied in some detail. The possibility now arises that the reduced permeability of the sandstones may be a general feature of the central area, when compared with the more marginal areas of the basin where, for example, the Ballymacilroy and Newmill boreholes are located. This would obviously have an important bearing on the geothermal potential of the aquifers.

Most of the information about the hydrochemistry of the groundwater in the Permo-Triassic sandstones has been derived from the formation fluids retrieved during drill-stem testing and by extraction from sandstone cores from the Ballymacilroy (Burgess, 1979) and Larne No. 2 (Downing and others, 1982b) boreholes. At Ballymacilroy the interstitial brines extracted from the Sherwood Sandstone core had total mineralisation averaging 104 000 mg/l. Brines which had been extracted during drill-stem tests on the Sherwood Sandstone and on the 'Permian' sandstone were chemically almost identical despite the fact that the two aquifers are separated by apparently tight strata: the total mineralisation was about 120 000 mg/l, roughly four times the salinity of seawater. The dominant dissolved mineral species were Na^+ and Cl^-. The piezometric surfaces calculated for each aquifer from the pressure data of the drill-stem tests were also identical, 110 m below ground level.

At Larne, brines extracted from Sherwood Sandstone cores were found to be contaminated by a lithium tracer which had been aded to the drilling mud, indicating that significant invasion of the core by mud filtrate had occurred. The chemical results are not, therefore, representative.

Reliable chemical results were obtained from brines which had flowed into the drill string during a drill-stem test on the Sherwood Sandstone. The total mineralisation was about 198 000 mg/l, considerably more saline that at Ballymacilroy, but overall comparison of the analyses suggests that the two brines are genetically related. The higher salinity at Larne is probably related to interaction with the halite deposits there, which are not present at Ballymacilroy.

An unexpected finding was that there were no significant differences between the ratios ^{18}O: ^{16}O and ^{2}H: ^{1}H of the brines and those of shallow groundwaters elsewhere in Northern Ireland. This indicates that the fluids saturating the deep aquifers are of meteoric origin and have not undergone significant isotopic exchange with aquifer materials. These waters have flushed out the aquifer system by natural flow which, from the evidence of the low temperature gradients, is probably still taking place.

Chapter 5 Devonian and Carboniferous Basins

D W Holliday

INTRODUCTION

Rocks of Palaeozoic age are widespread in the United Kingdom, cropping out over much of northern and western Britain and covered by younger strata in the south and east (Figure 5.1). Similarly in Northern Ireland Palaeozoic rocks are to be found both at outcrop and beneath a younger cover. Most of the rocks of Palaeozoic age are

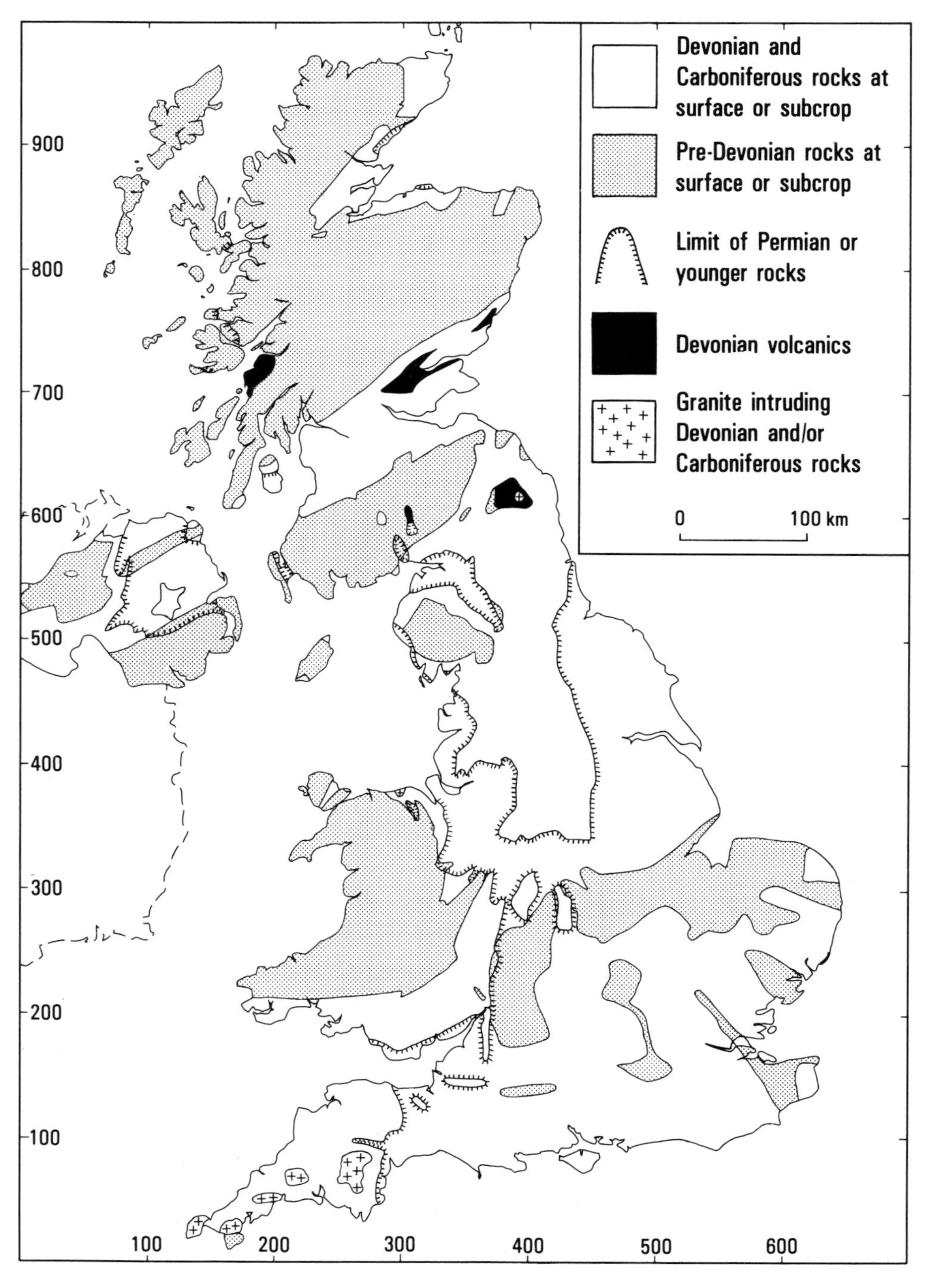

5.1 Distribution of Devonian and Carboniferous strata in the UK.

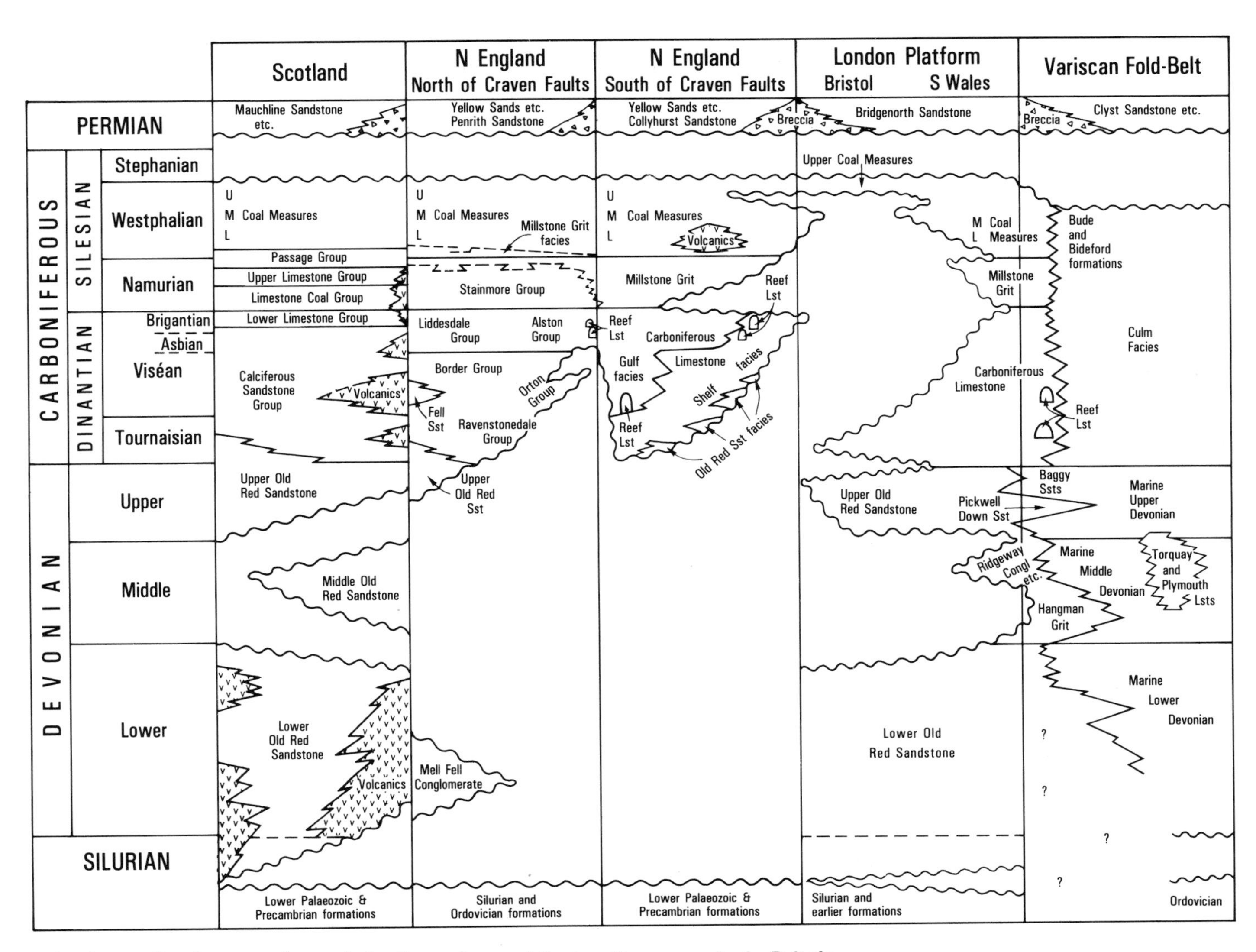

5.2 Generalised successions of the Devonian and Carboniferous rocks in Britain.

sedimentary in origin with a combined thickness commonly totalling several kilometres. Thus in many parts of the country strata of this age occur at depths where temperatures of geothermal interest are reached. This chapter considers the extent to which Devonian and Carboniferous rocks might contain aquifers of sufficient transmissivity to yield fluids of geothermal value. Rocks of Lower Palaeozoic age are not considered as their distribution and nature are not well known at depth. In addition they are not important aquifers at outcrop and there is no reason to suppose their performance will improve with depth. The geothermal potential of these rocks is more likely to be in the realm of hot dry rock. Permian aquifers have been considered in the previous chapter.

The distribution of Devonian and Carboniferous rocks in Britain is shown in Figure 5.1. The geological lines are simplified from published BGS maps, and some small areas, of no significance in the present context, have been omitted. Particularly in the south and east, Devonian and Carboniferous rocks are overlain by later strata and geological boundaries shown under this cover are uncertain. Figure 5.2 is a correlation chart showing the broad relationships of the main formations and facies mentioned in the text.

In this chapter the geothermal potential of the Upper Palaeozoic rocks is discussed in terms of the main geological formations rather than on a regional basis as in the previous chapter. If the regional approach had been adopted, the general similarity of the Upper Palaeozoic sequences in different areas would have entailed much repetition. To overcome any difficulty this may cause, the chapter concludes by summarising the main issues on a regional basis.

PALAEOGEOGRAPHY AND GEOLOGICAL HISTORY

The Devonian and Carboniferous rocks of Britain were laid down in the period of time between the orogenic movements caused by the final closing of the Iapetus Ocean and resultant plate collision at the culmination of the Caledonian Orogeny, and the folding and intrusive episodes at the height of the Variscan Orogeny of southern Britain. Though of less intensity than the main orogenic phases, tectonic activity, locally accompanied by vulcanism, continued sporadically and locally throughout much of Devonian and Carboniferous time. In response to these events, and to climatic and sea-level changes, a wide range of sedimentary rocks were laid down which, except in southern Britain, are for the most part now only weakly to moderately deformed. There is an extensive literature relating to the Devonian and Carboniferous rocks of the United Kingdom. Their biostratigraphy is summarised in a number of Special Reports of the Geological Society of London (George and others, 1976; House and others, 1977; Ramsbottom and others, 1978) which contain extensive bibliographies. Useful reviews of the stratigraphy and palaeogeography, more detailed than can be attempted here, are to be found in a number of standard texts (e.g. Rayner, 1981; Anderton and others, 1979; Johnson, 1982). The Devonian and Carboniferous rocks of Britain have been viewed in the light of modern plate tectonic theory and crustal stretching models of basin formation by Johnson (1982), Dewey (1982) and Leeder (1982).

Devonian

Towards the close of Silurian times (Ludlovian) much of England, Wales, Ireland and southern Scotland was covered by sea. Deep-water sediments were widespread, but deposits laid down in shallow seas are to be found in the English Midlands and South Wales. As a result of the major orogenic phase (Caledonian Orogeny), resulting from the closing of the Iapetus Ocean during late Silurian and early Devonian times, the Lower Palaeozoic rocks were uplifted into mountain ranges, commonly with a south-west to north-east trend. In the Welsh Borderlands and in some other areas, sedimentation was continuous from Silurian to Devonian times with a gradual change from marine to continental conditions. Elsewhere there is an unconformity between the rocks of the two systems which varies considerably from place to place in its angular magnitude and in the length of time of non-deposition represented.

Over much of the United Kingdom the Devonian rocks are largely non-marine and fluvial in origin, composed of red conglomerates, sandstones and shales, known as the Old Red Sandstone, derived from the erosion of contemporary areas of uplift. The prevailing climate seems to have been semi-arid. In Scotland some of these rocks were laid down in areas of internal drainage and include lacustrine deposits. Periods of folding and faulting, representing late phases of the Caledonian Orogeny, are revealed by unconformities within the Old Red Sandstone. In northern Britain and Ireland calc-alkaline volcanic rocks are commonly associated with the earlier parts of the Old Red Sandstone and are locally intruded by or closely associated with granitic plutonic intrusions. Other similar granites, probably of much the same age, cutting pre-Devonian rocks, are widespread in northern Britain and, on the basis of geophysical evidence, perhaps in central and eastern England also.

Marine sediments of Devonian age are found in the south and south-west of England and Wales. The most northerly marine occurrences are of thin intercalations of shallow-water sediments within the Old Red Sandstone of South Wales and central England. Southwards (north Devon and Somerset) these marine beds become dominant. Farther south (south Devon and Cornwall) deep-water sediments, including turbidites, are common, but locally there were areas of shallow-water, carbonate-bank deposition. Deep-water marine sedimentation continued throughout much of Carboniferous time (Culm facies) in south Devon and Cornwall.

Carboniferous

During Lower Carboniferous (Dinantian) times much of Britain was totally or partially inundated by the sea. A number of positive areas were not submerged until late Dinantian or even later times, or were only occasionally covered. The most important of these were an area stretching from Wales to East Anglia (commonly known as St George's Land), parts of the northern Pennines, the Lake District and the Southern Uplands of Scotland. The Scottish Highlands probably remained as land throughout, being part of an active source of terrigenous clastic material that was eroded to give rise to the Dinantian fluvial, coastal plain and deltaic sediments which are interbedded with marine limestones and shales in Scotland and northern England. These sediments are, for the most part, grey measures, suggesting a wetter climate than that prevailing in the Devonian. However, towards the base they include red beds of Old Red Sandstone aspect, and in many northern areas there is no clearly distinguished boundary between Devonian and Carboniferous rocks. Southwards these strata pass into marine beds, which are mainly shelf facies, shallow-water limestones on the positive areas (Carboniferous Limestone), but largely calcareous mudstones and argillaceous limestones of deeper-water origin in the rapidly-subsiding basinal areas of the central Pennines (gulf facies). Reef limestones occur commonly at the hinge zones between gulf and shelf and are also known in the extreme south of Wales and in Somerset at the boundary of the shelf limestones and the Culm facies to the south. Alkaline-basaltic volcanic rocks are commonly interbedded with Dinantian rocks in Scotland, and also are known locally in northern England, Derbyshire and Somerset. Volcanic rocks of spilitic character are associated with the Culm facies of Devon and Cornwall.

As a result of renewed uplift in late Dinantian and early Namurian times, rocks of Namurian age are absent over most of St George's Land and its southern margins. Attenuated sequences occur on its northern margin. The main depositional centres were in the Culm basin of south-west England and in the central and southern Pennines. In this latter area deposition continued, in the major basinal areas, without a break from Dinantian times, but commonly on the former shelf areas there is a disconformity or minor unconformity. In such areas a major karst system was developed within the limestone during this period of erosion, opening out joints and fractures and locally forming caverns. During Namurian times the area of sedimentation gradually increased to cover the shelves, so that younger Namurian rocks are to be found resting on Carboniferous Limestone or older rocks in the northern part of St George's Land. Similar relationships are found on the other side of St George's Land where Namurian rocks overlap against older rocks towards the edges of the

South Wales Coalfield. Deposition was also continuous from the Dinantian in the northern Pennines and in Scotland, where the dominant lithologies are sandstones and shales of shallow marine, deltaic and fluvial origin. Widespread sheets of fluvial coarse-grained sandstone were laid down in the later part of the Namurian. Limestones, indicative of open marine incursions, are common, particularly in the early part of the Namurian. Reduced thicknesses occur over positive areas and locally there is evidence of contemporaneous earth movements affecting sedimentation. Similar phenomena are observed in the central Pennines and adjacent areas. In early Namurian times these areas were occupied by a deep water basin in which black shales were deposited, but later a series of delta lobes, each locally preceded by turbidites, built out into the basin and gradually infilled it. The fluvial channels associated with the lobes are commonly composed of coarse-grained, pebbly sandstone from which has been derived the term Millstone Grit, now applied to the combination of facies found in the Namurian of this part of the Pennines. Rocks of Millstone Grit facies also occur in the South Wales Coalfield.

In early Westphalian times tropical coastal swamp conditions were established over much of the country. Marine deposition still persisted to the south in the Culm basin, while St George's Land and the Scottish Highlands were largely areas of non-deposition. The sandstones, shales and coals of the Westphalian, known as Coal Measures, overlapped older rocks around the edges of St George's Land. During this period conditions of deposition were similar throughout the country and there was little local structural control on sedimentation. Volcanic rocks and associated intrusions, of alkaline affinities, are locally associated with the Coal Measures in Scotland, and the east and south Midlands.

Leeder (1982) has attempted to apply McKenzie's (1978) model of lithospheric stretching to British Carboniferous basin formation in a continental lithosphere rendered inhomogeneous in part by the presence of Caledonian (including early Devonian) granitic plutons in the upper crust. The driving force for the stretching is believed to be plate motions and stresses caused by subduction to the south of Britain. These caused the lithosphere north of St George's Land to become stretched during the Dinantian, causing regional tension and rift formation. Dewey (1982) and Johnson (1982) in addition have suggested that dextral strike slip motion on major fault lines has led to the formation of 'pull-apart' basins. The major change in tectonic behaviour during Namurian times, from fault-bounded rifting of the Dinantian to a general crustal sagging, centred near Manchester, has been ascribed to athenospheric cooling.

Towards the end of Carboniferous times, as a result of thrusting from the south, the Culm basin and parts of the adjacent shelf area to the north were intensely deformed, uplifted and eroded during the Variscan Orogeny, and a new mountain range formed. Granites were intruded into the cleaved rocks of Cornwall and south Devon. This highland area extended from central Europe, across southern England, into Ireland. To the north of the fold-belt, the deformation generally was less strong, but locally there was considerable uplift and erosion. The boundary between the intensely deformed strata of the south, and the more gently deformed beds to the north, is known as the Variscan Front. The nature of the Variscan Front and its position under the Mesozoic cover of southern England is uncertain. In late Westphalian times sediments, derived from the new highlands thrown up by the earliest phases of the Variscan Orogeny, were carried northwards and spread out over St George's Land where commonly they are unconformable on earlier rocks. In the north of England and Scotland, contemporaneous strata (Upper Coal Measures), conformable with earlier strata, are found in most coalfields and in many areas include red beds indicative of a return of more arid climates.

Post-Carboniferous

All the Devonian and Carboniferous rocks were subjected to the later phases of the Variscan Orogeny, and Britain entered a major period of subaerial erosion. North of the Variscan Front all or part of the Upper Carboniferous succession was removed, and in areas of major uplift, Dinantian or older rocks were exposed (Wills, 1973), allowing the development of karst phenomena in the Carboniferous Limestone. In the Variscan fold-belt erosion was particularly intense and over large areas Carboniferous rocks were completely removed. Progressively during Permian and Mesozoic times, the Palaeozoic land surface was covered by renewed sedimentation. This sedimentary cover was formerly more widespread than at present, and in many areas Devonian and Carboniferous rocks were once buried to depths up to 3 km greater than today. During Mesozoic and Tertiary times the country was subjected to several periods of earth movement. Locally, particularly in southern England, the Mesozoic rocks are intensely deformed but otherwise these movements were less strong than those of the Variscan Orogeny and largely resulted in block faulting of the Devonian and Carboniferous rocks.

ESTIMATION OF SUB-SURFACE TEMPERATURES

Of the large number of subsurface temperature measurements in Britain (Burley and others, 1984), only a limited proportion are reliable equilibrium determinations, the bulk of the data being derived from bottom hole temperature readings taken soon after drilling. The methods employed to estimate temperatures and the results obtained are discussed in more detail in the individual reports on each basin published by BGS (see Appendix 4) and are summarised in the figures accompanying this chapter. However, a number of points of more general importance have emerged and can with convenience be discussed here. It is noticeable that almost everywhere, geothermal gradients in the Lower and Middle Coal Measures are close to or above the national average value. This is because the Coal Measures are essentially an argillaceous formation and have low thermal conductivity (up to 2 W/mK). Thus, even in areas where the heat flow is below the national average of 54 mW/m^2, geothermal gradients as high as 25°C/km can be anticipated through the Coal Measures. If such rocks are covered by Mesozoic rocks, as in the East Midlands where the Jurassic and upper part of the Triassic are also largely argillaceous, temperatures well in excess of 60°C can be expected at depths less than 2 km, even where heat flow is below average. Conversely formations such as the Pennant Measures, Old Red Sandstone or Carboniferous Limestone, which have high thermal conductivities (around 2.5 to 3 W/mK) commonly are characterised by relatively low geothermal gradients even where heat flow is above average. Where there is a thick cover of argillaceous Coal Measures and Mesozoic strata, this is relatively unimportant but, where high conductivity formations are at or close to the surface in areas with below-average heat flow, then depths well in excess of 2 km, where few open fractures can be anticipated, are required for temperatures to reach 60°C.

GENERAL GEOLOGY AND HYDROGEOLOGY

Throughout much of Devonian and Carboniferous times, shallow marine or continental deposits, commonly of substantial thickness with a large proportion of arenaceous sediment, were laid down in Britain. In Lower Carboniferous times widespread shelf limestones were deposited. Apart from those areas intensely folded during the Variscan Orogeny, the Carboniferous and Devonian rocks are generally little deformed. This suggests that aquifers with good intergranular permeability could be widespread in rocks of this age. However, although at outcrop in many areas sandstones and limestones of this age yield significant quantities of water, the bulk of the yield is from fissures and fractures. Most sandstones also have some intergranular permeability but this, with local exceptions, is generally low. This appears to result from a number of causes, principally from the fine-grained nature of many of the rocks, from poor sorting resulting from high silt and clay contents, from the products of diagenesis blocking pore spaces and from compaction during later deep burial and earth movements. Except locally, little is known of the hydrogeological properties of deeply buried British Devonian and Carboniferous rocks, but it is unlikely that these will be more favourable than those of strata at shallow depths.

The extent to which open fracture systems occur at depth in British Devonian and Carboniferous rocks is unknown. However, there is some evidence that there is movement of water through the Carboniferous Limestone at depths of around 3 km (Bullard and Niblett, 1951; Burgess and others, 1980), probably by way of open fractures. Hot springs occur at several localities and provide further evidence of deep circulation of groundwater, probably by way of fractures. Deeply buried palaeokarstic fissures and caverns have been shown to play an important role in allowing passage of fluids through the limestone at depth in Belgium and northern France, yielding economically important quantities of hot water (Delmer and others, 1980). Palaeokarst systems of various ages have been demonstrated in the Carboniferous Limestone at outcrop and shallow depths in many parts of Britain.

Thus, before the basin studies described below were undertaken, it was realised that any successful exploitation of geothermal energy from Devonian and Carboniferous strata in Britain will depend either on drilling into natural or artificial fissure systems or on locating formations with hydraulic properties considerably in excess of those normal for rocks of that age. In view of the wide extent of Devonian and Carboniferous rocks in Britain, a preliminary appraisal was carried out (Holliday and Smith, 1981) in order to ascertain which areas and formations had the most promise and for which there was sufficient geological, geophysical and/or hydrogeological data to support a major desk study. In all, five major regions were studied: Nottinghamshire-Lincolnshire-South Humberside (Gale and others, 1984c), South Wales (Thomas and others, 1983), the West Pennines (Smith and others, 1984), the Midland Valley of Scotland (Browne and others, 1985) and the Fell Sandstones of the Northumberland Coalfield (Cradock-Hartopp and Holliday, 1984). In addition the thermal springs of the Bath-Bristol area have been studied in some detail (Burgess and others, 1980). The following reviews of the low enthalpy geothermal potential of the major subdivisions of the Devonian and Carboniferous, taken in reverse order from the youngest to oldest, are intended to include the whole country, but naturally most detail is available for those regions studied in detail. In particular, the Nottingham-Lincolnshire-South Humberside area, which has by far the most data as a result of many decades of hydrocarbon and coal exploration and exploitation, provides the main yardstick by which other regions can be evaluated.

COAL MEASURES

The Coal Measures (Westphalian) of Britain contain the vast majority of the productive coal seams which form the country's greatest single resource of fossil fuel. The main productive measures are dominantly grey in colour and include a relatively small number of rock types which are repeated many times over in a full succession — non-marine mudstones and siltstones, sandstones, seatearths and coal with sporadic marine bands. These latter beds, which are commonly widespread, occurring in most coalfields, provide the means for the main lithostratigraphical and chronostratigraphical divisions of the succession. The Upper Coal Measures differ from the Lower and Middle divisions by generally containing fewer and thinner coals, by the presence of thicker and more widespread sandstones (especially in the south) and by the widespread presence at some levels of red beds which are only very locally developed in the underlying strata.

As a result of post-Carboniferous earth movements, the coalfields of Britain are now widely scattered, but originally Coal Measures probably covered much of the country. The area between St George's Land and the Southern Uplands of Scotland, extending out to the southern North Sea, formed one large basin during this time. South of St George's Land Coal Measures were probably continuous in the area north of the Variscan Front from Pembrokeshire to Kent, although there is evidence in this belt of more than one centre of deposition. Upper Coal Measures are well developed hereabouts and overlap onto St George's Land. It is likely that Coal Measures were formerly present over most of the Midland Valley of Scotland, and perhaps original thickness changes, there is much variation in total thickness of Coal Measures from place to place. The maximum amounts, locally exceeding 2 km, are to be found in South Wales, North Staffordshire and Lancashire.

The environments of deposition indicated by the Lower and Middle Coal Measures include fossil soil and forest-swamp (seatearths and coal), shallow lake (mudstones with non-marine bivalves), inter-distributary bay, tidal inlet and shallow sea (mudstones with *Lingula* and/or marine shells), prograding deltas (upward-coarsening mudstone-siltstone-sandstone sequences), delta distributary channel and river channel (upward-fining sequences) and flood-plain (rooty and planty mudstones). Almost everywhere deposition took place close to or below the water table, except locally around positive areas or where elevated ground was built up by local volcanic activity. In such instances the grey measures pass into red beds with the coals dying out though seatearth horizons may continue. In the northern coalfields the Upper Coal Measures show broad similarity to the underlying beds, but fluvial environments predominate, there being no immediate marine influence. The sediments were mainly deposited by both low sinuosity braided streams and by meandering rivers flowing across an extensive flood-plain. The contemporary water table was evidently fluctuating, red beds forming when it was low and grey measures with thin coals forming when it was high. Fluviatile sedimentation also dominated in the southern coalfields but proximity to the rising Variscides in the south resulted in the deposition of successions either dominated by or containing thick units of coarse-grained, pebbly sandstone with only subordinate mudstones and coals (e.g. the Pennant Measures of South Wales, and the Arenaceous Coal Group of Oxfordshire). Secondary reddening of any part of the Coal Measures may occur where these are close to the sub-Permo-Triassic surface.

The Lower and Middle Coal Measures are dominantly argillaceous and silty, sandstones generally forming no more than, and commonly less than, 30% of the total thickness. The coarsest sandstones occur in channel fills and form narrow, mostly discrete meandering bodies, generally up to 20 m thick, and account for less than 1% of the total sequence. However, where channels are stacked upon each other thicknesses of up to 100 m are attained locally. The sandstones are quartz rich, with minor feldspar and accessory mica and vary between very fine and coarse-grained. Sandstones also occur thinly interbedded with siltstones in levée, distributary mouth bar and interdistributary bay deposits. These also have a limited lateral extent and are generally more argillaceous than the channel sandstones.

Coalfields east of the Pennines

The Coal Measures of the eastern Pennine region in Yorkshire, Derbyshire and Nottinghamshire and their easterly concealed extension in many ways can be regarded as a standard for the country as a whole. Generally the Coal Measures succeed the underlying Millstone Grit conformably but around the margin of St George's Land they overstep onto Carboniferous Limestone or older rocks. The total thickness ranges up to 1500 m, but generally no more than 25% of the succession is sandstone. Isopachytes, simplified structure contours and estimated temperatures on the top and bottom of the Coal Measures are shown in Figure 5.3. These maps illustrate that over much of the region, all or part of the Coal Measures are at temperatures in excess of 40°C and, in the north-east around Grimsby, above 60°C.

The hydrogeology of Coal Measures at outcrop in this region has been studied by Gray and others (1969) and Downing and others (1970). Yields from boreholes in the outcrop and adjacent areas are very variable and generally not more than a few litres per second. Upper Coal Measures sandstones tend to provide greater yields and mine shafts can yield 20 to 50 l/s (Rodda and others, 1976). Although the sandstones have a porosity of 12 to 15% the permeability decreases with depth due to compaction, weight of overburden and more widespread cementation. The lateral extent of the sandstones is seldom greater than 250 km^2 and thinning from outcrop in an easterly direction is common. The aquifers are further isolated and reduced in size by block faulting. This reduces the area of recharge, and high initial yields soon decline substantially due to abstraction exceeding recharge. Although jointing in the sandstones may be present, interconnection seems to be poor and sandstones at depth can be dry despite the presence of joints. Folding apparently has a greater control on permeability; joints which open in anticlinal regions allow water to infiltrate and collect in synclinal regions resulting in collieries in synclines being wetter than those in anticlines. The mining of coal has had a major effect on the aquifers; firstly dewatering alters the local flow regime and secondly the tensional and compressional forces generated by the subsequent collapse of the mine workings result in changes in the secondary permeability of the overlying strata. Ineson (1953) attempted to calculate the permeability of the Coal Measures strata from mine pumping data, estimates of hydraulic gradient and cross-sectional area of the working face from which the water is derived. The values of transmissivity obtained must be regarded as order-of-magnitude estimates but range from 0.2 to 20 m^2/d and estimates of permeability range from 0.1 to 34 m/d (about 0.1 to 40 mD), decreasing away from outcrop areas.

Away from outcrop the Coal Measures sandstones form the major reservoirs of the East Midlands oilfields and consequently their reservoir/aquifer characteristics have been studied in detail. Porosity values are determined both from geophysical logs and core analysis, whereas permeability is determined from core analysis and DSTs. Where both have been determined the DST value is regarded as being the more representative as it measures the bulk permeability of the sandstone, including fracture permeability. Transmissivity values can be determined for the section of core tested or the particular sandstone on which a DST was carried out. However, they have little meaning in any formation-wide comparison due to the variability of both individual and cumulative sandstone thicknesses.

Because of the varied nature of the sandstone bodies and the different methods of aquifer property determination no trends can be derived. Mean porosity in the Lower and Middle Coal Measures sandstone units ranges from 7 to

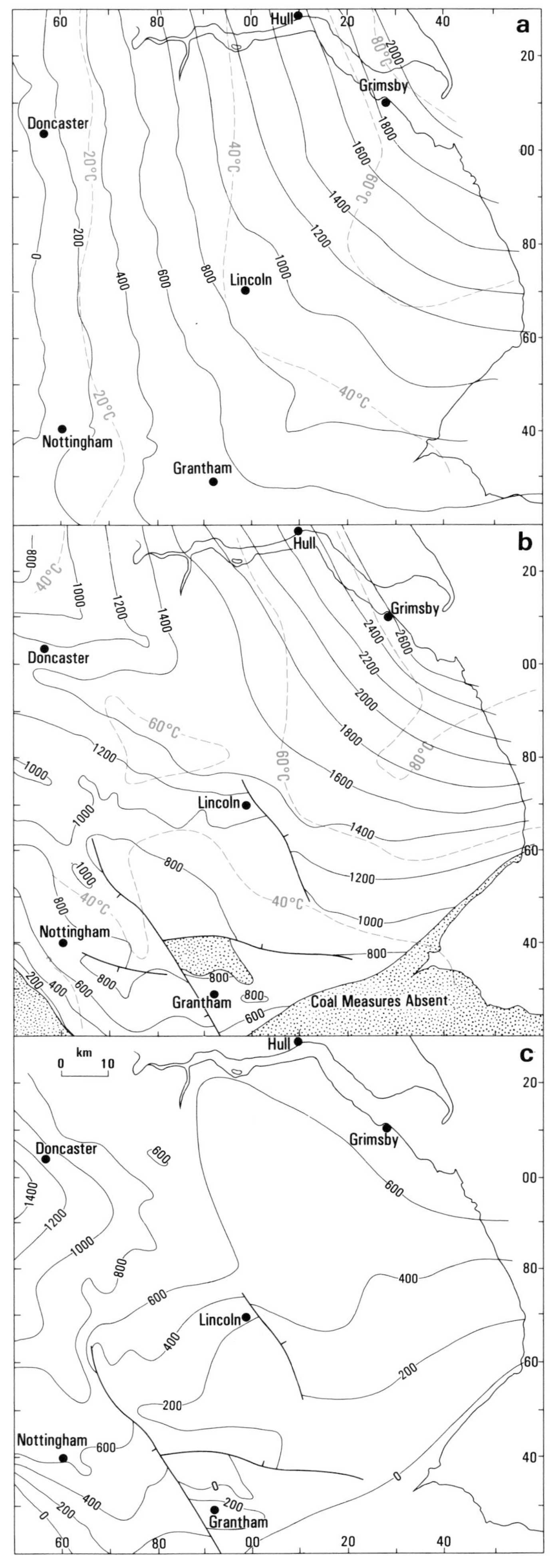

5.3 East Midlands:

a) Structure contours (metres below sea level) and estimated tempertures (°C) at the base of the Permo-Triassic.

b) Structure contours (metres below sea level) and estimated temperatures (°C) at the base of the Coal Measures.

c) Isopachytes of the Coal Measures (metres).

20%, the majority of values being around 12%. In contrast, the Upper Coal Measures sandstones range in porosity from 12 to 19% and have a mean value of approximately 15%. Mean permeability values show a similar variability, the Lower and Middle Coal Measures sandstones typically ranging from 0.06 to 37 mD, the mean value being approximately 13 mD. Locally exceptionally high values with average permeability up to 400 mD have been recorded, but these are apparently of limited extent. The Upper Coal Measures sandstones range from 2 to 160 mD, the mean value being approximately 60 mD. It must be remembered that only the sandstones with a potentially high permeability are likely to be tested, a large proportion of sandstones being disregarded as having low permeability on the basis of geophysical log analysis.

Of the boreholes where the percentage of sandstone was calculated, the cumulative thickness ranged between 7 and 210 m. Assuming a mean cumulative thickness of 100 m of sandstone, a mean permeability of 50 mD would be required to give a transmissivity of 5 D m and to justify the Coal Measures sandstones being called a geothermal resource. From the permeability values given above, this value seems unlikely to be obtained, even in the Upper Coal Measures, except locally.

The pressure data available from Westphalian sandstones are few. Although no flow regime can be defined the data indicate that mine dewatering has reduced the pressure, particularly in the south near Nottingham. The inferred flow regime is an easterly flow from outcrop and a westerly flow from the Gainsborough-Lincoln area towards the intervening low pressure zone caused by coal mine dewatering. In order to retain an elevated pressure in the east of the region, upward movement of groundwater from the underlying Namurian and Dinantian formations must be inferred. This has been found to be the case at some boreholes.

The chemistry of the Coal Measures groundwaters was discussed in detail by Downing and Howitt (1969) and few additional data have been gathered since that time. The most noteworthy aspects of these groundwaters are that they have a markedly higher concentration than the underlying Namurian and Dinantian groundwaters. The contours shown on Figure 5.4 indicate an increase in salinity in a northerly direction and the approximate location of the chloride concentration of present-day seawater. The areal distribution of salinity is similar to the Millstone Grit and Carboniferous Limestone groundwaters in that in the south the groundwater is less concentrated than seawater, indicating recharge from some source, whereas in the north the brines are highly saline, even at relatively short distances (about 20 km) from outcrop.

Coalfields west of the Pennines

The thickest Westphalian sequences in Britain occur in the Lancashire and North Staffordshire coalfields. A total thickness of 2.5 km is believed to occur in the area south of Manchester (Figure 5.5). Structure contour and estimated temperature maps for the top and bottom of the Coal Measures are shown in Figure 5.5 and identify a possible geothermal target area south and south-west of Manchester, under the Permo-Triassic rocks of the Cheshire Basin. Around 25% of the measures are sandstones.

Compared with the eastern Pennine region, there are few data from deeply buried water-bearing Coal Measures sandstones. An overall impression can be derived from experience of coal mining and from water supply data. Generally the sandstones are hard compact rocks that yield water principally from fissures. The porosities tend to be low because of secondary cementation. Limited data from public and private water supply wells in the Upper Coal Measures indicate that specific capacities are very variable. Overall, the values (average 11 m^3/d m for the Keele Beds) are indicative of low permeabilities, even at shallow depths. A value of only 23 m^3/d m has been obtained from a Middle Coal Measures sandstone near Manchester. With the exception of two sites with specific capacities of 400 and 250 m^3/d m, all other sites in the Lower Coal Measures have specific capacities of less than 25 m^3/d m. Drill-stem and Repeat Formation tests have been carried out on a limited scale in boreholes but the resultant data are of poor quality. Laboratory tests on core from three NCB boreholes gave very low porosities (less than 15%) and permeabilities (less than 1.6 mD). Thus yields from water boreholes are unlikely to exceed 12 l/s. Even in the more arenaceous Upper Coal Measures not more than about 5 l/s can be anticipated from a borehole and, indeed, few mine shafts in the region can yield higher flow rates. Thus it would seem unlikely that yields high enough for geothermal purposes can be obtained from the Coal Measures in this region.

Coalfields of northern Britain

For the most part the more northerly coalfields contain thinner successions than those previously discussed and have no, or thin, Permo-Triassic cover. Thus the Coal Measures in Cumbria or Northumberland and Durham have no geothermal potential, not being sufficiently deeply buried. Depth contours and estimated temperatures at the base of the Coal Measures in the Midland Valley of Scotland are shown in Figure 5.6. These suggest some potential in the 40°C range.

However, aquifer properties are poor and this potential is unlikely to be realised. Borehole yields range from 1 l/s to 15 l/s, principally from fissure flow. Permeability values of 1 mD to 600 mD, with a mean value of 10 mD, have been obtained from laboratory analysis of core samples from depths less than 100 m. Values of 7 mD to 70 mD have been obtained from core samples from Seafield Colliery. The corresponding porosity values range from 4 to 24% (near surface samples) and average 16% for the Seafield samples. Up to 150 l/s have been pumped from mine shafts but, because of the large effective areas drained, such values have little relevance to geothermal developments.

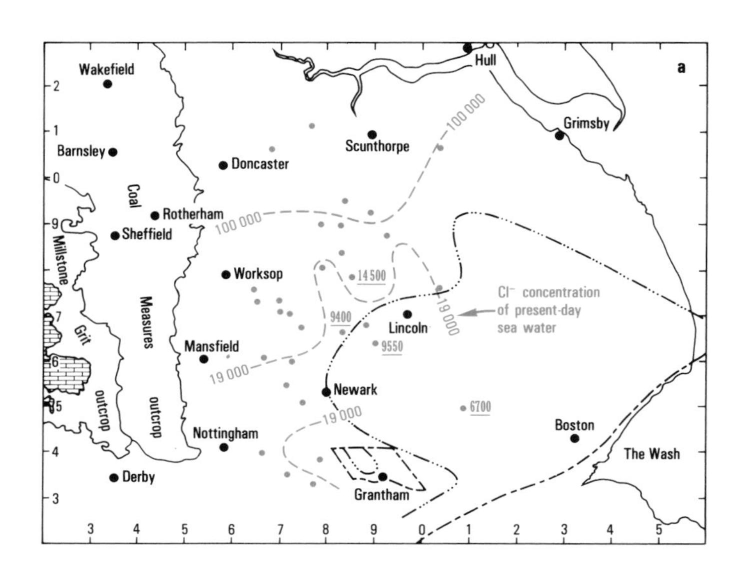

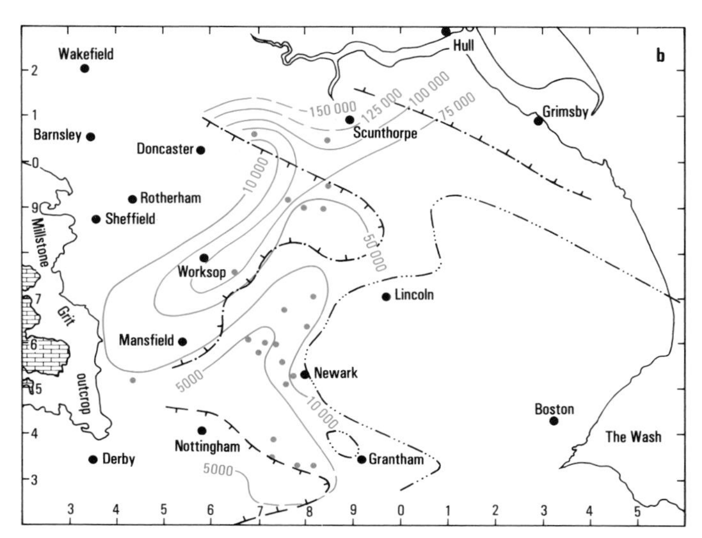

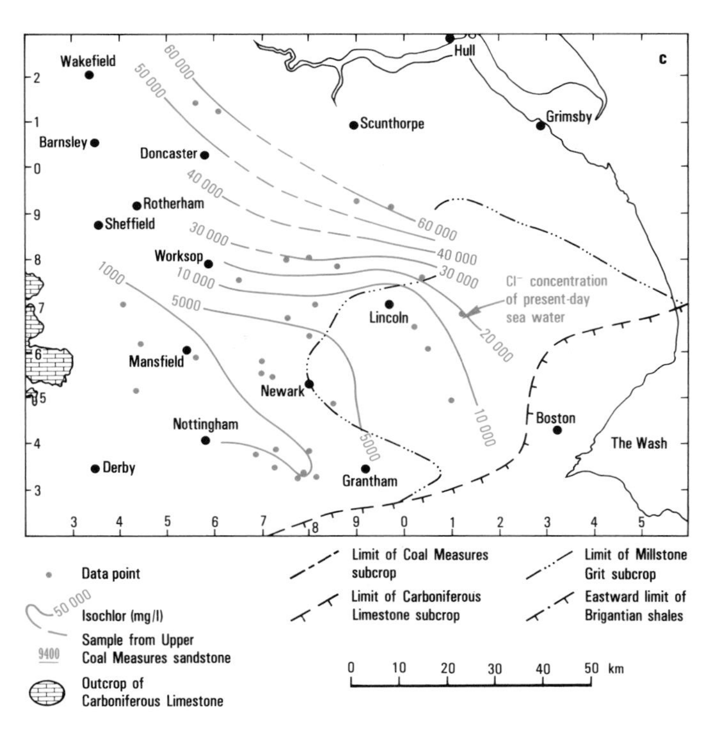

5.4 Chloride concentration of groundwaters from the Carboniferous of the East Midlands a) Coal Measures b) Millstone Grit and c) Carboniferous Limestone

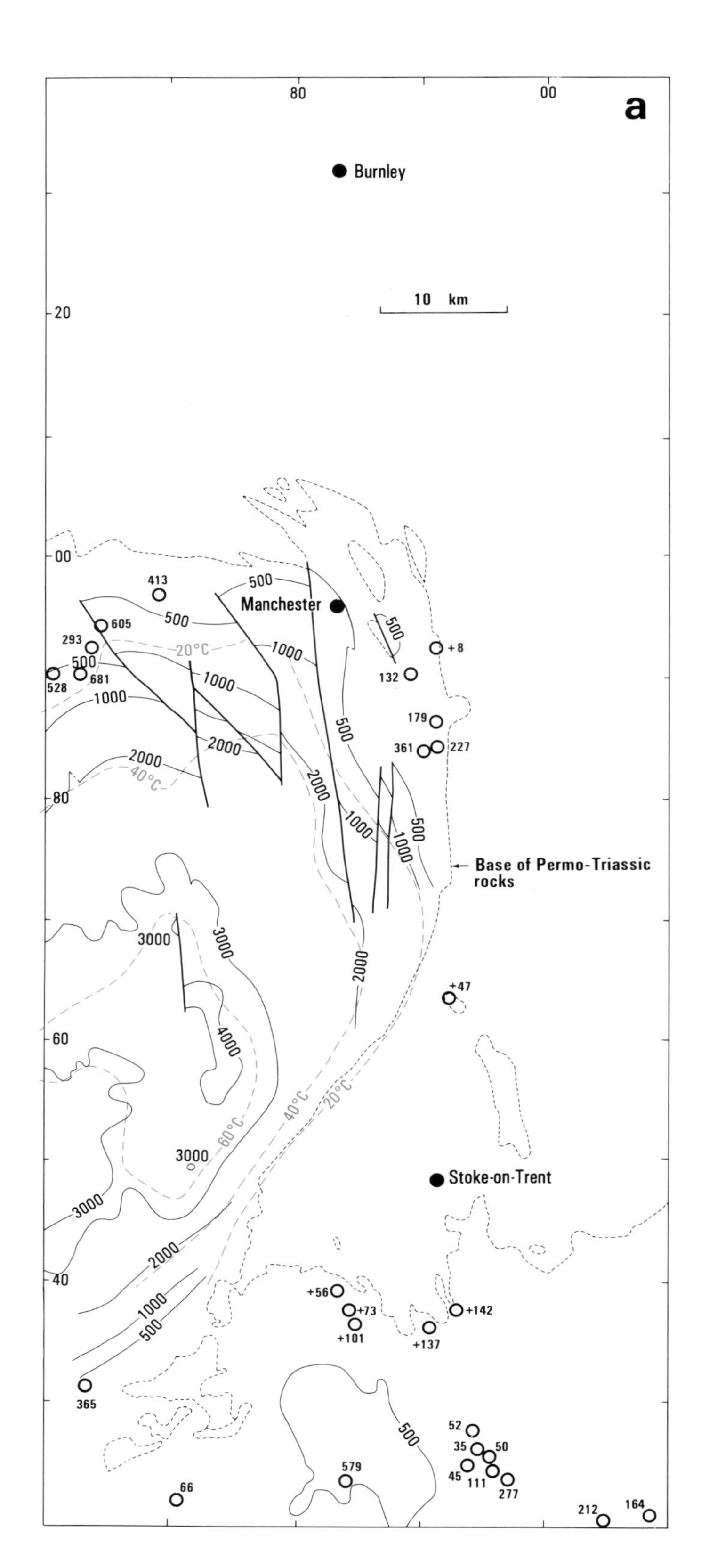

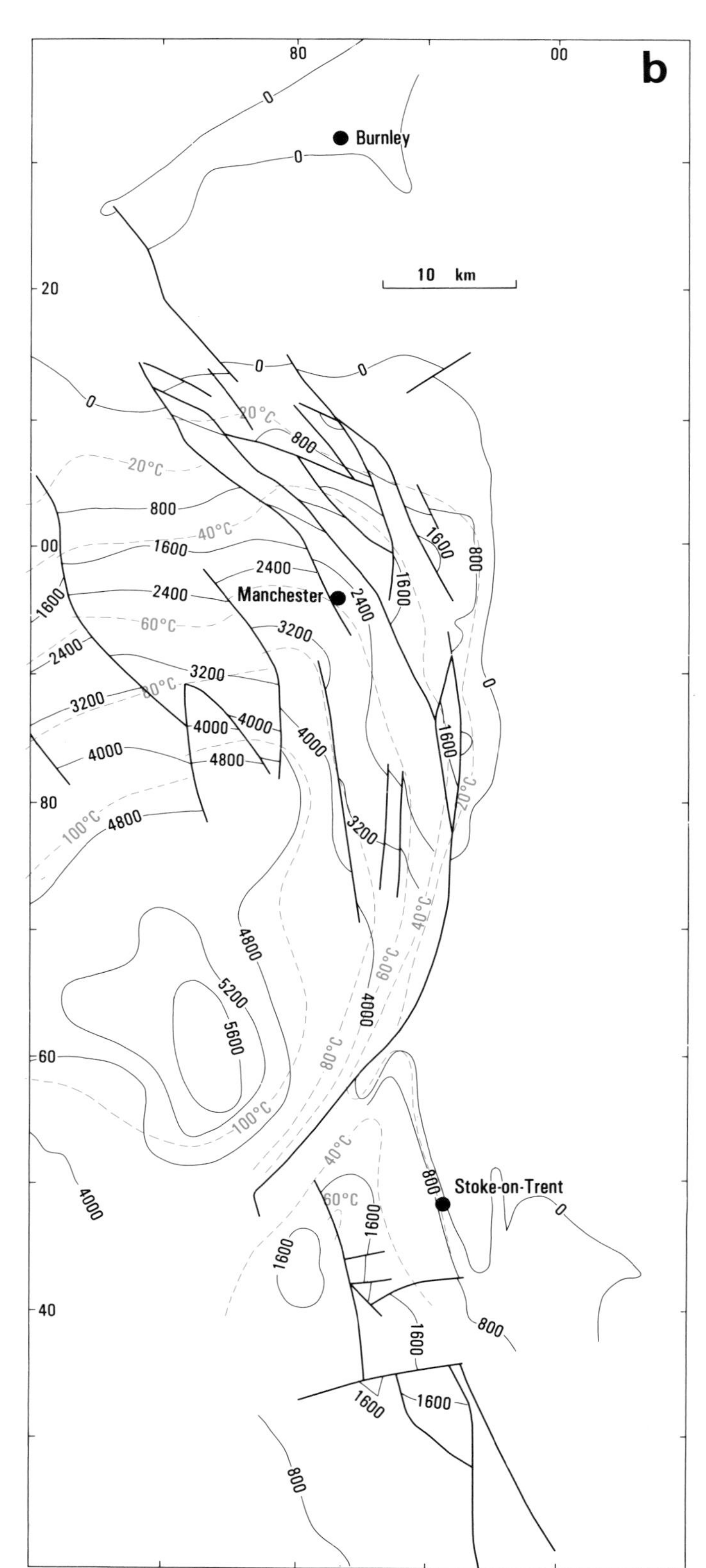

5.5 North-west Midlands:

a) Structure contours (metres below sea level) and estimated temperatures (°C) at the base of the Permo-Triassic.

b) Structure contours (metres below sea level) and estimated temperatures (°C) at the base of the Coal Measures.

c) Isopachytes of the Coal Measures (metres)

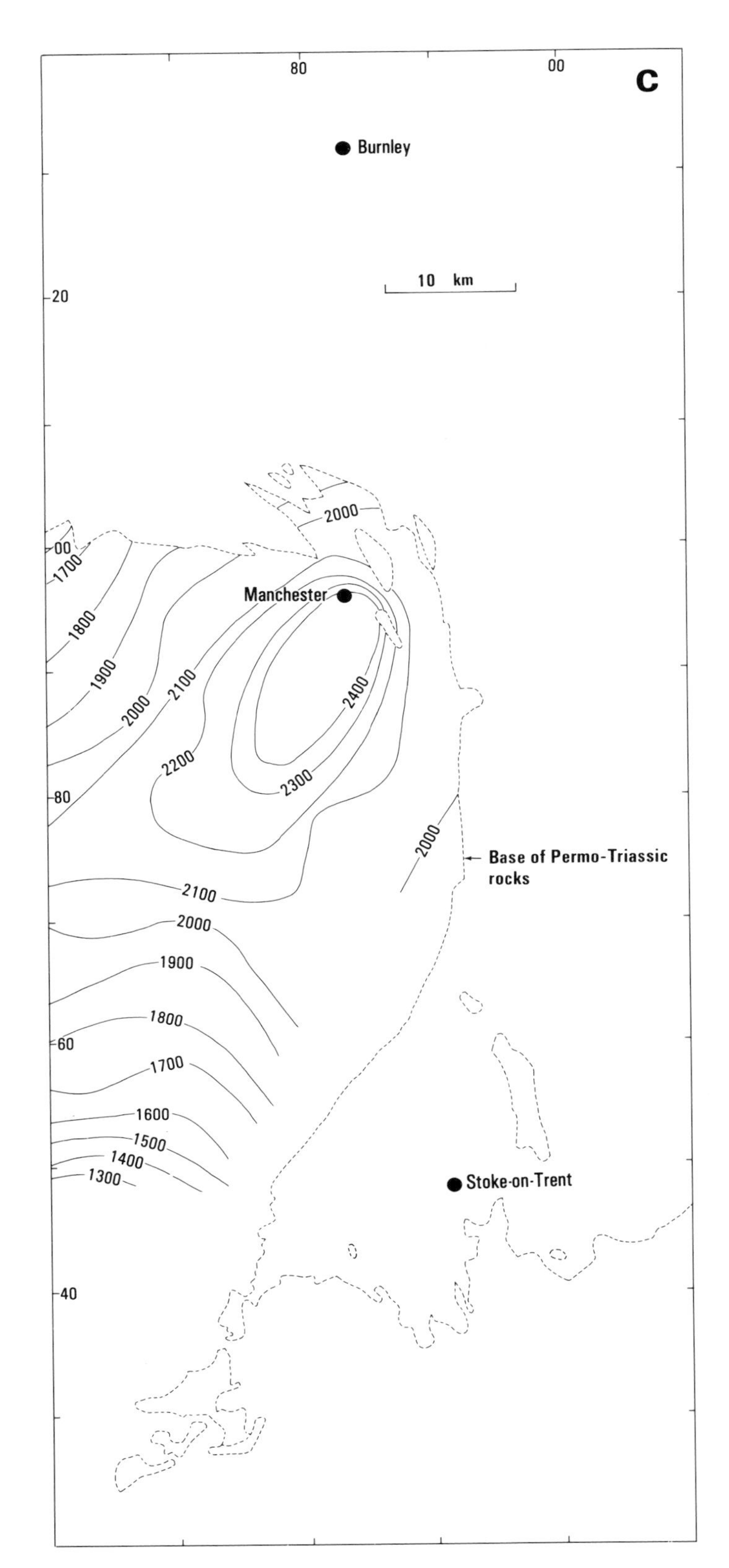

Coalfields south of St George's Land

A number of Coal Measures basins lie south of St George's Land but only South Wales has been studied in detail. Here, the Upper Coal Measures (or Pennant Measures) consist predominantly of thick, massive, feldspathic and micaceous sandstones, some of which are coarse-grained and conglomeratic. The total thickness of the sequence ranges from about 700 m in the east of the coalfield to over 2000 m in the west (Figure 5.7); the thickness of actual sandstones is about 900 m in the west, reducing to 600 m in the Taff Valley and 240 m in the east. The sandstones of the Pennant Measures form the high ground in the centre of the coalfield and their base is relatively close to the surface over the northern and eastern parts of the coalfield. However, between Swansea and Llanelli it is over 1500 m below sea level, where temperatures in excess of 40°C have been inferred (Figure 5.7).

5.6 Midland Valley of Scotland:

a) Structure contours (metres below sea level) at the base of the Coal Measures (top of Passage Group)

b) Estimated temperatures (°C) at the base of the Coal Measures (top of Passage Group)

c) Isopachytes (metres) of the Passage Group.

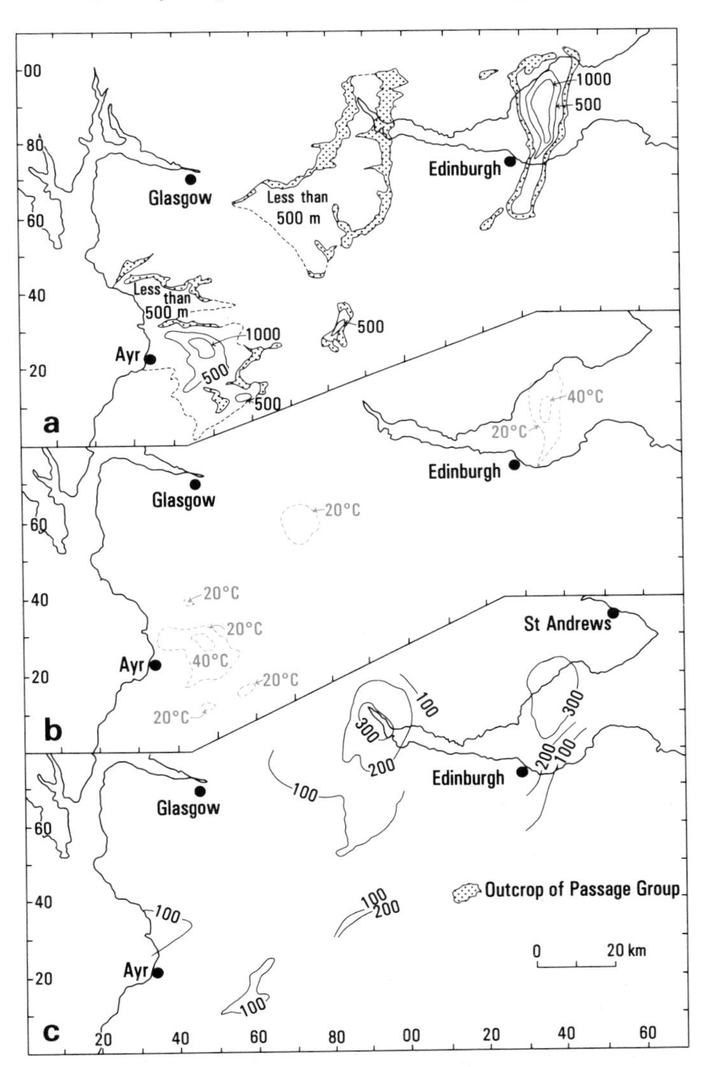

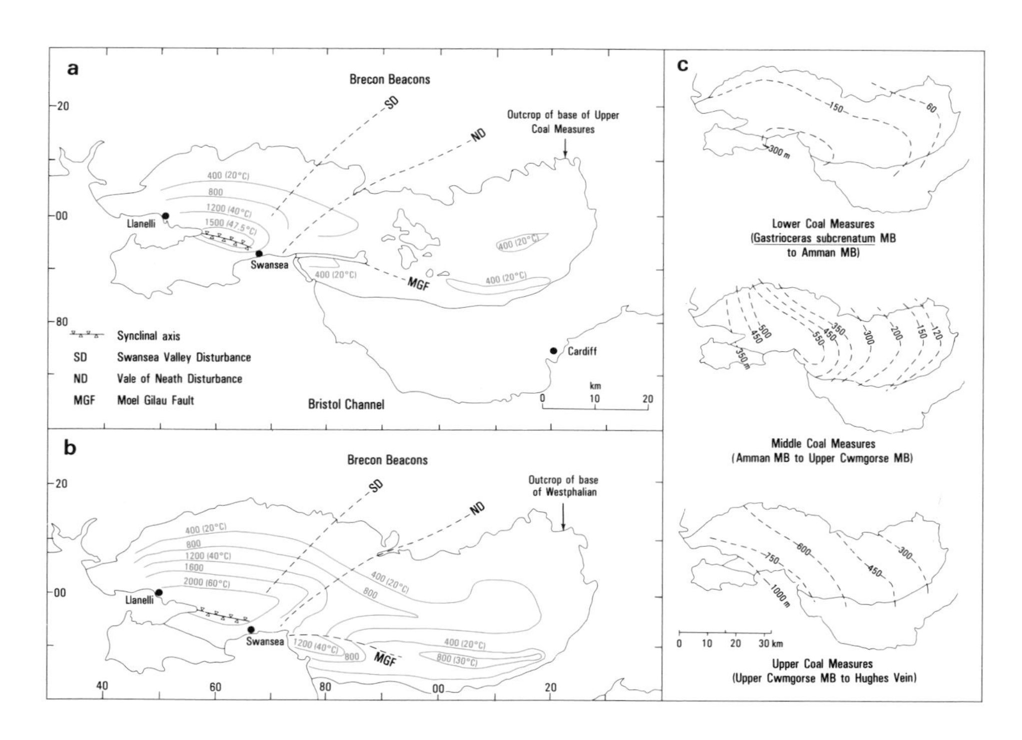

5.7 South Wales:

a) Structure contours (metres below sea level) and estimated temperatures (°C) at the base of the Upper Coal Measures.

b) Structure contours (metres below sea level) and estimated temperatures (°C) at the base of the Coal Measures.

c) Isopachytes (metres) of the Coal Measures.

The Lower and Middle Coal Measures can be considered, from the hydrogeological point of view, as a single group of predominantly argillaceous measures which include the main coal-bearing horizons of the coalfield. The sequence does contain a number of sandstones, some of wide lateral extent, although others are impersistent. The maximum thickness of the combined Lower and Middle Coal Measures is about 1000 m near Swansea, reducing to less than 250 m in the east. Individually some sandstones attain 50 m in thickness but their total thickness is not significant and is invariably less than 100 m. They lie at depths of more than 2000 m below sea level in the deeper parts of the coalfield near Swansea, where temperatures around 60°C have been inferred (Figure 5.7).

The sandstones in both the Pennant and Lower and Middle divisions are extremely hard, dense rocks that are divided into blocks by two vertical joint systems more or less at right angles. Porosities and matrix permeabilities are very low because of secondary cementation. The actual permeability of the sandstones is directly related to the distribution of fissures which can be several centimetres wide at outcrop and at relatively shallow depths. The transmissivity of a Coal Measures sandstone generally ranges from less than 1 to about 20 Dm depending on thickness. Yields from water boreholes are unlikely to exceed a few litres per second, a figure as high as 8 l/s being exceptional for drawdowns of up to a few tens of metres (Ineson 1967), which are typical for water supply boreholes in the outcrop zone. The permeability of sandstones in the Pennant Measures is generally higher than those in the Lower and Middle Coal Measures. The permeability of the sandstones is closely related to artificial changes caused by mining subsidence. Subsidence produces irregular zones of tensional and compressional strains. The rate of groundwater flow in tensional zones can exceed 500 m/d implying very high fissure permeability compared with the low intergranular permeability of the sandstones of about 10^{-5} darcys (Mather and others, 1969).

South Wales is one of the 'wettest' coalfields in the UK. In the past, mines pumped an average of about 8 m^3 of water per tonne of coal mined, compared with an average drainage-output ratio for mines in England and Wales of 2 m^3 of water per tonne of coal mined (Rodda and others, 1976). But to put this in context, about 30% of mines in South Wales pump less than 4 l/s and a further 40% less than 20 l/s; only about 15% abstract more than 50 l/s (Rae, 1978). This gives a good indication of the low yields that can be anticipated from individual wells, for these figures account for drainage from quite thick vertical sequences draining into very extensive mine workings. Workings in the deeper seams in the coalfield tend to be drier than those at shallow depth as the permeability is reduced at depth by the increasing thickness of overburden and the closure of discontinuities. The structure of the Coal Measures appears to control the overall flow of water through the sequence for, as in Nottinghamshire, there is a tendency for mine-drainage to be greater from collieries working in synclines than anticlines (Ineson, 1967). In this context conditions in the deep mines of the west of the coalfield are particularly relevant. Generally the workings in these mines are dry (for example Cynheidre Colliery which is 700 m deep); any water that does enter is usually derived from the Upper Coal Measures and penetrates into the workings down faults.

During the drilling of an underground borehole at Cynheidre Colliery, an opportunity was taken to study the groundwater conditions in the Coal Measures (Ineson, 1967). The borehole began at a depth of 365 m below sea level and continued for 400 m to 765 m below sea level. It penetrated the Cwm Berem and the Twelve-Feet sandstones and the Basal Grit. Water flows were small and the transmissivity of the Cwm Berem Sandstone, which is about 9 m thick, was only 0.05 Dm as derived from a pressure recovery test, giving a mean permeability of about 6 mD. Results

Depth (m)	Permeability (mD)	Porosity (%)
483	0.076	2.30
492	0.044	2.26
493	0.034	1.96
509	0.037	2.21
515	0.017	2.31
518	0.280	2.28

Table 5.1 Aquifer properties of a typical sandstone in the Middle Coal Measures of South Wales

Borehole	Depth (m)	Permeability (mD)	Porosity (%)
Chacombe	563	2.11	16.38
	618	0.59	10.43
	654	0.53	13.15
	660	0.84	14.01
	665	1.13	13.20
	670	0.11	7.86
	675	1.00	12.37
	680	2.47	15.26
Warkworth	532	1.91	15.88
	613	4.70	17.82
	634	27.50	21.64
	638	7.40	18.36
	639	1.86	16.87
	647	4.81	16.92
Shutford	730	101.80	18.75
	750	0.50	14.24
	764	6.03	16.21
	813	0.64	13.28
	828	1.46	13.85
	842	7.47	15.84
	872	0.82	13.35
	931	4.10	13.92
	944	3.37	14.02

Table 5.2 Aquifer properties of typical sandstones in the Upper Coal Measures of the Oxford Coalfield

of laboratory measurements of permeability and porosity for a sandstone above the Big Vein in the Middle Coal Measures at Cynheidre are given in Table 5.1. The low intrinsic permeability of this typical sandstone is obvious. These data may be compared with those in Table 5.2 which are for sandstones from the Upper Coal Measures in Oxfordshire at depths of 500 to 1000 m and they are probably similar to those for Upper Coal Measures sandstones at comparable depths in South Wales. Values are an order of magnitude greater than for the Middle Coal Measures but even so the intrinsic permeabilities are invariably less than 10 mD and porosities are about 15%. Some intervals of very limited thickness have permeabilities of as much as 100 mD, but this is exceptional.

The chemical nature of groundwaters in the Coal Measures of South Wales has been studied in detail by Ineson (1967). The salinity of the waters is very low compared with groundwaters from the Coal Measures in other parts of the country. The majority has a salinity of less than 1000 mg/l and few exceed 10 000 mg/l. Waters with a high salinity tend to be sodium-sulphate waters; sodium-chloride waters, common in deep Carboniferous strata in other coalfields, do not occur. Waters in the Upper Coal Measures are usually of the calcium bicarbonate type. In the more argillaceous Lower and Middle Coal Measures, calcium- and magnesium-rich waters tend to be restricted to the periphery of the coalfield while sodium-rich types (associated with both bicarbonate and sulphate) predominate in the central syncline as a consequence of ion exchange.

Similar rocks to those of South Wales are to be found in the Forest of Dean and Bristol-Somerset coalfields, but the total thickness is somewhat less. There is no reason to believe the rocks in these coalfields have significantly different hydraulic properties from those of South Wales and likely transmissivities are low. In addition limited temperature and heat flow data suggest (Holliday and Smith, 1981) that inadequate temperatures will be achieved

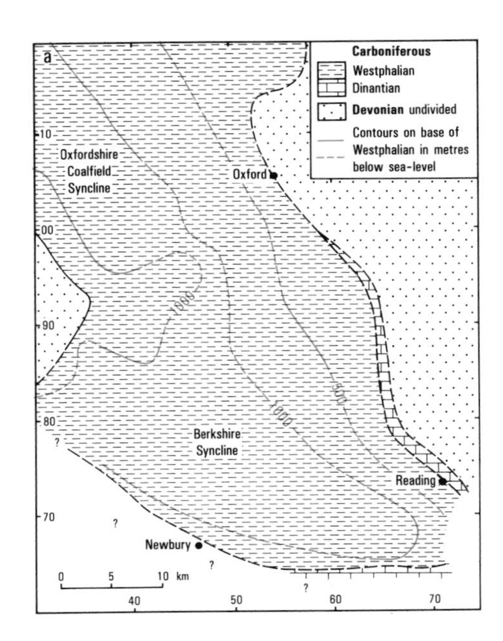

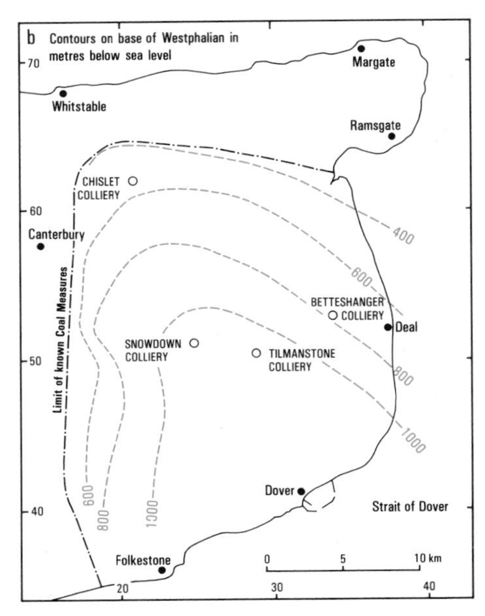

5.8 a) Palaeogeographical map of the floor beneath Mesozoic formations and structure contours on the base of the Coal Measures in Berkshire and south Oxfordshire.
b) Structure contours on the base of the Coal Measures in the Kent Coalfield. Simplified from maps by Allsop and others (1982) and Mitchell (1981).

in these coalfields. A further drawback to the Bristol-Somerset area is its location close to the Variscan Front and the resultant disturbed state of the strata, which makes accurate prediction of the subsurface structure impossible with present exploration techniques.

The Coal Measures of the Oxfordshire and Berkshire areas mainly belong to the Upper division. The Arenaceous Coal Group of Oxfordshire, at the base of the Upper Coal Measures, is very similar to the Pennant Measures of South Wales. Some porosity and permeability values are given in Table 5.2. The strata are buried to depths greater than 1000 m only locally (Figure 5.8) and thus temperatures of 40°C will rarely be reached in these rocks. Similarly the Coal Measures of the Kent Coalfield are only locally buried deeper than 1000 m (Figure 5.8). Since, here, the lowest and more deeply buried beds belong to the argillaceous Lower and Middle Coal Measures, there seems little likelihood of a geothermal resource in these rocks.

MILLSTONE GRIT

This section considers all rocks of Namurian age, not only the typical Millstone Grit of the central and southern Pennines, but also the comparable facies in South Wales and the rocks of similar age in northern England and the Midland Valley of Scotland. In the classic Millstone Grit country of the central and southern Pennines and adjacent areas where the rocks are concealed, the Namurian rocks consist of a basal (E_1 to H_2) argillaceous succession overlain by fine- to coarse-grained sandstones (grits) with intervening shales and rare coals (R_1 to G_1). Northwards the sandstones are found at lower levels, while north of the Craven Fault limestones and coals, though proportionally still small, become more important and are used for stratigraphical subdivision. Particularly in Scotland the coals have economic significance. In the more northerly areas fine- to coarse-grained sandstones continue to dominate the succession of R_1 to G_1 age, and in Durham, Northumberland and the Midland Valley of Scotland persist into the early Westphalian and are the lateral equivalents of the early part of the Lower Coal Measures of more southerly areas.

Namurian rocks are widespread throughout Britain, occurring either beneath or around the margins of most coalfields. Around the margins of St George's Land the Coal Measures locally overstep Namurian strata and the Millstone Grit is absent. This is particularly the case in Lincolnshire and around the Wash. South of St George's Land the only significant occurrence of Namurian rocks, north of the Variscan Front, is in South Wales. In parts of South Wales and many areas on the northern margin of St George's Land there is a break in sedimentation between Namurian and underlying Dinantian rocks. However, in basinal areas of the south and central Pennines, and more generally to the north, there is no depositional break at this time. The greatest thickness of Namurian rocks observed in Britain (up to 2 km) is in east Lancashire.

The environments of deposition indicated by the Namurian rocks are for the most part similar to those of the Coal Measures, with a greater emphasis on marine or deltaic sediments. The sandstones that formed in distributary and river channels are generally more sheet-like than the string-sandstones of similar facies in the Coal Measures. In the central and southern Pennines, the early Namurian shales are believed to be deep-water marine in origin and, preceding the main fluvio-deltaic beds of late Namurian age, are overlain by poorly sorted feldspathic sandstones and shales deposited in submarine fans and channels, and on the delta slope, by turbidity currents.

Individual channel-sandstone bodies may range up to about 60 m thick; the cumulative total may exceed 150 m but is commonly less than 100 m. The sandstones are commonly pebbly and mostly feldspathic, the coarser fraction being subrounded to rounded, the finer fraction subrounded and subangular. Pressure solution of quartz grains, kaolinisation and sericitisation of the feldspars, partial replacement by iron oxides and some secondary silicification combined with the poorly sorted nature of these sandstones has led, in general, to very low intergranular porosities and permeabilities. Fracturing of the sandstones occurs locally at depths exceeding 1 km and has been detected in some boreholes by core examination or by the analysis of geophysical logs. The turbiditic sandstones form fans up to 150 m thick of limited lateral extent. The grain-size of these sandstones is very variable from very fine to very coarse, the coarsest varieties occurring within the submarine channels and the finest in the outer fan and between channels. They are commonly very 'dirty', poorly sorted and feldspathic and, outside the channels, thinly interbedded with shales. Porosities and permeabilities, therefore, are likely to be very low, and it is thought unlikely that this facies could form a viable deep aquifer.

East Midlands

Rocks of Namurian age underlie the coalfields of the eastern Pennines and extend eastwards to the coast. The thickness of strata varies laterally (Figure 5.9), ranging up to around 1000 m. The upper more widespread part of the Namurian ranges up to 300 m thick, and contains the main sandstones. The maximum proportion of sandstone in this higher division is 50%. In many parts of the region it is buried to depths where 60°C and above can be expected (Figures 5.3 and 5.9).

The outcrop of Millstone Grit (Namurian) in the region to the south of the Humber and to the east of the Derbyshire Dome, forms a minor aquifer which is exploited locally. It has been calculated by Gray and others (1969) and Downing and others (1970) that in this area there exists a potable groundwater resource of approximately 6×10^5 m^3/d, of which only 3×10^4 m^3/d was used in 1964.

The sandstones of the Millstone Grit normally act as discrete aquifers due to the presence of intervening mudstones and shales. Flow is either intergranular or through fissures but the latter dominates even where the formation is at shallow depths. The flow in the aquifer tends to decrease rapidly with depth and it is found that sandstones which are soft and decalcified at outcrop generally become hard and compact down-dip. Boreholes in the sandstones of the Millstone Grit yield from 0.5 l/s to 50 l/s and are commonly artesian. Pumping, however,

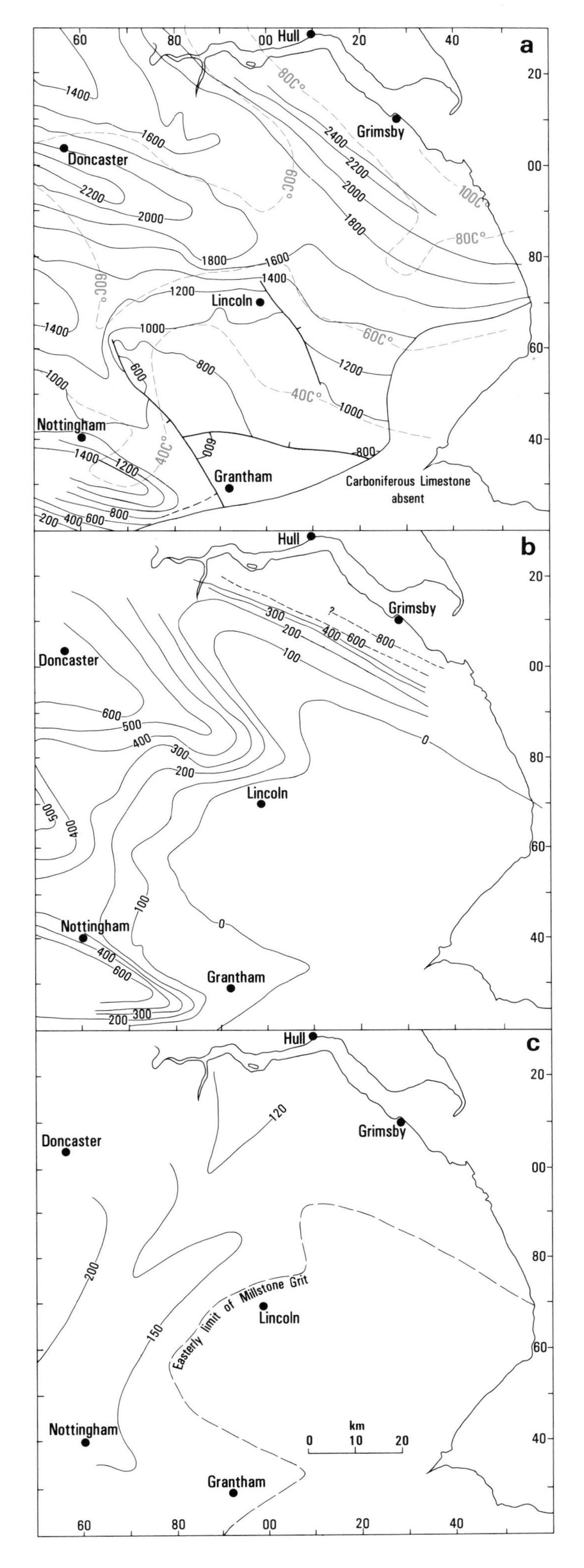

rapidly reduces the flow as abstraction exceeds recharge. Explosive fracturing of the rock has been attempted to increase yields but has met with little success. The most prolific boreholes that sustain their yield are those completed in several sandstones, thus increasing the transmissivity and the recharge area.

At depth the Namurian sandstones have proved to be oil and gas reservoirs and, therefore, have been extensively drilled and tested in the East Midlands oilfields. The results of over a hundred drill-stem tests and many core analyses have been available for study. The sandstone reservoirs of the Millstone Grit are numerous and variable in extent, thickness and facies type with porosities ranging up to 20% (mean about 12%). Permeabilities are also variable but only average a few tens of millidarcys. In addition, log analysis and core examination show that the Namurian sandstones are locally extensively fractured. The majority of tests have been carried out in the most favourable sandstone sequences where aquifer properties would be expected to be highest, not only because of the grain size, presence of fractures and lack of cement but also because the presence of hydrocarbons may have restricted the diagenetic processes that tend to reduce permeability (Hawkins, 1978). There appears to be a general trend of decrease in aquifer properties in a northerly direction, the porosity decreasing from 16% near Newark to between 7 and 11% in the Gainsborough area. Similarly values of permeability decrease from approximately 30 mD to the north of Newark to around 1 mD in the northern oilfields area. The mean permeability is 14 mD.

Taking the mean value for the permeability, a thickness of approximately 360 m of sandstone would be required to meet the minimum transmissivity of 5 D m considered necessary to form a geothermal resource. Such thickness of sandstone has not yet been encountered, even in the gulf areas of thicker sedimentation where around 100 m is the maximum value. Even if the permeability were enhanced by natural fracturing it is unlikely that a sufficiently high transmissivity value would be reached. The area to the north-east of Newark has given the most promising permeability results but the total thickness of the Namurian rocks reaches only 200 m in this area. For example, at Eakring (Edwards, 1967), of the 180 m total thickness, only 60 m is composed of sandstone. Assuming that, on average, these sandstones have a permeability of 30 mD then the maximum transmissivity would be only 1.8 D m. Applying the same optimistic permeability to the thicker basinal sections still only produces a transmissivity of 3 D m.

Despite the small pressure gradient estimated (0.8 to 2.1 m/km) across the area it is possible to contour the piezometric surface expressed as fresh water head (Figure 5.9). These contours imply flow to the east, away from the

5.9 East Midlands:

a) Structure contours (in metres below sea level) and estimated temperatures (°C) at the top of the Carboniferous Limestone.

b) Isopachytes (metres) of the Millstone Grit.

c) Calculated fresh-water head of groundwater (in metres above sea level) in the Millstone Grit.

outcrop where the head is more than 300 m near Sheffield. Superimposed on this general trend are two lobes of higher pressure. These lobes could be explained by variations in transmissivity, a view which is supported to a certain extent by the porosity and permeability data to the north and north-east of Newark. An alternative explanation is that the lobes occur in the zone between the easterly limit of the underlying Brigantian and low Namurian shales and the easterly limit of the Millstone Grit. In this area the upward movement of Carboniferous Limestone waters would be less restricted, resulting in slightly higher pressures (Figure 5.9).

The high head lobes described above are reflected in the distribution in concentration of the groundwater in the Millstone Grit. Figure 5.4 shows this distribution expressed as the chloride concentration of the groundwater, which in all areas exceeds the chloride concentration of the groundwaters in the underlying Carboniferous Limestone. The lobes of fresher groundwater can be explained by the same mechanism as the lobes of higher pressure. Zones of greater transmissivity would permit faster flow of water and thus more rapid dilution of the original water or brine in the sandstone. Upward movement of water, from the Carboniferous Limestone where not restricted by Brigantian/low Namurian shales, would also dilute the formation water of the Millstone Grit which would have a higher concentration than the Dinantian water due to residual formation water and/or the solution of additional material. The waters in the Millstone Grit overlying the Brigantian Shale in the Worksop/Mansfield area generally have salinities greater than present-day seawater. This highly saline water is located very close to the Namurian outcrop, perhaps indicating very little movement of groundwater.

North-west England

The maximum development of Millstone Grit occurs in this area. Isopachytes are shown in Figure 5.10 and inferred temperatures and structure contour maps on top and bottom in Figures 5.5 and 5.10 respectively. The maximum thickness (more than 1800 m) occurs around Burnley. Sandstones and grits account for about 35% of the thickness in the south, increasing to over 60% in the north. However, not all arenaceous members are everywhere fully developed and the proportion of sandstones varies considerably.

Although the Millstone Grit sandstones have been used widely as a water supply source on a local scale, they appear to be of little value for large-scale supplies. Yields from wells are only a few litres per second. The range of specific capacity values is great; the mean is less than 25m^3/d m. The construction of the Bowland Forest Tunnel afforded an opportunity to study the aquifer characteristics of Millstone Grit (Earp, 1955). High flow rates encountered initially from fissured zones declined with time and, below approximately 200 m, inflow into the tunnel was generally insignificant. No information is available from core material about aquifer properties at depth, but the permeability is likely to be low and comparable to those of Coal Measures sandstones at similar depths. This view is supported by limited Drill-stem Test and Repeat Formation Test data which suggest that permeabilities of only a few millidarcys can be anticipated.

Northern England

Although Namurian rocks are widespread in northern England, they are for the most part at relatively shallow depths. Where deeply buried in north Yorkshire, the limited data to hand suggest that sandstones form a relatively small part of the succession and that they have very low porosities resulting from two periods of deep burial in their history.

Midland Valley of Scotland

Namurian rocks are also well developed in the Midland Valley of Scotland, where three main lithological divisions are recognised. The highest of these, the Passage Group, up to 380 m thick (Figure 5.6), is dominantly composed of fine- to coarse-grained sandstone. Few data are available on the hydrogeological properties of the group, although the deposits are considered to be a reasonable aquifer. Yields of up to 30 l/s have been recorded and specific capacities, calculated from pumping tests, range from 7 m^3/d m to 50 m^3/d m. No porosity or permeability data are available but values of 20% and 200 mD are possible. However, as the group only locally is deeper than 1000 m (Figure 5.6), it is unlikely that the temperature will anywhere exceed 35°C.

The underlying Upper Limestone Group and Limestone Coal Group contain channel sandstones that are only locally up to 35 m thick and in which the grain size is up to medium-grained. For the most part these are of limited extent and only in Ayrshire are they sufficiently deeply buried to approach 40°C. They, therefore, have limited geothermal potential.

South Wales

The Millstone Grit in South Wales comprises a variable group of sandstones and shales, up to 600 m thick (Figure 5.11), that display rapid lateral changes in lithology. Few arenaceous horizons can be traced throughout the region. Along the north crop of the Carboniferous syncline, the Millstone Grit can be divided into an upper sandstone unit, referred to as the Farewell Rock, a middle shale group and a lower sandstone unit, the Basal Grit. Towards the south the lithology becomes finer-grained and only the sandstones at the base retain prominence and even then not everywhere. On the north crop, north-west of Merthyr Tydfil, the Millstone Grit includes some 150 m of sandstone in the sequence, but along the southern outcrop the thickness of sandstones is about 100 m in the south-west and no more than 20 m in the east. These sandstones lie at depths of over 1500 m below considerable areas in the south-western part of the coalfield, where temperatures may exceed 60°C (Figures 5.7 and 5.11).

The sandstones in the Millstone Grit are invariably hard, massive and quartzitic with very low porosities and inter-

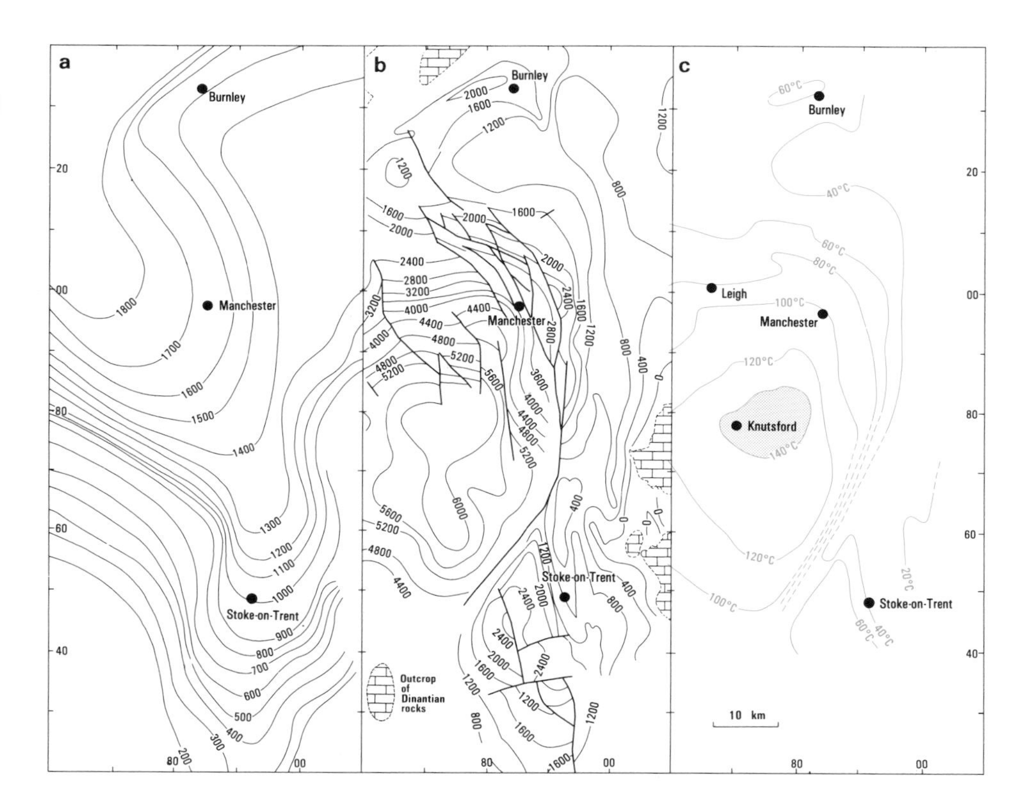

5.10 North-west Midlands:

a) Isopachytes (metres) of the Millstone Grit.

b) Structure contours (metres below sea level) on the top of the Carboniferous Limestone.

c) Estimated temperatures (°C) at the top of the Carboniferous Limestone.

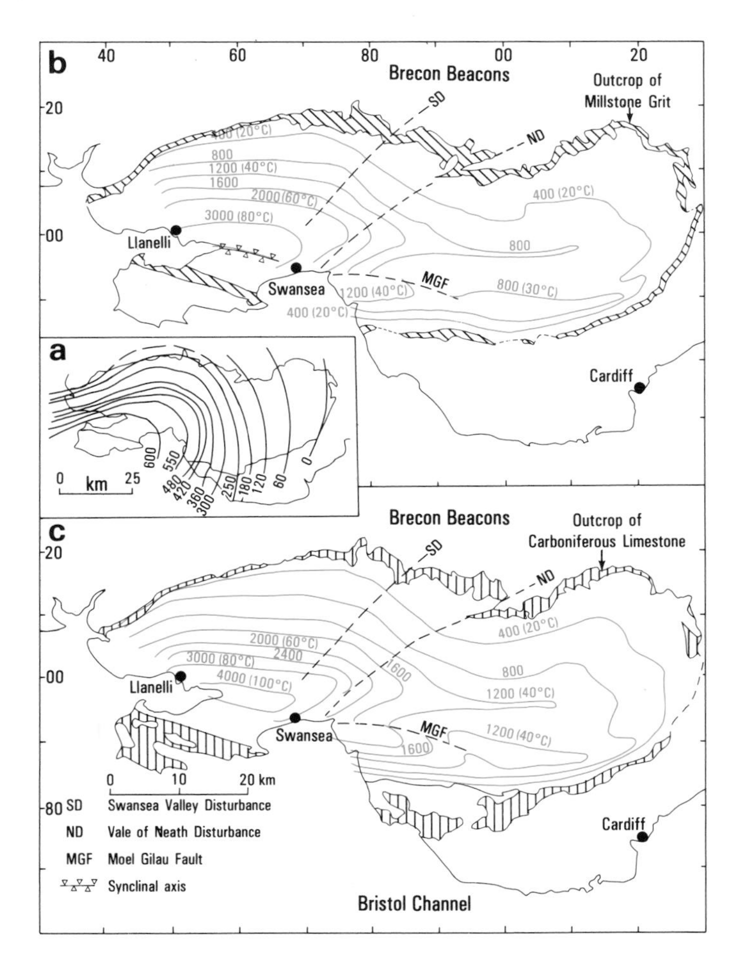

5.11 South Wales:

a) Isopachytes (metres) of the Millstone Grit.

b) Structure contours (metres below sea level) and estimated temperatures (°C) at the base of the Millstone Grit.

c) Structure contours (metres below sea level) and estimated temperatures (°C) at the base of the Carboniferous Limestone.

granular permeabilities. As with Coal Measures sandstones, any water movement is through fractures and fissures which are more prevalent at outcrop. Nevertheless, even near the outcrop, the sequence is of little value for water supply. Yields from wells are only a few litres per second and although information is not available about aquifer properties at depth, the permeability must be very low and certainly no better than for Coal Measures sandstones at similar depths.

CARBONIFEROUS LIMESTONE

The Carboniferous Limestone comprises the shallow water shelf carbonate and related deposits laid down in Britain in Dinantian times. The maximum proven thickness of more than 1800 m is in Derbyshire; elsewhere there is considerable variation. Thin or absent on St George's Land, the limestone thickens southwards towards the Culm Basin. Similarly, it broadly thickens northwards away from St George's Land, although locally it passes into argillaceous strata in basinal areas. Farther north it passes into strata, considered in more detail below, of which limestone forms only a small proportion. The stratigraphy and the depositional and diagenetic histories of the limestone are complex, the details being beyond the scope of this account. Although some of this detail has hydrogeological significance, it is sufficient, for the most part, here to regard the Carboniferous Limestone as one broadly uniform hydrogeological unit.

Prior to lithification, many of the deposits that now make up the Carboniferous Limestone are likely to have had high porosities and permeabilities. However, such are the results of diagenesis on these rocks that the limestones have uniformly very low values for hydraulic properties and intergranular flow is negligible both at outcrop and in the sub-surface. Total replacement of calcite by dolomite results in a maximum volume reduction of 12% so that dolomitic rocks can be expected to have a greater porosity than their parent limestones. At outcrop this is commonly found to be the case where coarse sucrosic dolomite replaces limestones of several different kinds, obliterating earlier fabrics. Locally dolomitisation is relatively superficial and relates to the former presence of nearby unconformably overlying Permian rocks. Elsewhere more extensive dolomitisation occurs perhaps resulting from the passage of saline fluids during diagenesis. Petrographical work and chemical analysis shows that dolomitisation is rarely complete and, given the very low porosity of the parent limestone, the maximum porosity of dolomitised rocks is, therefore, likely to be in the range 10 to 12%. Taking into account the patchy irregular nature of the dolomitisation, it is possible that, locally, aquifers with limited storage and with intergranular permeability produced in this way could exist at depth. It is doubtful, however, whether aquifers with extensive storage, such as are required for geothermal development, could be formed solely by dolomitisation.

Groundwater flow in the limestones at shallow depths follows fissures and fractures such as joints and bedding planes which are enlarged by solution. Because of the massive nature of the limestone, such fractures tend to be widely spaced, except locally, for example, close to faults. Core examination and log analysis suggest that sporadic sub-vertical fractures are present in more deeply buried limestones. Where these are present it is probable that the permeability is locally enhanced.

Limestone rocks are susceptible to solution in even mildly acid water. The effects of limestone dissolution are extensively displayed in limestone outcrops. Similar features are displayed to a lesser extent in limestone cores obtained from the sub-surface. The most prevalent solution phenomena seen are stylolites and vugs, but nowhere are these sufficiently numerous and interconnected as to increase significantly the permeability of the limestone. In limestone country much of the drainage is underground. Extensive solution and widening of joints and bedding planes in the zone of fluctuation of the water table and above have produced extensive cave and passage systems through which very large quantities of shallow groundwater flow. This karst system is probably of relatively recent origin, dating from the exposure of the Carboniferous Limestone in Tertiary times. It is unlikely that this modern karstifaction has had any effects on the concealed limestones of Britain. The buried limestones, however, have been exposed previously at several earlier periods in their history, notably during the Dinantian itself, during early Namurian times and, particularly in the south, in Permian and Mesozoic times. Palaeokarst phenomena, attributable to these periods, can be detected at outcrop and in the subsurface rocks. Walkden (1974) has described, from outcrop examples, the development of mamillated (so-called 'pot-holed') surfaces as a result of contemporaneous subaerial solution of the limestone, and the deposition of laminated caliche-type crusts. Similar features can be seen in borehole cores, both within the limestone and at the junction with the overlying Namurian rocks. Cavern and fissure formation at this boundary has been reported from Derbyshire outcrops by Simpson and Broadhurst (1969), but the cavities so formed are infilled by calcite and fluorite. Particularly in southern Britain extensive cave and fissure systems were developed in the limestone during Permian and Mesozoic times, but generally these have infillings of later sediments. Thus, while the possibility of a widespread open palaeokarst system in the buried Carboniferous Limestone remains, the evidence to hand suggests that such is unlikely. The presence of such an open system in rocks of the same age and lithology in Belgium and northern France (Delmer and others, 1980; Bernard and others, 1980) can be related to the longer period of emergence (until late Cretaceous) and thinner total cover. The presence of thick anhydrite deposits in Belgium and the collapse of overlying limestone brought about by their solution is a contributing factor. Known evaporite occurrences in the British Dinantian are thin and there is little evidence that they have undergone solution or that overlying beds have collapsed.

East Midlands

The carbonate deposits of this region were laid down on the East Midlands shelf. In the late Dinantian, in basinal areas, the limestones pass into shales with limestone and sandstone turbidites. The exposed part of the shelf, in Derbyshire, is known in considerable detail, but elsewhere knowledge is limited and largely restricted to borehole data. Although around a hundred boreholes prove the top of the Carboniferous Limestone in this region, only six penetrate its full thickness. However, it seems likely that the main lithological divisions of the Carboniferous Limestone at outcrop can be recognised in the sub-surface. From the limited data to hand, the Carboniferous Limestone shows a broad overall thinning from north-west to south-east, being absent in the extreme south and south-east (Figure 5.9). In detail the stratigraphy is complex and the thickness is locally variable due to fault movements and block tilting during deposition. A wide range of lithologies is to be found in the basal beds of the Carboniferous Limestone. Apart from limestones and dolomites, and local occurrences of anhydrite, there are commonly important developments of conglomerates, sandstones and shales. Such rocks may persist high in the succession close to basement prominences. There are insufficient data to allow the preparation of detailed isopachyte maps, but around 1800 m is the greatest thickness proved. Structure contours and likely temperatures at the top of the limestone are shown in Figure 5.9.

The hydrogeology of the Carboniferous Limestone outcrop of Derbyshire has been summarised by Edmunds (1971). In addition to precipitation, run-off from the Millstone Grit in the north-west contributes to the recharge of the aquifer. Discharge from the aquifer is mainly eastwards, with smaller discharges to the north and south-west. The rivers derive their water almost entirely from groundwater discharge as the karstic nature of the limestone allows for little surface run-off. Such streams as do flow are either at the level of the general water table or indicate the presence of a perched water table. A detailed study of the geology and its relationship to the hydraulics of the aquifer reveals a very complex system. The hydraulic properties of the Dinantian aquifers vary considerably depending on the type of limestone, the extent of fracturing, the shale content (which increases towards the top of the sequence) and the occurrence of igneous rock suites (both intrusive and contemporaneous lavas and tuffs). Apart from such natural complexities, the area has been extensively mineralised and consequently mined, necessitating artificial drainage of parts of the aquifers and resulting in locally very complex hydrogeology. However, the local variability is not regarded as being significant from the overall basin-wide aspect and from the hydrogeological viewpoint the limestone is considered as a single lithological unit, the outcrop area being the source of recharge and providing the driving head of the system.

The aquifer properties of the Dinantian limestones are difficult to define. Since fissure flow predominates, porosities and permeabilities determined from core analysis are largely meaningless in relation to the flow characteristics. Yields from boreholes for water supply, as would be expected, are very variable depending on how many fissures are intersected. Of the four water supply boreholes mentioned by Downing and others (1970) the specific capacity range from 0.02 l/sm to 0.42 l/sm, whereas a mine was pumped at 380 l/s. Where the limestone is penetrated at depth, by hydrocarbon exploration boreholes, great variability in flow rates was noted. Drill-stem tests were carried out at approximately 30 sites. The majority of these produced only a small amount of water, indicating low permeability and providing insufficient data for a quantitative estimate. In only a few instances were 'high flow rates' reported. Zones of mud loss were also recorded indicating the presence of open fractures. It was also noted that the limestone is commonly fractured just below the contact with the Millstone Grit but the extent of these fractures does not appear to cause any appreciable permeability. Where the test data were adequate for detailed analysis, permeabilities and transmissivities were low, up to 3 mD and 0.1 D m respectively. These values accord well with the core analyses, which show for the most part porosities of less than 10% and permeabilities up to 5.5 mD. No trends in the areal distribution of permeability and porosity can be identified.

Despite the negative features described above, evidence of groundwater flow at depth in the outcrop area is given by the occurrence of thermal springs. The spring waters are meteoric in origin (Edmunds 1971), the four main discharges being at Buxton (27.7°C), Stoney Middleton (17.7°C), Bakewell (15.5°C) and Matlock (20.0°C). Tritium has been used to identify the recent component of the spring discharges, and dilution of the thermal waters both in ionic concentration and temperature, is indicated. Geothermometry (Burley and others, 1984) indicates that temperatures greatly in excess of those measured at the surface have not been attained in the thermal history of the waters. The thermal springs are all located around the periphery of the Carboniferous Limestone outcrop, indicating that the contact with the overlying Millstone Grit is the dominant factor controlling surface discharge. Trace element chemistry varies over the region but very similar suites are found in non-thermal groundwaters in the same area, differing only in amounts (Edmunds, 1971). The high manganese and arsenic contents of some of the springs suggests that some of the water may be derived from the Millstone Grit area to the west (Nichol and others, 1970).

Deep groundwater movement in the Carboniferous Limestone is also indicated by the negative temperature gradient between 450 and 575 m, recorded in the Eyam Borehole. The anomalous temperature measurements in this borehole are attributed to relatively rapid movement of recharge water to considerable depths. To the east, upward movement of thermal water in the western limb of the Eakring Anticline has been invoked to explain the high heat flow and geothermal gradients observed thereabouts (Bullard and Niblett, 1951; Burley and others, 1984). The observed temperatures at Eakring are consistent with those anticipated from the present depth of burial of the Carboniferous Limestone to the west of the anticlinal structure.

As will be discussed below, major groundwater circulation at depth in the Carboniferous Limestone has been demonstrated in the Bath-Bristol-Mendip Hills area, where thermal water discharges at temperatures up to 46.5°C. In Derbyshire the circulation system is either not as deep as in the Bath area, or more heat is lost through mixing. Recharge is presumably within the area of limestone outcrop, although recharge via the Millstone Grit is also possible. Deeper circulation beneath Carboniferous and Mesozoic cover has no obvious localised outlet but a general upward movement into the Millstone Grit in certain areas is postulated, principally from pressure and geochemical data. In all twelve pressure measurements were available for analysis from the Carboniferous Limestone in deep hydrocarbon boreholes. They show no strong trend in hydraulic gradient and are insufficient in number to identify any local anomalies in the pressure gradients. They generally indicate a flow from outcrop in an easterly direction with a head gradient of 0.7 m/km.

Water chemistry data from oil exploration boreholes were studied in detail by Downing (1967) and Downing and Howitt (1969). The few new data do not alter the pattern of isoconcentration contours greatly and merely serve to confirm the original pattern. From the outcrop area the groundwater concentration increases gradually to the east and much more rapidly to the north as shown in Figure 5.4, which gives the contours of the chloride concentration (which are proportional to the total concentration). One notable feature is that the chloride concentration increases gradually to between 7000 and 10 000 mg/l and then the increase in concentration is very rapid to over 60 000 mg/l in 15 to 20 km. The 19 000 mg/l isochlor (the chloride concentration of present-day seawater) runs from Worksop eastwards to the north of Lincoln and then to the south-east. To the south of this line the original seawater, in which the limestone was deposited, has been flushed out at some subsequent stage or diluted by fresh water. To the north of this line flushing could also have occurred but the net result is a highly concentrated brine, the possible origins of which are discussed by Downing and Howitt (1969).

In the east, the Dinantian limestones are unconformably overlain by Upper Carboniferous and locally by Permian rocks. These unconformable contacts are likely to represent periods of subaerial erosion and hence periods of flushing of the original pore water by meteoric water. The present pattern of isoconcentration lines, however, does not reflect the areas of unconformity and it is therefore thought that the effects of these periods of flushing are no longer detectable as a result of subsequent marine transgression and water movement within the aquifer. The present location of the outcrop and the position of the lobe of fresher water suggest flushing by meteoric waters in an essentially easterly direction, the water presumably discharging upwards into the overlying Carboniferous and Permian formations. To the north of the line representing present-day seawater the concentration increases rapidly by a factor of more than two and a half times, to over 90 000 mg/l. The postulated mechanisms by which such concentrations are attained were discussed by Downing (1967).

South Lancashire-Cheshire-North Staffordshire

Very little is known of the Carboniferous Limestone west of the Derbyshire Dome, it having been proved in only a few boreholes. Northwards the limestones pass into the more argillaceous facies of the Craven Basin. Estimated structure contours and temperatures at the top of the limestone are shown in Figure 5.10.

Where proven, the limestones are massive, crystalline with very limited porosity and intergranular permeability, and locally dolomitised. A 21 m section of dolomite at a depth of around 1.5 km in a hydrocarbon borehole was tested and produced brine, with a salinity of 37 000 mg/l, at a rate of approximately 3 l/s. A 23 m section of dolomite at a depth of 1810 to 1833 m near Burnley had an average permeability of around 240 mD. Tests on limestone have produced only a little water, even where fractured. By comparison with the East Midlands, it is likely that there is groundwater flow in the limestone westwards, away from the Derbyshire outcrop, and a more gradual upward movement into overlying formations. There are at present too few data to test this hypothesis. Maximum temperatures of water in the limestone are estimated to exceed 140°C (Figure 5.10).

South of St George's Land

Carboniferous Limestone occurs extensively south of St George's Land, in South Wales, the Forest of Dean, around Bristol and in the sub-surface discontinuously to the Kent Coalfield. South of the Variscan Front, highly disturbed Carboniferous Limestone occurs in a belt of country north of the Culm Basin, extending eastwards into the sub-surface of southern England from the Somerset coast and the Mendips. The limestone thickens to the south, but because of erosion the original thickness (possibly up to 2 km) is not now preserved. However, locally more than 1 km is present.

The Carboniferous Limestone of South Wales can be divided into a main sequence of limestones overlain and underlain by the Upper and Lower Limestone Shales respectively. The limestones in the middle part of the sequence are about 1000 m thick in the south-west, in Gower and south Dyfed, but only about 300 m along the north crop and much less along the east crop. Near Swansea, the base of the limestone is over 3000 m below sea level and over extensive areas below the south-western part of the coalfield it is over 1500 m below sea level (Figure 5.11).

The limestones are compact, well-jointed rocks with low porosities. Permeabilities are invariably less than 10 mD and yields of water from the body of the rock are negligible. Groundwater flow is along fissures and joints that have been enlarged by solution. The karstic landforms so apparent in South Wales are of relatively recent origin and confined to the outcrop and the zone immediately below the overlying Millstone Grit. The presence of such features enhances the permeability at shallow depths, but has little relevance in the context of developing any hot water that may exist at depth. Palaeokarst features have been recognised in the Carboniferous Limestone of South Wales but there is no evidence to date that significantly enhanced permeability has resulted.

In some areas the limestone sequence contains appreciable thicknesses of dolomite of both primary and secondary origin. Commonly secondary dolomites are cavernous, containing vugs and cavities, that may be open or infilled with calcite and silica. Dolomites are common around, and probably below, the eastern part of the coalfield. North of Merthyr Tydfil they occur in the upper part of the limestone immediately below the Millstone Grit, where the dolomitised limestone is brecciated with a significant density of fractures and fissures (Oxley, 1981), leading to enhanced permeabilities. There is some evidence that fractures exist in the limestone at depth. As well as the possible presence of palaeokarst already referred to, a thermal spring issuing from the Coal Measures at Taffs Well in the Taff Valley could originate in the Carboniferous Limestone. The temperature varies from 13 to 22°C, no doubt because the flow includes a variable component of cold recent meteoric water (Squirrell and Downing, 1969). If a geothermal gradient of 20°C/km and a mean surface temperature of 10°C are assumed, the maximum temperature of the spring implies a source for the warm component at a depth of about 600 m. As the spring water clearly contains a significant recent meteoric component, the hot water source is likely to have a higher temperature than 22°C and therefore probably originates from a greater depth than 600 m. The most probable source rock is the Carboniferous Limestone which is at a depth of about 700 m below the spring. The thermal water has probably circulated in this formation before rising up a fissure, or more likely a fault plane, in the Taff Valley. Chemical analysis of the major elements in the water indicates that it is not mineralised and has a composition similar to groundwater from the Carboniferous Limestone. The water has also been analysed for helium-4 and dissolved inert gases (Burgess and others, 1980). The dissolved inert gases show that the temperature at which it infiltrates the ground is about 5°C, lower than would be anticipated from its position on the southern outcrop of the coalfield. This implies that the source of the water is at an elevation about 500 m higher than the spring outlet and points to a possible origin along the northern outcrop of the limestone, north of the coalfield, the water flowing to the spring through the limestone below the coalfield, a distance of some 25 km. However, the water has a high helium-4 content and it is unlikely that this could have been generated during passage through the Carboniferous Limestone only, and a component originating from greater depths is indicated (Burgess and others, 1980).

The extent to which water flows below the coalfield from limestone outcrops to the north of the coalfield is unknown, as is the extent of fissure systems in the rocks at depth. One piece of evidence for the presence of flow channels is the economic occurrence of haematite in the limestone along the southern outcrop. One view (Williams quoted by Squirrell and Downing, 1969) is that the source of the iron is from hydrothermal solutions circulating through the limestone possibly in Tertiary times, a circulation that was associated with folding and faulting at that time. These earth movements could have fractured the limestone over extensive areas below the coalfield.

Further evidence for the presence of water-bearing channels in the Carboniferous Limestone, where under cover of younger rocks, is provided in the Bath-Bristol-Mendips area. Groundwater under artesian pressures from the limestone was encountered during the construction of the Severn Tunnel in 1879. The flow was about 90 Ml/d and pumping is necessary to the present day. At two localities in

5.12 Distribution of Palaeozoic rocks, major structural features and position of thermal springs in the Bath-Bristol area.

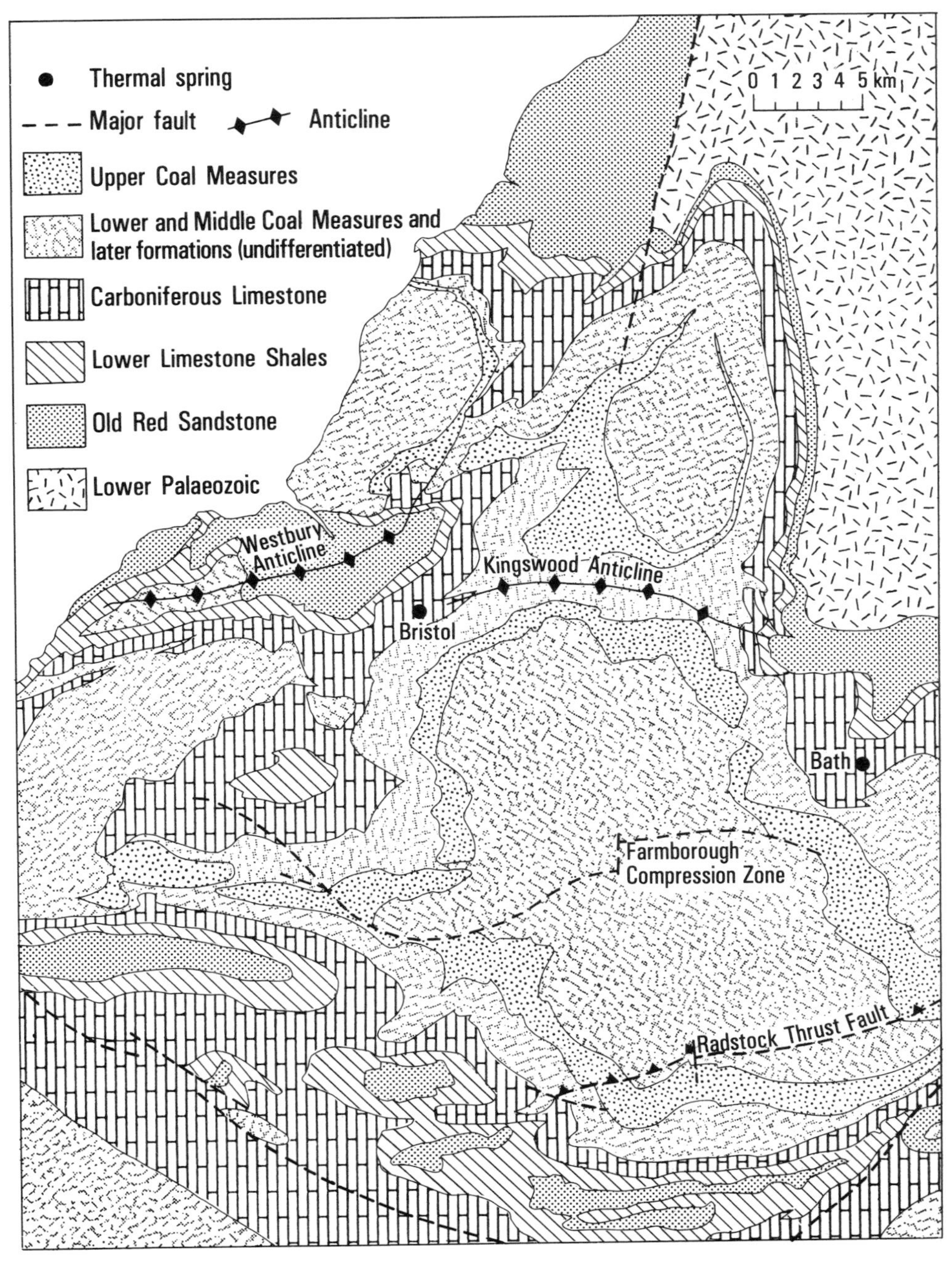

the Bath-Bristol area thermal water discharges naturally at the surface, and historical records refer to the interception of tepid water in boreholes. The major discharge area is in the city of Bath (Figure 5.12) where three thermal springs have been used for balneological purposes at least since Roman times, the largest spring discharging from Lower Jurassic rocks at a rate of nearly 10 l/s and a temperature of 46.5°C. At Hotwells, Bristol (Figure 5.12), a thermal spring at 24.3°C discharges from the Carboniferous Limestone at an approximate rate of 0.2 l/s. A detailed geochemical and hydrogeological study of these occurrences has been undertaken by Burgess and others (1980); their results are summarised in Chapter 6.

The thermal waters are demonstrated to be meteoric in origin and to have fallen as rain at a temperature of about 9°C during climatic conditions similar to the present day. Hydraulic evidence suggests that the flow of the thermal water is controlled by the piezometric head in the Carboniferous Limestone and Old Red Sandstone of the Mendip periclines and further that the Triassic sequence in the vicinity of Bath is itself recharged by leakage of thermal water radially from the Bath thermal springs.

Groundwater flow in the Carboniferous Limestone is confined to fissures which have been opened by solution along pre-existing planes of weakness such as joints, faults, or bedding planes. Whilst the piezometric head difference between the Mendip recharge area and the major thermal discharge area at Bath provides the potential for groundwater flow, it is considered that the southerly dipping Variscan thrust fault planes provide the necessary hydraulic connection between the deeply circulating groundwater and the sub-Mesozoic surface (Figure 5.12). The thermal water is stored at depth in the fissure system of the Carboniferous Limestone, whose precise geometry and limits cannot be assessed on present evidence. The pattern of flow of the thermal waters that have recharged the Triassic and Liassic at Bath is also controlled by a local east-west fault system. The distribution of thermal water within the Triassic rocks is predominantly intergranular, allowing cooling of the water and slight hydrochemical modification radially away from Bath.

Farther east Carboniferous Limestone has been proved in several deep boreholes both within and to the north of the Variscan Fold Belt (Figure 5.8), but little is known of its distribution and thickness. It seems possible that deep circulation of groundwaters, as in the Bath area, occurs in this region but so far remains undetected.

DINANTIAN CLASTIC ROCKS OF NORTHERN BRITAIN

This section considers the clastic strata of northern England and Scotland, that are the lateral equivalents of the Carboniferous Limestone, but in which shales and sandstones are the dominant rock types and limestones are of lesser importance. These rocks are dominantly of fluvio-deltaic and shallow marine origin. The sandstones generally occur as part of fining-upwards or coarsening-upwards sequences, of which the latter leads to thicker beds. For the most part the sandstones are less than 20 m thick and of limited extent, though there are many exceptions. Few porosity and permeability data are available but, as for the sandstones of the Coal Measures and Millstone Grit, most values obtained are low. Although these strata support many shallow wells, yields as a whole are not great and are derived from flow through joints and fissures. Though some sections may contain 40% and more of sandstone, it is unlikely that a transmissivity of 5 D m or above could be expected, except locally; thus, despite being buried in some areas to depths likely to be of geothermal interest, they have received little attention from the geothermal viewpoint.

Fell Sandstone Group of Northumberland

The Fell Sandstone Group, and its broad equivalent the Middle Border Group, form part of the infill of the Northumberland Trough, a graben trending east-south-east which occurs between the high ground of the Cheviots and Southern Uplands to the north and the Pennines and Lake District to the south (Figure 5.13). Sandstones occur in all the major divisions of the Carboniferous succession in the Trough, but the Fell Sandstone is the only thick persistent unit dominated by rocks of this kind (upwards of 60% of the thickness is sandstone). The dominantly south-easterly or easterly dip suggests, though presently confirmation from drilling is lacking, that the Fell Sandstone or its lateral equivalent should be present at depths of geothermal interest under a cover of younger Carboniferous rocks in the southern part of the Northumberland Coalfield and the large urban area of Tyneside (Figure 5.13). However, the Fell Sandstone Group passes in a south-westerly direction into the broadly equivalent Middle Border Group in which sandstones are much less abundant (only representing around 25% of the total thickness). It is not known which facies underlies the urban area. The Fell Sandstone/Middle Border Group ranges from 200 to 450 m thick, being around 300 m thick over much of the outcrop. A similar figure is likely under Tyneside.

The Fell Sandstone Group is largely composed of fine- to medium-grained sandstone with rounded to subangular grains, moderately well-sorted. At outcrop it has extremely variable but commonly good aquifer properties. Porosities up to 33% have been recorded, with a mean around 14%. Permeabilities in excess of 100 mD are common. Middle Border Group sandstones give lower values: the average porosity is about 7% and permeability is up to 70 mD. The analysis of pumping test data gave a range of transmissivity values from 9 to 189 m^2/d (equivalent to 6 to 122 Dm at 10°C) for five wells in the Fell Sandstone, and a value of 60 m^2/d (40 D m) for the Middle Border Group at Spadeadam. Hydraulic conductivities calculated from rough estimates of saturated thickness range from 0.05 to 0.6 m/d for the Fell Sandstone. These values are at least double those obtained for intrinsic permeability in the laboratory, indicating that there is a significant contribution from fissures. Indeed, plots of drawdown against time from pumped wells are commonly stepped, indicating that as the water level is lowered, fissures are dewatered. The yield of a well invariably depends on the number of fissures intersected. The geothermal potential of the Fell Sandstone thus depends on whether it can sustain a sufficiently large fracture system at depth. Any water the sandstone would contain at 2 km is likely to be saline.

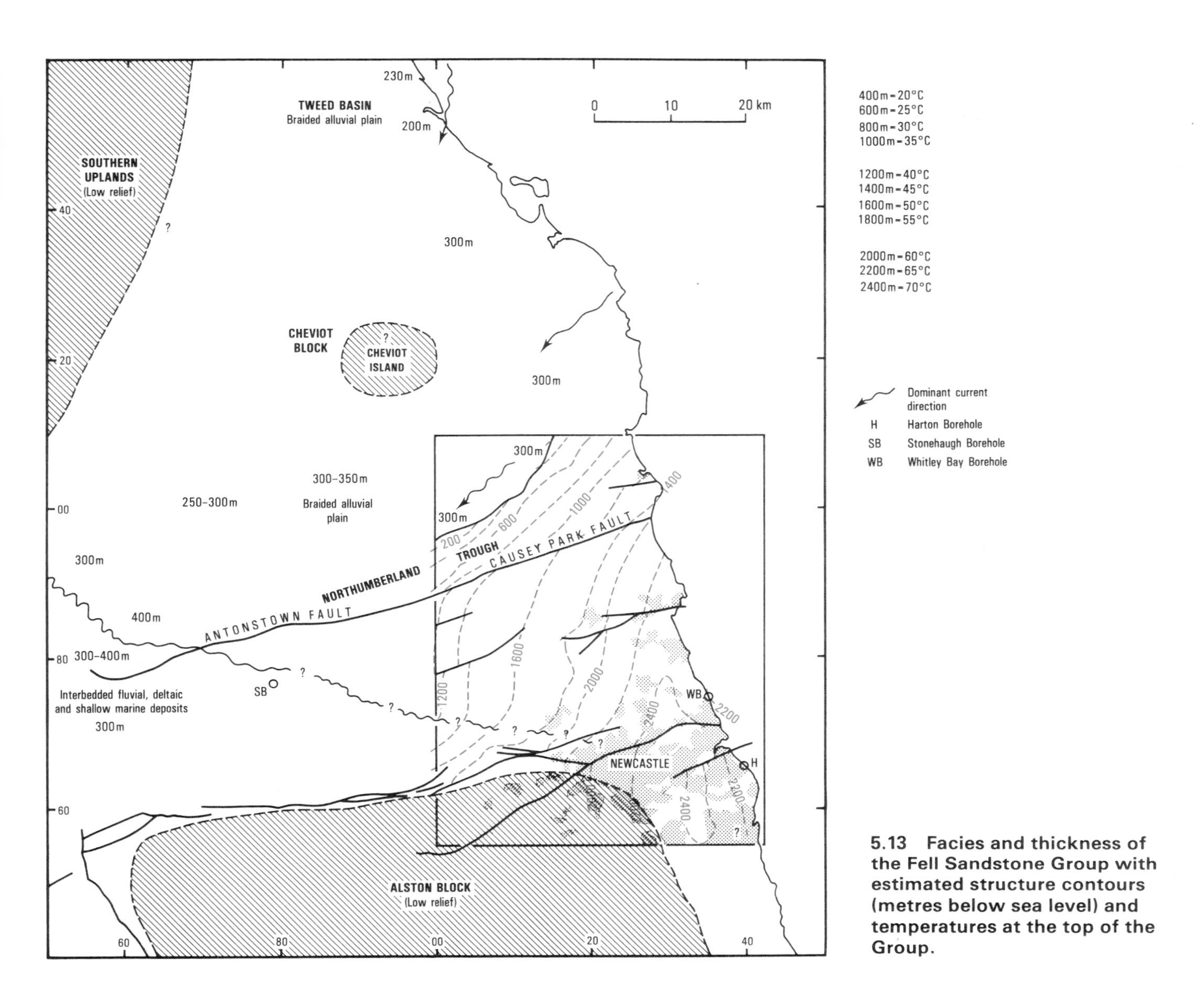

5.13 Facies and thickness of the Fell Sandstone Group with estimated structure contours (metres below sea level) and temperatures at the top of the Group.

Midland Valley of Scotland

Little is known about the geographical distribution of channel sandstones in the Dinantian of Scotland and on the whole sandstones are scarce and argillaceous rocks predominate. Dinantian rocks in Scotland, therefore, are not considered to be important regional aquifers, yields from wells being less than 10 l/s. The general level of porosity and permeability of the sandstones is very low, though some exceptional rocks, with porosities of more than 20% and permeabilities greater than 1000 mD, are known. Thus, although these rocks are buried to depths where attractive temperatures can be expected (Figure 5.14), they cannot be considered to have significant geothermal potential because of their limited properties as an aquifer.

OLD RED SANDSTONE

The Old Red Sandstone is predominantly a continental red-bed sequence deposited as a post-orogenic molasse as a consequence of the uplift of the Caledonian mountain chains. Sandstones, shales and conglomerates are the dominant lithologies. Though principally of Devonian age, the lowest strata are commonly of Silurian age, while some of the highest rocks may belong to the Carboniferous.

There were three main areas of Old Red Sandstone deposition in Britain. In the south, rocks of this facies, commonly in excess of 2 km thick, crop out in South Wales and the Welsh Borderlands and again in the Mendip Hills. Eastward extensions in the subsurface have been proved in southern England, the south Midlands and in East Anglia. Much of the sediment belongs to the Lower Old Red Sandstone. The middle division is only locally represented, this seemingly being a period of erosion. The more persistent Upper Old Red Sandstone is generally less than 200 m thick, resting unconformably on lower beds. To the south the continental deposits pass into the marine strata of the Variscan Fold Belt in Devon and Cornwall. In north Devon the marine beds are divided by a thick Old Red Sandstone development at the level of the Middle Old Red Sandstone. In the Midland Valley of Scotland and the Scottish Border area, Middle Old Red Sandstone is again missing and the upper, thinner division rests unconformably on the thicker lower unit. The total thickness is variable but is thought to be up to 9 km. Volcanic rocks are extensively associated with the Lower Old Red Sandstone. The extensive outcrops of Old Red Sandstone in north-east Scotland, the Orkneys and Shetlands, belong mainly to the Middle division and are between 5 and 10 km thick.

Depositional environments that can be recognised within the Old Red Sandstone include meandering and braided alluvial plains, alluvial fans, and freshwater lakes. Various lines of evidence suggest that in England and Wales streams drained from the north, over a broad alluvial plain stretching across much of southern Wales and England, into the sea which lay farther south. In Scotland and the Borders drainage seems to have been into interior basins. The nature of the sediments, their red colour and the occurrence of pedogenic carbonate deposits, all point to a semi-arid climate during deposition.

South Wales

The Upper Old Red Sandstone of South Wales is predominantly an arenaceous deposit. In the east much of the sequence is represented by the Quartz Conglomerate which comprises red and brown sandstone, quartzites and coarse conglomerates up to some 50 to 80 m thick. The conglomerates tend to occur in lenticular beds, particularly near the base, and pass upwards into soft, poorly cemented fine- to coarse-grained sandstones. Farther west they pass laterally into a series of sandstones, quartzites and conglomerates. The upper part of the Lower Old Red Sandstone is primarily an arenaceous deposit about 700 m thick, variable in nature and including soft fine-grained argillaceous sandstones, hard coarse-grained sandstones and sometimes conglomerates. However, more than half the Lower Old Red Sandstone is represented by the Red Marls or their equivalents which, as the name implies, are predominantly argillaceous and essentially impermeable. Thus any water potential in the Old Red Sandtone is in the upper part of the Lower Old Red Sandstone and in the Upper Old Red Sandstone. The rocks of the Old Red Sandstone tend to be well-cemented and indurated with low porosities and permeabilities. Where the sandstones crop out, yields from shallow wells are typically less than a litre per second. Water movement is primarily through small fractures but below about 50 to 100 m these are unlikely to be extensive and those that do exist are probably reduced in size by mineral deposits such as silica and calcite. A study of boreholes in the Old Red Sandstone of the Welsh Borderland (Monkouse, 1982), which included the area around the eastern part of the South Wales Coalfield, indicated that the mean specific capacity was about 4.8 m^3/d m from 150 to 250 mm diameter boreholes about 50 m deep with only a 10% probability of obtaining more than 25 m^3/d m. These values imply very low permeabilities for aquifers even at shallow depths and they will be much reduced at the greater depths that would be required for geothermal boreholes.

The Upper Old Red Sandstone and the upper part of the Lower Old Red Sandstone attain a total thickness of some 700 m. If the average permeability is 1 mD, the transmissivity would be less than 1 D m but on the other hand it would be 7 D m (a value near the lower limit for economic development) if the mean permeability were 10 mD. This indicates the difficulty of assessing the potential for geothermal development of an area where the rock properties are known imperfectly. On balance, however, it must be concluded that even a value of 1 D m for the transmissivity must be optimistic because the total thickness is commonly less than 700 m, and the sequence includes a significant proportion of fine-grained argillaceous layers; as a consequence the average permeability is likely to be near 1 mD. In these circumstances, the prospects of obtaining adequate yields from the Old Red Sandstone in South Wales at depths of 2000 m must be remote.

Southern England

Old Red Sandstone similar to that of South Wales, but with significant marine intercalations, has been proved in a number of boreholes beneath cover of younger rocks in southern England and the south Midlands. Up to 2 km of sediment may be present but little detail is available.

A limited number of samples from late Devonian deltaic sandstones in the Steeple Aston Borehole in Oxfordshire (Poole, 1977, in his table 6), taken from depths between 837 and 877 m, had intergranular porosities of up to 20% and permeabilities up to 130 mD. The mean values were 16% and 21 mD (or 10 mD if one very high value is discounted) respectively. These particular samples are not sufficiently deeply buried to be of geothermal interest at present; nor is it likely that they are representative of the main bulk of Devonian sandstones in the region. However, they do suggest that some geothermal potential is possible in similar sandstones elsewhere. Nearly 1000 m of Old Red Sandstone were penetrated below a depth of 1725 m in the Marchwood Well, near Southampton. Two cores cut in the formation revealed porosities of less than 3% and an average permeability of less than 0.2 mD with individual values never more than 1 mD (Burgess and others, 1981). This example probably gives a better indication of the order of magnitude of the aquifer properties that can be expected at the depth of about 2000 m necessary for geothermal boreholes.

Cheviot Hills and Scottish Border

Upper Old Red Sandstone rocks, around 200 m thick, made up of sandstones, conglomerates and marls, crop out in the Border Country and around the Cheviot Hills, where they rest uncomformably on older rocks. Locally some of these sandstones appear to have useful aquifer properties. However, the presence of concealed continuations of the Upper Old Red Sandstone under the Carboniferous rocks of Northumberland is uncertain (Leeder, 1973). Lower Old Red Sandstone lavas and sediments are exposed in the Cheviot Hills and adjoining coastal areas, and in the former are intruded by granite. The Cheviot volcanics are probably more than 1000 m thick, but near the coast are represented by about 600 m of sandstones and marls associated with volcanics. The sandstones are not thought to have much aquifer potential because of the associated argillaceous and volcanic material. The extent of these rocks under cover of later rocks is unknown and they are not considered to be of geothermal interest.

Midland Valley of Scotland

Old Red Sandstone and associated volcanic rocks form around one-third of the land surface of the Midland Valley and occur in the sub-surface extensively beneath the Carboniferous cover.

The Upper Old Red Sandstone consists predominantly of sandstone with subordinate mudstone and locally with coarse conglomerate at the base, but also interbedded into the sequence on the Clyde coast. The sequence is probably up to 1000 m thick in the west and usually over 500 m elsewhere. Maps showing isopachytes of the Upper Old Red Sandstone and structure contours and estimated temperatures at the top of the formation are included in Figure 5.14.

The Upper Devonian sequence of central Fife is an important fresh water aquifer and production wells and observation boreholes have been examined in detail (Foster and others, 1976). The sequence established in Fife (Chisholm and Dean, 1974) is as follows:

Formation	Thickness	Group
Kinnesswood Formation	40–130 m	
Knox Pulpit Formation	170 m	Statheden Group
Dura Glen Formation	0–40 m	Statheden Group
Glenvale Formation	450 m	Statheden Group
Burnside Formation	150 m	Statheden Group

The most important aquifer is the Knox Pulpit Formation but significant yields have been achieved locally from other formations. The Knox Pulpit Formation is composed entirely of fine- to coarse-grained cross-bedded sandstone of possible aeolian origin which, for the most part, is not cemented or only weakly cemented at outcrop. Laboratory analysis of samples taken from a depth of less than 80 m give a porosity greater than 20% and permeability greater than 600 mD. Jointing and fissure systems are, however, the dominant groundwater flow systems. Investigations have shown that only 30% of the overall transmissivity is represented by intergranular permeability, and fissure flow is the major contributor to transmissivity and contributes 60% of the yield. Specific yields of over 12% and possibly over 15% are likely from units with porosity greater than 20%. Yields of 40 l/s are common with specific capacities of around 130 m^3/dm. Laboratory analyses of cores from the Glenburn Borehole give porosity values between 10% and 19% and permeability values between less than 1 mD and 215 mD for strata between depths of 128 and 275 m (Table 5.3). Laboratory analyses of shallow samples from other formations of the Stratheden Group also yield promising results, with porosities of 10 to 20% and permeabilities of about 240 mD. However, boreholes yield less than 10 l/s with an average specific capacity of around 40 m^3/dm. Tests on some shallow samples from the Kinnesswood Formation were less promising, porosity and permeability being around 15% and less than 10 mD respectively. The extent of the Knox Pulpit Formation beyond Fife is uncertain, but the lithological characteristics of sandstones in exposures at numerous localities suggest it may be widespread.

Tests on samples from the Upper Old Red Sandstone of Ayrshire, taken from cores extending only to depths of 100 m, gave values of porosity and permeability ranging from 3 to 22% and 10 to 47 mD respectively. However, investigations have shown that groundwater movement is dominated by fissure flow which contributes up to 90% of total transmissivity.

The Upper Old Red Sandstone in general, and the Knox Pulpit Formation in particular, on the basis of their properties at shallow depths, appear to have good potential

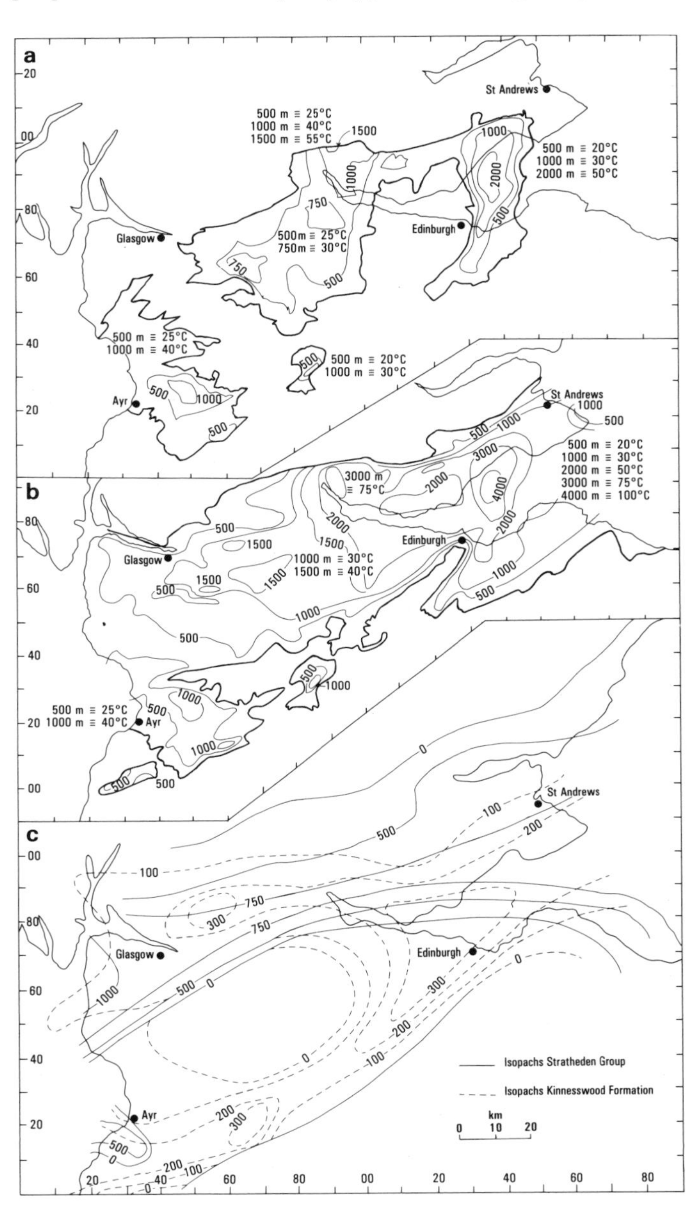

5.14 **Midland Valley of Scotland:**

a) Structure contours (metres below sea level) and estimated temperatures (°C) at the top of the Dinantian.

b) Structure contours (metres below sea level) and estimated temperatures (°C) at the base of the Dinantian.

c) Isopachytes (metres) of the Upper Old Red Sandstone.

as geothermal aquifers. However, their character at depth remains unproven and success will be dependent to a large degree on the extent and number of open fissures. Taking thickness, depth of burial and likely temperatures into account (Figure 5.14), the Fife area would seem to offer the best chances of success.

The most extensive and best known area of outcrop of Lower Old Red Sandstone is in Strathmore, adjacent to the Highland Boundary Fault. This sequence is around 9 km thick but probably decreases southwards under Dundee and south-westwards towards the Clyde coast to around 2 km. The sequence comprises thick units of conglomerate, sandstone and muddy siltstone. The conglomerate facies is most common at the north-eastern end of the 'Strathmore Basin' around Stonehaven and also along the Highland Boundary Fault. This distribution reflects the main direction of transport of sediment from the Highlands in the north. Similarly, northward transport is indicated for the sediments in the Edinburgh area, indicating that the Southern Uplands formed a bounding hinterland to the south.

The Lower Devonian conglomerates usually contain an abundant fine-grained matrix and, where rarely of openwork character, the interstices are filled usually with carbonate minerals (mainly calcite). The sandstones, which are probably the dominant lithology within the Lower Devonian successions, also commonly contain a matrix of clay minerals and 'mica' or alternatively are well-cemented by diagenetic minerals (mainly carbonate such as calcite). The potential of the Lower Old Red Sandstone is, therefore, restricted as the porosity and intergranular permeability is low. The permeability distribution is heterogeneous as it relies on the occurrence of fissures which are only locally common. Geophysical logs and drilling indicate that fractures seen at the surface are closed at depth because of increasing pressure. Step tests and geophysical logs show major water-bearing fissures down to depths of only 125 or 150 m.

Data on the aquifer properties of the Lower Old Red Sandstone are limited and can be misleading in interpretation. Transmissivity and storativity values derived from pumping tests are relevant only to the borehole tested and cannot be applied to the aquifer as a whole because of the heterogeneous nature of the rocks. Yields depend upon the extent of the fissures system intercepted and hence vary enormously. Values ranging from 20 to 2500 m^3/d have been recorded. A statistical study (Clark, 1981) showed that within 95% confidence limits over 33% of wells have a specific capacity less than 1 m^3/d m and less than 5% have specific capacity over 10 m^3/d m. Data from BGS Well Records suggest a mean specific capacity of 30 m^3/d m but over 80% of wells have a specific capacity of less than 5 m^3/d m. Wells with low specific capacities probably do not intercept water-bearing fissures and no water-bearing weathered zone is present. Intermediate wells with specific capacities of around 40 m^3/d m probably take water from a weathered zone whilst the few wells with high specific capacities intercept major water-bearing fissures. Calculations of transmissivity range from 9 m^2/d to 450 m^2/d. However, the transmissivity in these cases is a measure of the capacity of a fissure system encountered by a borehole to supply water to that specific borehole and is not representative of the aquifer as a whole. Similarly, calculated storativity values range from 0.01 to 0.42. The actual storage available is the finite volume of water held in the fissure system intercepted by a borehole, plus leakage from superficial deposits, which cannot be accurately calculated from pumping test data.

Laboratory analyses of sandstone cores taken from a borehole near Forfar give porosity values of around 14% and permeabilities of less than 4 mD. These results were obtained from cores down to 150 m only (Table 5.3). At depths required for geothermal potential the permeability will probably be less than 1 mD and the porosity could be as low as 3%.

Target depths for water at 60°C will be about 2200 m. The Lower Old Red Sandstone occurs at depths greater than this within the Midland Valley. However, at this depth, even with a thickness of potential water-bearing sediments in excess of 2200 m, the likely average permeability of 1 mD and resultant transmissivity of about 2.5 D m would be too low for economic development in the absence of fissures.

Table 5.3 Aquifer properties of typical sandstones in the Old Red Sandstone of the Midland Valley of Scotland

Depth (m)	Porosity (%)	Permeability (mD)	
		Horizontal	Vertical
Upper Old Red Sandstone (Glenburn Borehole)			
Kinnesswood Formation			
62.0	16.5	6.8	6.4
77.4	14.0	2.7	2.0
90.3	14.4	32.6	7.8
Knox Pulpit Formation			
128.0	17.7	19.3	34.0
142.4	13.0	5.0	1.2
161.3	18.1	215.0	131.0
216.3	12.3	0.6	0.4
273.0	11.6	1.4	0.7
Lower Old Red Sandstone (Forfar Borehole)			
6.75	17.5	—	0.3
6.85	16.1	0.5	—
7.25	13.2	—	1.9
7.45	17.6	4.1	1.0
7.85	17.5	1.3	—
48.00	7.4	—	1.7
48.25	10.0	3.9	—
49.45	7.0	—	1.4
49.50	8.3	3.9	—
100.25	20.9	—	1.4
100.50	21.3	1.6	—
146.10	12.1	—	2.5
146.50	12.2	3.6	—

Orcadian Basin

Old Red Sandstone deposits occur in Caithness and around the shores of the Moray Firth, extending to the Orkneys and Shetlands. Little is known of their structure and nature at depth. Their great thickness, around 4 km, suggests a possible geothermal potential. However, heat flow values hereabouts are low and, taking into account the probably high thermal conductivity of the beds, low geothermal gradients are likely. Because of this and the limited heat loads in the area no detailed study of the geothermal possibilities of these sandstones has been carried out.

VARISCAN FOLD BELT

The Old Red Sandstone and Carboniferous Limestone of the northern part of this fold belt, as they occur in north Devon, the Mendips and eastwards beneath Mesozoic cover, have already been considered.

In Cornwall and south Devon, the Devonian and Carboniferous rocks have a complex structure and are highly deformed and indurated. Several major granite plutons have been intruded into and have locally metamorphosed these rocks. Although sandstones and greywackes are not uncommon, the strata are dominantly argillaceous with strongly developed slaty cleavage.

Important exceptions are the thick developments of shallow-water limestone around Plymouth and Torquay. None of these rocks has other than very low permeability, and any water occurs in fractures and fissures. No major regional near surface aquifers are known and none is expected to occur in the sub-surface. This is likely also to be the situation in the easterly continuation of these rocks beneath Mesozoic cover. Higher than average heat flow is found in much of Cornwall and south Devon, though most measurements have been made in the granites which are generally relatively enriched with uranium. The high heat flow combined with the likely low thermal conductivity of the dominantly argillaceous succession suggest geothermal gradients will be high. However, the general lack of aquifers and the improbability that accurate predictions can be made of the structure and disposition of beds at depth, all suggest little low enthalpy geothermal potential.

NORTHERN IRELAND

Carboniferous rocks, about 2 km thick, overlying Old Red Sandstone, occur at outcrop in the west of the province. Structurally these rocks can be regarded as largely lying in a south-westerly extension of the Midland Valley of Scotland, and it is with this latter region that the rocks are comparable both in lithology and facies. Given the disappointing assessments of geothermal potential of Carboniferous rocks in Scotland, similar conclusions, therefore, are likely to be reached in Ulster. Water is obtained from Carboniferous rocks in Northern Ireland but from fissures at shallow depths. Geothermal gradients are below average, however, 21° to 24°C/km, and are too low for even the basal Carboniferous beds to reach 60°C in the outcrop area. The Old Red Sandstone, generally, has poor hydraulic properties where encountered at the surface.

Eastwards the Carboniferous and Devonian rocks are overlain by up to 3 km of Permian, Mesozoic and Tertiary rocks. Geothermal gradients and heat flow are believed to be higher here. However, the Devonian and Carboniferous rocks are known only from a few boreholes and little is known of their detailed distribution. In addition it is thought unlikely that sufficiently extensive open fracture systems can be expected at such great depths.

REGIONAL PROSPECTS IN UPPER PALAEOZOIC ROCKS

The rocks that comprise the Upper Palaeozoic sequences are hard and compact with low porosities and permeabilities. Water flows through them mainly in fractures and fissures. The principal areas where thick sequences could have potential are the East Midlands, South Wales, the West Pennines and Cheshire, the Bath-Bristol area, Northumberland and the Midland Valley of Scotland. The sandstones and limestones that form large parts of the Upper Palaeozoic sequences in these areas are generally poor aquifers and drilling deep wells in the hope of obtaining adequate yields of water would be a very speculative venture. The exceptions to this generalisation could be the Fell Sandstone below south Northumberland and Tyneside, and the Upper Devonian sandstones of the Midland Valley of Scotland, in particular the Knox Pulpit Formation which has properties approaching those of the Permo-Triassic sandstones of England. The nature and properties of these two Upper Palaeozoic formations at depths where temperatures would be suitable for geothermal development are unknown but their properties at outcrop indicate that potential could exist.

The Knox Pulpit Formation is believed to be deeply buried throughout central and west Fife, in the Clackmannan and Falkirk area and westwards into Glasgow. The thickness is about 170 m. Throughout Fife and around Falkirk and Glasgow, the temperature at the top of the group should exceed 40°C and where the group is as deep as 3000 to 4000 m temperatures of more than 75°C could be expected.

The aquifer properties of the Fell Sandstone are attractive where it crops out in Northumberland and if these persist at greater depths the aquifer could have a transmissivity of more than 5 Dm beneath the Tyneside urban area. In this region the sandstone may be at depths of more than 2 km with temperatures in excess of 60°C.

The hydrothermal systems centred on Bath and Bristol and also found in Derbyshire and South Wales have been referred to and are also discussed in Chapters 6 and 10. Buried hydrothermal systems not yet recognised undoubtedly exist and may in the future be detected by measuring heat flow in shallow boreholes. These systems are invariably in the Carboniferous Limestone, a formation believed to contain zones of palaeokarst at appreciable depths resulting from the solution of the limestone during the geological past. The problem is that the positions of these zones are unknown and they are only likely to be discovered by drilling for purposes other than hot water. This was the case in Belgium where exploration drilling for gas storage sites found a prolific fracture zone in the Carboniferous Limestone at depths of 1500 to 2500 m; this zone is now being considered as a possible geothermal reservoir. The geothermal prospects for the Upper Palaeozoic rocks of the UK could be increased if more elegant exploration methods were developed allowing identification of fractures at depth.

Chapter 6 Geochemistry of Geothermal Waters in the UK

W. M. Edmunds

THE ROLE OF GEOCHEMISTRY IN GEOTHERMAL EVALUATION

Interest in the chemical composition of mineral and thermal waters of the UK has aroused the curiosity of philosophers, scientists and laymen for centuries. Although Britain has not as many thermal springs as several other countries in continental Europe, saline waters or waters with unusual chemistry having alleged therapeutic value occur in several parts of the country (Edmunds and others, 1969). The exploration for geothermal resources since 1977 has provided a new stimulus for understanding the origin and chemistry of traditional sources, such as Bath, and new information has been obtained about much deeper waters in the sedimentary basins.

The possible use of geochemistry in evaluating the low enthalpy geothermal resources of the UK are summarised in Table 6.1. Helping to meet exploration objectives is a primary concern although it has also been possible to assess geochemical methods in the course of testing some geothermal wells. In high enthalpy geothermal areas, geochemistry is usually the principal exploration tool and, where hot springs occur, can give valuable information about the nature of the resource at depth, especially the likely maximum temperature (Ellis and Mahon, 1977). In low enthalpy areas such as Britain, geochemistry cannot be used in many of the traditional ways. Geothermometers, for example, generally do not work in sedimentary basin environments at low temperatures; in such areas the geothermal gradients are more predictable and can be inferred from relatively shallow drilling for heat flow measurements. During the present series of studies, the possibilities have been explored of using geochemistry to determine such parameters as the age of the water, the origin of the water and of its solutes, and the recharge characteristics. The approach has differed in those areas which are part of hydrologically active systems at the present day, such as the Bath-Bristol area, and those areas, such as the Wessex Basin, where the fluids are at present moving extremely slowly, if at all. Thermal waters occurring in tin mines in Cornwall were also investigated in detail.

The thermal waters in the Bath-Bristol area were studied initially so that all the potential geochemical techniques could be developed and then assessed. The parameters investigated and the approach adopted is illustrated in Table 6.2. This approach has been followed in studies of other areas.

Six broad lines of investigation have been considered: (1) major elements, (2) minor and trace elements, (3) physico-chemical parameters, (4) stable isotope ratios, (5) radioelement contents and radioisotope ratios, (6) dissolved gases. Each of these gives information on one or more characteristics of the water and a comprehensive picture of the thermal and hydrological history can be built up using all of them. In the case of Bath, the estimation of groundwater age is inferred from the balance of evidence of several different indicators and the results are considered in more detail below.

Drilling in the sedimentary basins has necessitated the development of new geochemical methods or adaptation of existing techniques. Some data, mainly major element analyses, already existed from drill-stem tests carried out on boreholes drilled for hydrocarbon exploration. Drill-stem tests, however, are designed mainly to obtain reservoir characteristics rather than high quality geochemical information and most of the geochemical data has come from new boreholes. As new data points within any basin are so few and often a considerable distance from any other site, it was necessary to ensure that the maximum information was obtained at each site. Accordingly, one or more sets of multi-element analyses were accumulated for each borehole, and an outline picture of the geochemistry of

Table 6.1 Application of geochemistry in the evaluation of low enthalpy resources.

A	Exploration
	Geothermometry
	Origin of the water
	Origin of the solutes
	Source of the heat (radiothermal or other source)
	Residence time of the fluids
	Local reservoir characteristics
	Extent of mixing with shallow waters
	Assessment of fluid chemistry during drilling
	Extent of formation damage (using tracers)
B	Exploitation
	Fluid characteristics
	Corrosion and scaling potential
	Reinjection assessment
	Economic recovery of constituents of brines
	Environmental aspects e.g. disposal of brines

Table 6.2 Chemical parameters and their relevance to geothermal investigations as illustrated by reference to the Bath-Bristol area.

Parameter	Aspects of study	Conclusions relevant to Bath-Bristol thermal springs
Na, K, Ca, Mg, HCO_3, pH, SO_4	Cation ratio geothermometry; Mineral equilibria; Definition of major hydrochemical type	Na/K/Ca geothermometer invalid in the sedimentary aquifers of the region. Major ion chemistry constant at Bath over 12 months and calcite, dolomite and gypsum equilibria established. Thermal water at Bristol mixed 1:2.3 with shallow limestone groundwater. Geochemical relationship between Weston and Bath denotes that thermal water is leaking into Triassic.
SiO_2	Dissolved silica geothermometry	Chalcedony gives 64°C and quartz 96°C max equilibration temperature. 64°C used as minimum reservoir temperature.
NO_3	Indication of near surface (pollutant) contamination	Nitrate (1.2± 0.1 mg/l) remains constant during monitoring period and shallow pollution must be negligible.
Sr, Li, Ba, F, Ni, Cu, Fe, Mn	Possible indication of source aquifer	Li supports theory of leakage into Triassic. Fluorite equilibrium attained. High Sr consistent with evolution within Carboniferous Limestone, but $^{87}Sr/^{86}Sr$ (0.71075± 6) suggests Triassic origin and derivation of Sr from secondary mineralisation.
pH, Temperature, Eh, dissolved O_2	Definition of controlling mineral equilibria and redox system	Temperature constant (46.5± 0.1°C) but pH, Eh and O_2 varied during monitoring period. Sulphide absent. Possibility of contribution of small amount of groundwater at intermediate depth.
Tritium, ^3H	Indication of post-1953 recharge	No contribution of recent (post 1950s) groundwater at Bath
$\delta^{14}C$, $\delta^{13}C$	Indication of groundwater age and carbon isotope evolution	$\delta^{13}C$ (−1.6‰) diagnostic of marine carbonate equilibration/evolution suggesting the Carboniferous Limestone is the main thermal reservoir. Radiocarbon results (mean 4.5% modern) denote *maximum* age of 19,200± years
$\delta^{18}O$, δ^2H	Indication of origin of groundwaters (juvenile; meteoric; marine)	Meteoric origin indicated at Bath with δ values similar to local present day groundwaters (−7.4‰, −46‰). Recharge during glacial or periglacial conditions therefore ruled out.
$^{234}U/^{238}U$, ^{238}U content	Investigation of system's potential to indicate source aquifer and evidence for mobilization of uranium	^{238}U content at Bath (0.06± 0.002) extremely low, but activity ratio (2.62± 0.23) varied during monitoring, which implies up to 10% mixing with intermediate depth groundwater.
^{226}Ra, ^{222}Rn, ^4He	Indication of residence time, flow history and radiochemical characteristics	^{222}Rn and ^4He levels varied during monitoring reflecting a) 1976 drought, b) possible seasonal head change in confining strata. Anomalously high ^4He content (1432 cm^3 STP/cm^3 H_2O) at Bath implies a long residence time or an enriched radiogenic source
Inert gases Ne, Ar, Kr, Xe	Estimation of recharge temperature	Recharge temperature of Bath thermal water (8.7°C) comparable with that of modern groundwaters.

saline groundwaters in UK sedimentary basins has been obtained during the programme, although a comprehensive view will require much more data. The various drilling, sampling and analytical methods used in the study have been discussed elsewhere (Edmunds and others, 1981, 1983, 1984) and are briefly summarised below, especially where new or improved techniques have arisen in the course of the programme.

Interstitial waters. Drilling programmes were designed to obtain core material representative of potential geothermal reservoirs. Pore fluids were then extracted from these cores so as to obtain information about fluid chemistry in advance of well completion. The method used was originally developed for shallow boreholes (Edmunds and Bath, 1976) and involved centrifugation of 5 to 20 ml of water. During the study an additional centrifuge scheme involving immiscible fluid displacement was developed (Kinniburgh and Miles, 1983).

A tracer (LiCl or KCl) was generally added to the drilling mud to assess the extent that the mud invaded the cores, but the mud chemistry, being distinctive, could also be used as a tracer. Mud invasion was only found to be significant in medium- to coarse-grained lithologies but even in formations which overall are very permeable, such as the Sherwood Sandstone, pore fluids from minor fine-grained units gave reproducible results, indicating that fluid invasion was negligible. Using this technique it has been possible to determine salinity and chemical gradients both within and between formations (Bath and Edmunds, 1981). At several sites where further testing was not carried out, interstitial fluids provided the only information on formation fluids.

Drill-stem tests. Two types of samples have been obtained from drill-stem tests (DSTs): (1) samples at formation pressure from the base of the DST string, (2) samples from the fluid column in the drill pipe. In the wells that were tested by this method the water did not overflow at the surface and it was necessary to recover fluids as the drill pipe was raised. It was always necessary to monitor the change in quality of the fluid at each break of drill pipe. Typical profiles are shown in Figure 6.1 where the least contaminated samples are generally near the base of the fluid column. Even with this control it was very seldom that samples free of contamination and free from particulate matter were obtained.

Gas-lift testing. For the purpose of obtaining meaningful geochemical data, gas lift tests have not proved particularly useful because of the disturbance of the redox and carbonate systems. Important information has, however, been obtained from the major ion chemistry. This indicates, for example, from an analysis of tracers, such as potassium and lithium, the rate at which mud filtrate contamination reduces as the test continues.

6.1 Composition of fluids produced during four drill-stem tests on the Sherwood Sandstone in the Marchwood Well. Lithium (an added tracer) and pH were used to monitor the contamination of the formation brine. A stand is a section of drill-pipe about 27 m long. The stand number is the number of stands above the DST assembly in the well.

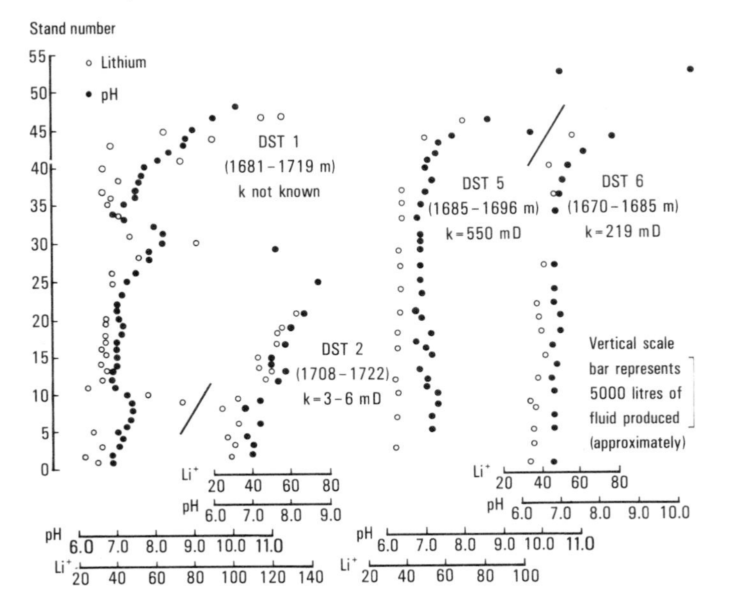

Production testing. Two production tests have been completed—at Marchwood and at Western Esplanade in Southampton. The Marchwood test provided an opportunity to determine the optimal sampling required for such tests as well as to check the validity of samples obtained by the above three methods. A complete sampling programme was carried out for major and trace inorganics, organic acids and physico-chemical parameters. Key variables were monitored at the well-head during the course of the test. A production test is able to provide the most comprehensive picture of the formation water chemistry, its residence time, likely source area and evolution, as well as providing essential information required during the production and development stages of the well. The work to date has indicated the type of data required from any future exploration or exploitation scheme. In particular, the comparison of earlier results with those of the production tests shows that interstitial water samples obtained at very little cost during coring can provide reliable information on formation water chemistry.

GEOTHERMAL FLUIDS IN THE SEDIMENTARY BASINS

Reliable geochemical data about formation waters in the sedimentary basins in the UK have been built up gradually during the programme. Detailed analyses of water from several of the most comprehensively examined sites are given in Table 6.3. The results for a number of major and minor constituents are plotted in Figure 6.2 as a series of graphs relative to chloride (which provides a measure of salinity). Relevant data, of similar reliability, about saline groundwaters from Carboniferous rocks in Northumberland and Durham (Edmunds, 1975), and the Chalk in East Anglia (Bath and Edmunds, 1981) are incorporated for comparison. Stable isotope data from Bath and Darling (1981) and other unpublished sources are also used in the discussion to illustrate similarities between formation waters (Figure 6.3). Radioelement and inert gas results from parallel geochemical studies by Andrews (1985) assist in creating a comprehensive picture.

Upper Palaeozoic basins

There are three known areas in Britain where thermal groundwaters discharge at the surface: (1) the Bath-Bristol area, (2) Taffs Well in the Taff Valley north of Cardiff, and (3) the Derbyshire Dome. Each of these sources is considered to be derived from deep circulation within Upper Palaeozoic rocks, notably the Carboniferous Limestone. The apparent relatively high permeability of the limestone at significant depths probably derives from several factors. Each thermal area is in Hercynian folded strata where faulting and jointing of the massive limestone has allowed opportunities for the development of secondary permeability. Mineralisation has occurred during one or more tectonic episodes which probably reactivated the fracture sets. Such fractures have been the principal vehicle for circulation of mineralising fluids. Changes in the

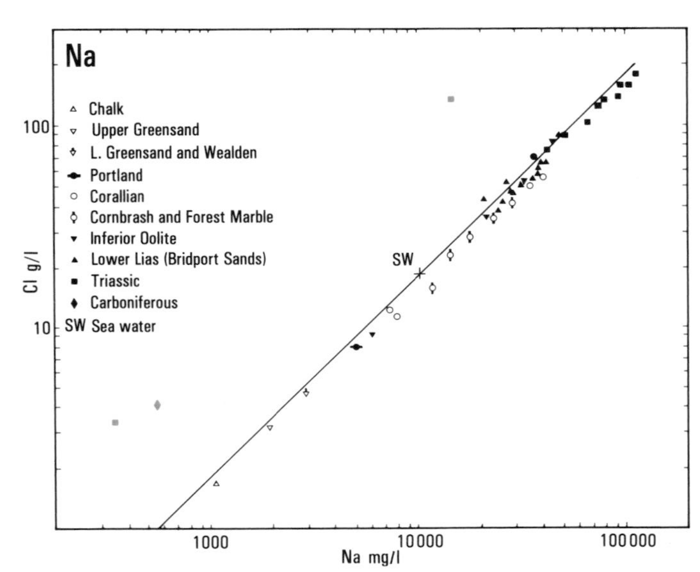

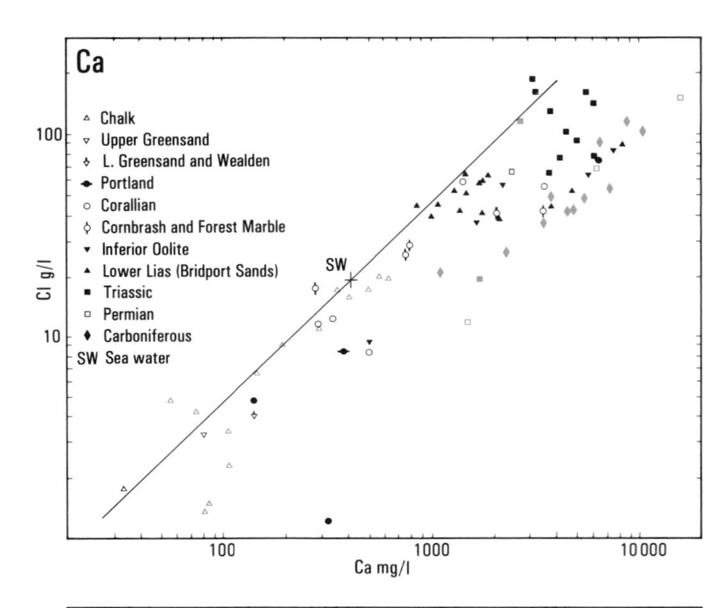

6.2 Variation relative to chloride of Na, Ca, Mg, K, Sr, Li, Br, I, B and Rb for formation waters in sedimentary basins in the UK. Chloride is used as an index of total mineralisation. Sites in the Wessex Basin are shown in black.

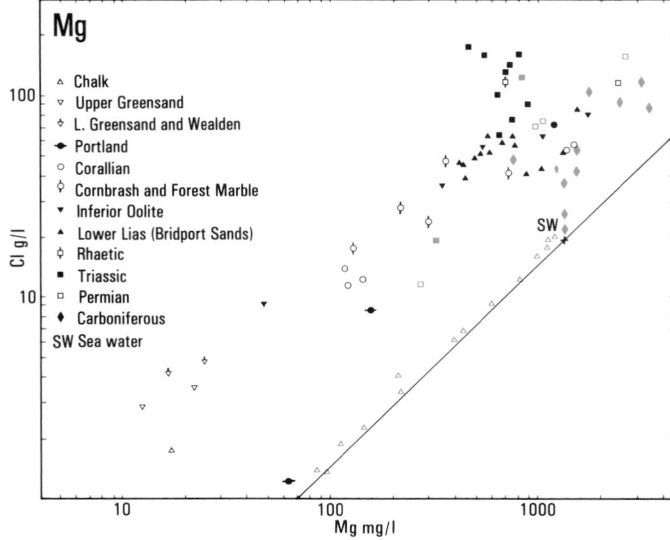

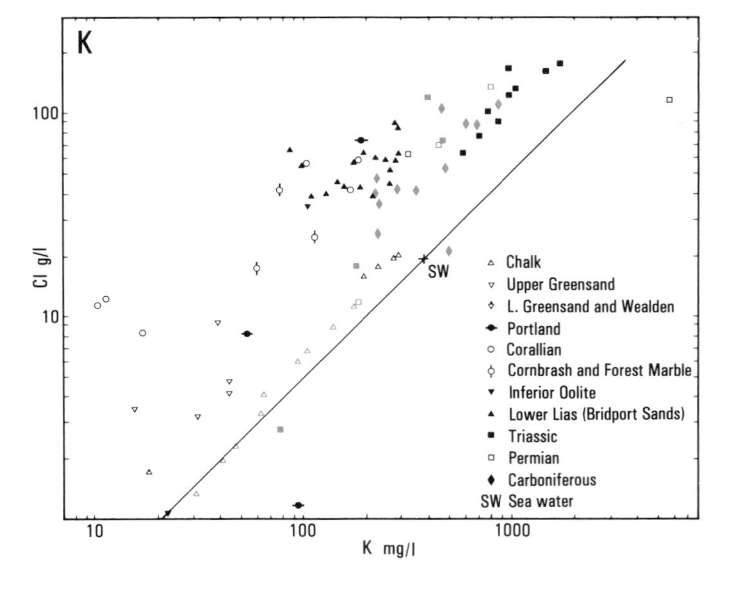

geothermal gradient over geological time have probably led to the solution of calcite and other minerals. During the Pleistocene glaciations this process was enhanced; karstification and solution in permafrost by cold groundwater circulation would have increased carbonate solubilities and developed the permeability; during removal of ice from the British landmass the joint systems would have relaxed. Hot brines occur widely in Upper Palaeozoic rocks where they lie at appreciable depths but details of their composition are only known in a few areas referred to below.

Thermal waters at Bath and Bristol

The Bath thermal waters, which have a temperature of 46.5°C, originate (Andrews and others, 1982) by deep circulation of rainwater falling on the Mendip Hills. The hydrogen and oxygen stable isotope compositions of the thermal water demonstrate that it is of meteoric origin (Figure 6.3) and the $\delta^{18}O$, δD values are close to those of present day local groundwaters in the vicinity. The very large piezometric head (some 10 m above that of the spring) and the contours on the piezometric surface lead to the conclusion that this is due to groundwater levels in the Carboniferous Limestone of the Mendip Hills some 15 km away (Figure 6.4).

The gas composition of the springs can be explained by the exsolution of atmospheric gases, dissolved at recharge, modified (a) by geochemical reaction of oxygen in the aquifer, (b) by addition of radiogenic ^4He due to radioelement decay, and (c) addition of hydrocarbon gases. About half of this gas mixture exsolves at the spring in response to the lower solubilities at 46°C. The ratios of inert gases indicate that the recharge temperature must have been about 9°C.

The total mineralisation of the springs is around 2300 mg/l (Table 6.3). The dissolved ionic constituents provide information on the flow path through the aquifer, the depth of circulation, the maximum temperature, the residence time and the extent of mixing of the springs (see Table 6.2). The dissolved silica concentrations indicate that the maximum temperature reached by the groundwater was in the range 64 to 90°C and using the most likely geothermal gradient of 20°C/km, this places the depth of

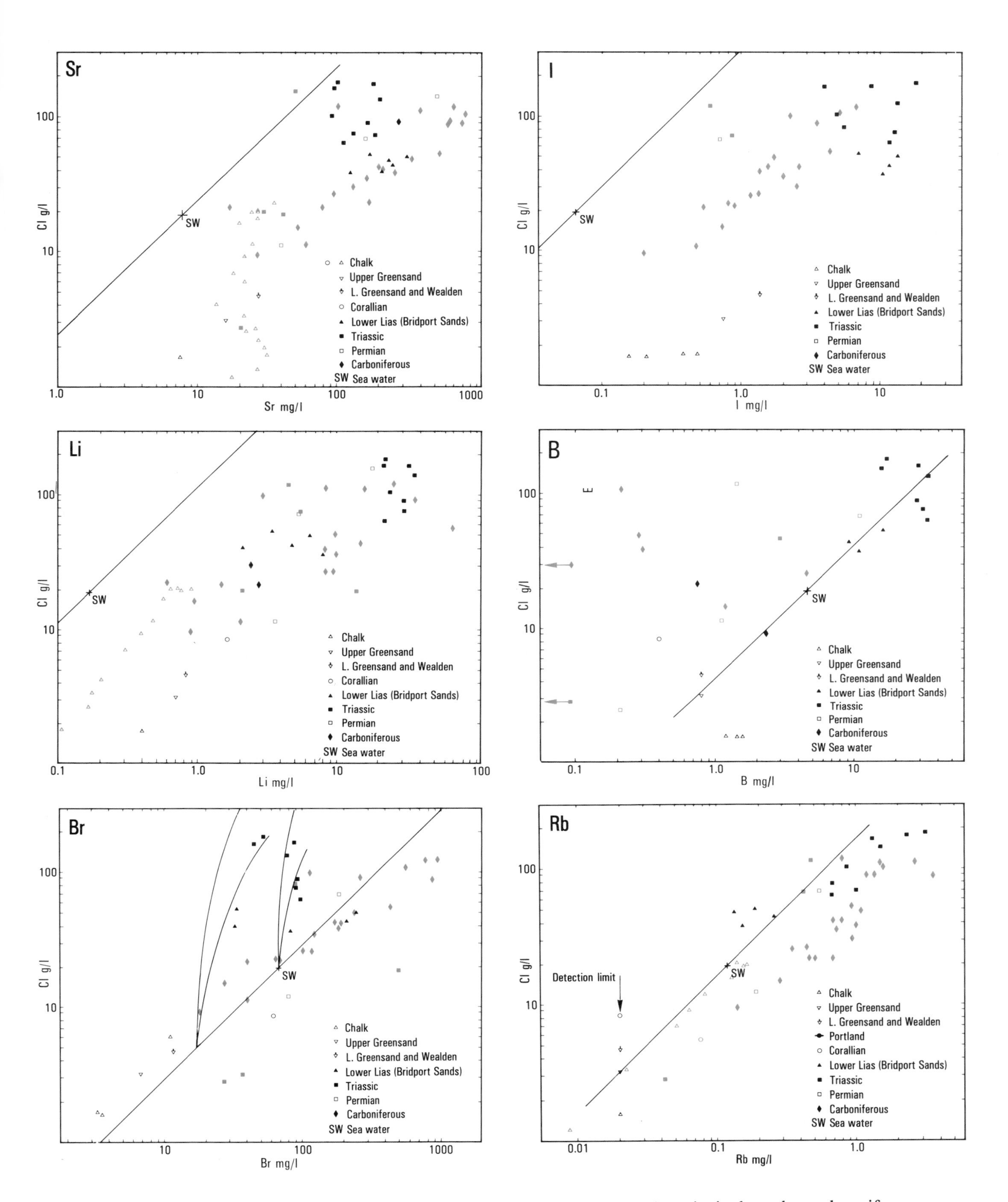

circulation of the water in the range 2700 to 4300 m implying that the Carboniferous Limestone and/or the Old Red Sandstone must be the main storage aquifer. The slightly negative $\delta^{13}C$ ($-1.5‰$) of the dissolved bicarbonate is strong evidence that the water has equilibrated with a marine limestone and thus the Carboniferous Limestone is almost certainly the principal geothermal aquifer.

The thermal spring contains very low tritium levels (1 to 2 TU) at certain times, and is thought, therefore, to contain a small component of recent (post-1950s) recharge. This is borne out by the presence of small amounts of NO_3, variations in dissolved O_2, the Eh and the $^{234}U/^{238}U$ activity

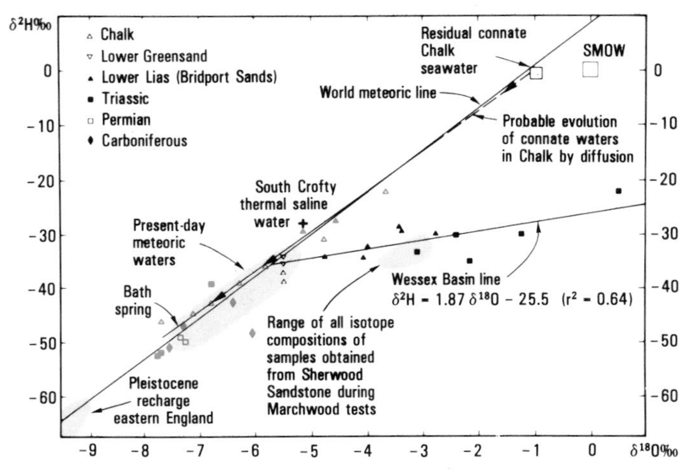

6.3 Isotope compositions ($\delta^{18}O$ and δD) for formation waters from the Wessex and other sedimentary basins in the UK. For key to stratigraphy see Figure 6.2. Sites in the Wessex Basin are shown in black.

ratio. The mix is likely to comprise recent seasonal inflow of water from the Lias or Trias and anoxic deeper thermal water.

The carbon-14 activity (4% modern) cannot be uniquely converted into a groundwater 'age' on account of possible mixing, some loss of ^{14}C by water-rock reactions and other uncertainties in the carbon geochemistry. Interpreted at its face value, the data indicate a maximum age of 20 000 years but the inert gas palaeotemperature implies that Pleistocene (glacial or periglacial) recharge conditions were not likely and therefore a mean residence time of less than 10 000 years is more probable. The ^{4}He contents are very high which suggest that a proportion of the water may be much older and derived from the Old Red Sandstone.

There is evidence that some thermal water is also recharging the Trias in the vicinity of Bath, since adjacent non-thermal groundwaters have a very similar chemistry. It is also likely that concealed thermal waters emerge elsewhere in the Avon valley. The Hotwells spring water at Bristol, which has a temperature of 24°C, is likely to be one such source and can be regarded as a Bath-type water mixed with some 40% of cold local groundwater.

The hot water issuing at Bath could possibly be used as an energy source but any further development by drilling could seriously affect the spring's natural discharge and the intrinsic historical and tourist value of the thermal springs.

Thermal waters in South Wales

Thermal water with a maximum temperature of 22°C discharges from the Coal Measures at Taffs Well, on the south crop of the South Wales Coalfield. Geochemical

Table 6.3 Representative analyses of thermal waters and brines (units are mg/l except where indicated)

	Bath	South Crofty	Wheal Jane	Marchwood	Southampton (Western Esplanade)	Winterborne Kingston	Larne No. 2	Ballymacilroy	Kempsey
Temperature °C	46.5	39.5	41.5	72.5	76.0	85.0	45.0	62.0	32
Formation	Carboniferous Limestone	Granite	Granite/ Devonian	Sherwood Sandstone	Sherwood Sandstone	Sherwood Sandstone	Sherwood Sandstone	Sherwood Sandstone	Sherwood Sandstone
Depth (m)	Spring	300	690	1662–1726	1725–1791	2380–2402	1347–1357	1534–1549	936
Flow (l/s)	15	10	3.5	29	20	DST	DST	DST	DST
pH	6.68	6.40	6.50	6.75	6.0	6.85	6.6	6.9	8.05
E_H				−300	−200				
Na	183	1250	4300	33240	41300	114000	75550	39700	1800
K	174	72.0	180	582	705	1820	311	490	80
Ca	382	835	2470	3670	4240	2920	2790	6200	340
Mg	53	22.0	73.0	658	752	450	866	1100	90
HCO_3	192	21.0	68	81	71		79		
SO_4	1032	148	145	1400	1230	2000	3075	2900	820
Cl	287	3300	11500	63815	75900	181000	115400	72000	2480
Sr	5.92	12.8	40.0	113	134	92	51.8	190	21
Ba	0.024			0.47	0.52	< 0.2	0.16	0.3	
Li	0.242	26.0	125	22.6	31.0	22.4	4.8	5.0	1.0
Rb				0.68				0.3	0.088
NH_4				35.0	36.0		2.2		
Br	1.32		43.0	97.0	91.0	57	171	150	14.8
I				11.8			0.63	0.9	0.014
F	2.6	3.3	2.7	0.46	0.37				
B		3.3	11	32.7	31.0		1.5		<0.1
Fe	0.879	22.4	4.75	4.2	4.1		4.6	<0.2	0.57
Mn	0.068	4.00	4.50	0.94	1.26	2.5	0.99	5.0	0.05
Cu	0.0021	0.005	0.023	<0.01	<0.05				n.d.
Zn				0.20	<0.1				n.d.
SiO_2	44.1	28.4	34.2	33.1			14.0	12.9	
Total mineralisation	2314	5747	19002	103370	124590	302360	198310	122000	5650
Acetic acid				43.0					
Natural uranium µg/kg	0.05	0.25	0.11	0.03					
$^{234}U/^{238}U$	2.5	1.67	1.59	1.76	2.09				
^{222}Rn pCi/kg	2350	12700	470	100	154				
^{226}Ra pCi/kg	11.0	677	83	34	39				
$\delta^{18}O$ ‰	−74	−5.7	−5.2	−3.1	−2.2	−4.1	−6.9	−7.8	
$\delta^{2}H$ ‰	−47.0	−33	−29	−33	−35	−28	−40	−53	
$\delta^{13}C$ ‰	−1.5		−17.4	−15.8	−15.5				

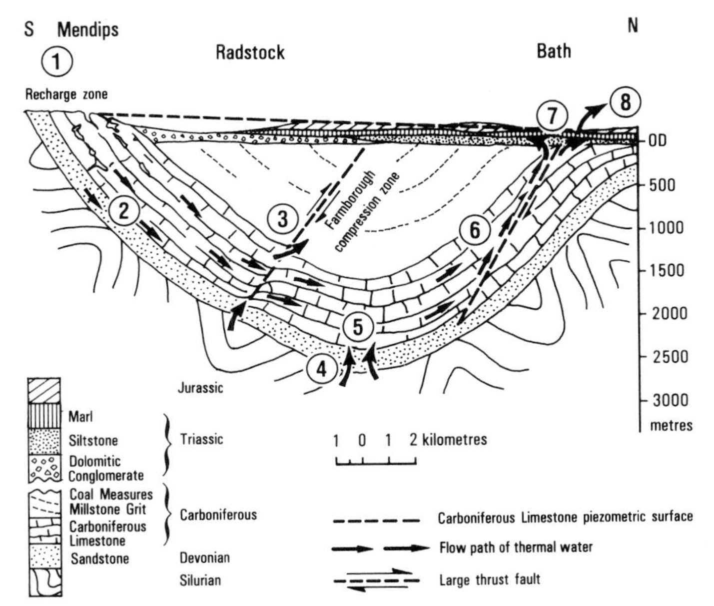

6.4 Conceptual model of the flow path of thermal water in the Bath-Bristol area. 1, recharge (9–10°C) at the Carboniferous Limestone/Old Red Sandstone outcrop; 2, down-dip and down-gradient flow, assisted by karstic features; 3, possible gain from Lower Palaeozoic and leakage to Upper Carboniferous via Farmborough compression zone; 4, some leakage of very old ^{4}He-bearing groundwater from Old Red Sandstone and possibly Lower Palaeozoic; 5, storage and chemical equilibration within the Carboniferous Limestone, at 64–96°C; 6, relatively rapid ascent along thrust fault system; 7, some recharge of Triassic by thermal water at Bath; 8, discharge of the thermal springs at Bath (46.5°C).

investigations of this spring demonstrate that, rather like the Hotwells Spring at Bristol, it contains both a deeper thermal and a shallower component. The thermal component is believed to be derived from recharge of the Carboniferous Limestone on the north outcrop of the coalfield. It then flows beneath the Carboniferous syncline to issue in the lower Taff Valley (Burgess and others, 1980; Thomas and others, 1983). Assuming a geothermal gradient of 20°C/km, a minimum depth of circulation of 600 m can be deduced. The most probable source rock is the Carboniferous Limestone which is at a depth of 700 m below the spring. Inert gas analyses demonstrate that mean recharge temperatures were some 5°C lower than those at the spring and this supports an origin along the high ground to the north of the coalfield. Despite mixing with shallow groundwaters the ^{4}He content is enriched indicating a very significant residence time.

Thermal springs of the Buxton and Matlock area

Eight thermal springs or groups of springs are located near the margins of the Carboniferous Limestone outcrop in the Derbyshire Dome (Edmunds, 1971), for example at Buxton (27.5°C) and Matlock (19.8°C). The geothermal gradient is higher than at Bath or in South Wales and the discharge temperature probably indicates a minimum circulation depth of 600 m. The structural setting of the springs is that they issue from the plunging limb of the dome and in this respect they differ from the other two areas just discussed where the flow is related to a syncline. A conceptual model for the origin of the warm waters has been given by Gale and others, 1984, (in their figure 9.2). The warmest springs are tritium-free but they have low ^{4}He contents (Burgess and others, 1980) and their mean residence time is likely to be over 20 but less than 100 years.

Formation brines in Palaeozoic strata

Thermal waters have also been found at depth in the Carboniferous and other Palaeozoic formations during coal mining, exploration drilling for hydrocarbons and during geothermal exploration drilling. These waters tend to be brines with salinities exceeding sea water and their occurrence in the Carboniferous of eastern England has been described by Downing (1967), Downing and Howitt (1969), Edmunds (1975), and Gale and others (1984c).

In the East Midlands, saline waters and brines are found at depth in three formations, the Coal Measures, the Millstone Grit and the Carboniferous Limestone. Generally, the salinity increases eastwards into the North Sea Basin with maximum salinities exceeding 200 g/l. There is also a sharp salinity gradient to the north of the region. The most notable feature of the Carboniferous brines is the decrease of salinity with depth, the lowest values (less than 10 g/l) occurring in the Carboniferous Limestone. Water fresher than sea water in all formations denotes the displacement of saline formation water by shallow, fresh groundwaters. All the Carboniferous brines have similar major ion trends and ionic ratios. Potassium is slightly depleted relative to sea water due probably to clay mineral uptake. Calcium is enriched and magnesium is depleted suggesting dolomitisation. The hydrogeochemical and piezometric evidence suggest upward movement from the Carboniferous Limestone to the Millstone Grit and to a lesser extent into the Coal Measures. Thus the saline formation waters are being modified by fresh water recharge into the outcrop areas of Carboniferous rocks in the Pennines, and the high concentrations downgradient are considered to be due to ultrafiltration through mudstone horizons.

In Durham and Northumberland, brines in the Coal Measures with salinities up to 230 g/l are considered to have evolved from the original formation water (Edmunds, 1975). Their compositions have also been modified by cation exchange, dolomitisation and ultrafiltration. These brines as well as others in the Coal Measures are often enriched in barium (up to 4100 mg/l) which has increased following the removal of sulphate by bacterial reduction. The reduction of sulphate waters, development of barium brines and barite mineralisation at moderately low temperatures is considered to have been a cyclic process and to have been important in the epigenetic mineralisation of the Pennines.

Only a limited number of Carboniferous brines have been analysed during the present phase of geothermal exploration and their chemistry is summarised in Figure 6.2. They include fluids from the concealed Oxfordshire and Nottinghamshire coalfields, which are compared with data from the Northumberland and Durham area. The salinities (expressed as Cl) range from 9 g/l to 118 g/l. They are depleted in K relative to sea water and depleted in

Mg. They are significantly enriched in Ca, Sr, Rb, Li, I, Br compared with sea water and also with respect to brines from most other formations. The enrichment is very uniform across the range of salinities and implies that this is a characteristic of the water-rock interaction with the Carboniferous formations whatever the salinity. The enrichment in strontium is very high (up to 900 mg/l) and the bromide enrichment is also important, being the result of diagenetic reactions in the Coal Measures with organic matter.

Mesozoic basins

As a result of the geothermal programme the data base on formation waters in the deep Mesozoic Basins has been improved although enough information has not yet been obtained to understand fully the relationship between these waters, either on a regional or on a stratigraphical basis.

In the Wessex Basin at least eight reliable sets of data have been obtained. These can be compared with the more scattered results from the other basins, including Worcester and Northern Ireland. The relevant results are included in Figure 6.2. As additional background to this discussion, geochemical information from a borehole at Trunch in Norfolk is included. This borehole, which fully penetrated the Cretaceous, provides a reference profile of pore-water chemistry which has been interpreted as showing that virtually unaltered 'sea water', approaching a true connate water, is present in the Lower Chalk (Bath and Edmunds, 1981). The stable isotope compositions (Figure 6.3) show a very slight modification, probably by diffusion, from a marine composition. They may be used (a) to demonstrate mixing trends in this case between fresh, meteoric water and sea water, (b) to show that formation waters may remain little changed over some 80 to 100 million years and (c) to show that the composition of sea water may have been different isotopically or chemically during the geological past.

The most saline water in the Cretaceous is found at Trunch (up to 19100 mg/l Cl). This is probably the result of a location near the western margin of the North Sea Basin, an area of subsidence since the Tertiary. In contrast, the formation water in the Chalk at Marchwood does not exceed 2000 mg/l Cl and generally the total mineralisation in the Wessex Basin Chalk should not exceed 10 000 mg/l. This suggests that significant exchange of fresh water for connate water has occurred since the Tertiary. The overall processes involved in early and late diagenesis of the Chalk by groundwater dilution have been discussed by Edmunds and others, 1973; Edmunds, 1976; and Bath and Edmunds, 1981. Both the Chalk and other Cretaceous groundwaters show depletion of Ca and Mg relative to sea water. The reaction with the Chalk has led to an enrichment of Sr, Li, and F in pore solutions. The upper limit of Sr in the Chalk pore solutions is controlled by celestite which has also been found infilling fissures at Trunch.

The Lower Greensand is an important water supply aquifer in East Sussex. Groundwaters with very low total mineralisation (340 mg/l) are developed at about 25°C from depths of over 400 m. The total mineralisation is still below 1000 mg/l at Portsdown where the depth is 530 m, although salinities over 7000 mg/l occur 200 m below this in the Wealden. Overall, the Cretaceous groundwaters are fairly uniform across the Wessex Basin, but, except for strontium, they are diagenetically modified in a different manner from samples from the Trunch Borehole; this is due to the greater flux of groundwater by flow or diffusion through aquifers in Wessex. Isotopically, Cretaceous waters in the Wessex Basin lie close to the meteoric water line.

Several horizons in the Jurassic of the Wessex Basin have yielded formation waters, either from DSTs or production tests or interstitial waters extracted from cores; little information is available on deep Jurassic formation waters from other basins although the saline groundwaters in the Inferior Oolite (Lincolnshire Limestone) of eastern England have been discussed by various authors (Edmunds, 1973; Downing and others, 1977; Andrews and Kay, 1983). In general in the Jurassic of the Wessex Basin the salinity increases with depth both stratigraphically and relative to sea level. Maximum salinities of 96 000 mg/l Cl are found in the Lower Lias. There are regional similarities, especially in the Bridport Sands; for example compare Ca, K concentrations (Figure 6.2). They are significantly enriched in iodide and depleted in rubidium compared with groundwaters in adjacent formations.

Brines have been obtained from Permian and/or Triassic formations in Northern Ireland (Larne and Ballymacilroy boreholes), and the Worcester Basin (Kempsey Borehole) as well as from nine sites in the Wessex Basin. In the Wessex Basin they are all of high salinity ranging from 104 g/l at Marchwood to 300 g/l at Winterborne Kingston; the regional variation of the salinity is shown in Figures 4.15 and 6.2. Very high salinities close to halite saturation ($SI_{HALITE} = -0.19$) occur near the centre of the basin, coinciding with the greatest depth and formation temperatures in the Sherwood Sandstone. High salinities are also found as edge waters to the Dorset oilfields. Despite the apparent uniformity of salinity gradient there are important local differences between waters in adjacent wells at the same depth so that the inter-relationships may not in fact be straightforward. For example, water from the Sherwood Sandstone had a salinity of 104 g/l at Marchwood and 125 g/l in the Western Esplanade Well in the centre of Southampton though the two sites are only 1.9 km apart. The brines in the Permo-Triassic formations in the Northern Ireland Basin are also highly saline, but the Sherwood Sandstone in the Worcester and the East Yorkshire and Lincolnshire basins contains water with a salinity lower than or equal to sea water.

There are important differences as well as similarities between brines in the Permo-Triassic formations which may be important genetic indicators. The K/Na ratios in all the brines, except one from the Permian of north-east England, are below 0.04 and are depleted with respect to sea water. The higher K/Na ratio may be associated with sylvite dissolution. In all the other high salinity Triassic formation waters the major and minor elements change significantly relative to chloride. This is most clearly demonstrated in the case of Br (discussed below) but a similar relationship is apparent for Ca, B, Mg, Li, Sr and I. The lower the element/Cl ratio, the more likely it is to have undergone 'dilution' by halite.

ORIGINS OF FORMATION WATERS IN THE SEDIMENTARY BASINS

There are two possible starting points for the origin of the formation waters and brines in the Palaeozoic and Mesozoic basins of the United Kingdom—sea water or meteoric water which entered the systems at various times during the geological history of the basins. It is quite possible that flushing of the systems at different periods has occurred and that the composition of the waters only records the most recent event. The origin of the water and its dissolved constituents must be considered separately.

The stable isotope ($\delta^{18}O$, δD) compositions can be used to summarise the history of the water (Figure 6.3). The isotope ratios of the geothermal fluids may be considered, relative to sea water (SMOW), in the context of the range of present-day meteoric waters and late Pleistocene waters which have more negative $\delta^{18}O$ values derived from cooler climates such as are to be found, for example, in the Triassic sandstones of the East Midlands (Bath and others, 1979).

It is apparent that almost all the thermal waters and formation brines with the exception of those from the Wessex Basin lie on or close to the world meteoric water line. All have $\delta^{18}O$ values between -5 and $-8‰$ which are very similar to present day groundwaters. The presence of Pleistocene recharge in any of the formations can be discounted and it is also clear that sea water, for which $\delta^{18}O = 0$, cannot be considered as the direct source of these brines.

This situation is rather ambiguous since most of the Jurassic and Cretaceous sediments must originally have contained sea water during early burial and diagenesis. The evidence from the Trunch Borehole in Norfolk (Bath and Edmunds, 1981) provides one likely solution. Here it is possible to follow the evolution of the interstitial water column in the Chalk from the almost unchanged marine composition (residual connate water) to recent groundwater. If a diffusion model is applied then the present isotopic profile can be shown to have developed by this process alone within 3×10^6 years. This process has probably taken place since the period of uplift and marine regression in the early to mid-Tertiary up to 40×10^6 years ago. Freshwater influx may not have commenced until the late Tertiary.

The Trunch profile therefore shows that replacement of sea water by fresh water can occur by diffusion but is likely to be speeded up in more permeable formations by groundwater flow. Groundwater flow will have taken place in the basins in response to tectonic movements, sea level changes and other hydrodynamic processes. Nearly all the sedimentary basins are thus now likely to be filled with meteoric water and the evolution of their salinity and chemistry must be considered mainly from this secondary starting point, possibly with 'initial' salinities in the range 1000 to 10 000 mg/l.

The main feature of the isotopic geochemistry is the different evolution of brines in the Wessex Basin relative to other areas. They are relatively enriched in the heavier oxygen isotope and are interlinked by a line with a slope of 1.87, with an intercept of $-5.5‰$ $\delta^{18}O$ on the meteoric line; the starting point of this evolution was meteoric water, similar to that of the present day, as in the other basins. It is also noted that deeper brines, in older formations are more enriched than those at shallower depths.

The best explanation for the trend is the fractional loss of lighter isotopes by ultrafiltration during compaction. This process would have to assume the preferential movement of water rather than solutes. The most likely hypothesis for the fractionation of the stable isotopes in the Wessex basin is that very high hydrostatic pressures during and after the Alpine tectonism drove the ultrafiltration whilst waters in Mesozoic basins to the north were little affected, if at all.

The only other possibility of accounting for oxygen isotope enrichment is considered to be the exchange of water with minerals in the aquifer (e.g. carbonates, gypsum, clay minerals). However this exchange is nowhere near equilibrium (for example with respect to carbonate minerals) and would not account for the unique behaviour of the Wessex Basin relative to other Mesozoic basins, although the explanation has been used to account for brine compositions with positive $\delta^{18}O$ values in the Western Canada sedimentary basin, for example Hitchon and Friedman (1969).

Experimental work (Coplen and Hanshaw, 1973) has indicated that oxygen enrichment by ultrafiltration produces a slope of 3.1 where adsorption is the controlling mechanism. Slopes as low as 0.5, may, however, occur if molecular diffusion is involved. The intermediate value for the Wessex Basin may therefore indicate that more than one process is responsible. Further data from the other basins in the UK are required before a conclusive explanation can be reached.

The strongest evidence for the origin and behaviour of solutes comes from the Br/Cl ratios of the fluids. Bromide and chloride, unlike all the other ionic constituents, do not participate in mineral formation at relatively low salinities and are unlikely to be affected by diagenesis, although some bromide enrichment may occur as organisms decay during diagenesis. The relative depletion of bromide in brines on the other hand almost invariably denotes some dissolution of halite evaporites with low Br/Cl ratios.

In Figure 6.2 the bromide data for the Chalk can be used as a model for 'dilution' of a marine water composition. The brines from Carboniferous strata all exhibit bromide enrichment, as do some of the Jurassic brines from the Middle Lias sands. Those in the Wessex Basin usually show varying degrees of depletion.

Halite generally contains bromide in the range 60 to 400 ppm. Lines of evolution are plotted on Figure 6.2 starting from (a) sea water, (b) groundwater with a total mineralisation of 5000 mg/l, and a Br/Cl ratio similar to diluted sea water dissolving halite with 65 and 200 ppm Br respectively. It can be seen that the Sherwood Sandstone brine at sites in and near Southampton could be explained by the dissolution of halite by a water with salinity close to sea water. In the western part of the basin, the low bromide but high salinity brines must originate from waters of much lower salinity. Diffusion of high salinity brines from the overlying Mercia Mudstone following tectonic activity in the Tertiary could account for the observed hydrochemistry, bearing in mind that evaporite deposits are thicker in the west of the Wessex Basin. It is also important to note that the Lower Lias (Bridport Sands) waters are also genetically related to the underlying brines in the Sherwood Sandstone rather than to other waters in the Lower Lias. An origin involving evaporite dissolution would also account for the observed differences in other ionic compositions. K and Rb dissolved from impurities in the halite would tend to be enriched whereas Mg, Ca, Sr and Li would not.

Although ultrafiltration is proposed for the isotope fractionation, it is unlikely that ionic ratios have been changed by this process in the Wessex Basin. Experimental studies (Kharaka and Berry, 1973) show that bromide should be retained and Br/Cl increased during ultrafiltration which is the opposite of that shown in the Wessex Basin. As already indicated the enrichment of calcium and depletion of magnesium and potassium in the UK brines generally can be accounted for by dolomite and K-feldspar formation. Exchange reactions and membrane filtration are unlikely to have modified ionic ratios since Na/Cl ratios are very close to 1 in most brines.

The residence time for fluids in the Wessex Basin, notably in the Triassic sandstones, has been examined using isotopic ratios of helium and the uranium isotope activity ratios, all other dating methods being inappropriate (Andrews, 1983; Edmunds and others, 1983). Helium-4 accumulates in an aquifer from the radioactive decay of uranium and thorium and, assuming this to be the only source, a residence time may be calculated using the equation:

$$t\,(\text{years}) = [\text{He}]\{\varrho\,(1.19 \times 10^{-13}\,[\text{U}] + 2.88 \times 10^{-14}\,[\text{Th}])\}^{-1}$$

where ϱ, [U] and [Th] are the density and the U and Th concentrations of the rock respectively. This assumes that the ^{4}He generated is distributed uniformly between the water and the mineral phases. Using the data from Marchwood an age of more than 15 million years (Ma) has been derived for the groundwater in the Triassic sandstones in the vicinity of Southampton. However, helium being very mobile could also have been exchanged by diffusion between the Trias and overlying and underlying confining formations. This is very difficult to quantify and at present there is a degree of uncertainty over the residence time depending upon which model is used. Uranium isotope activity ratios (^{234}U/^{238}U) confirm that the water must be in excess of 1.25 Ma. However if the most pessimistic model for helium loss is used, a maximum age of 85 Ma would be possible.

The median age for the brines, of about 15 Ma (which would be Miocene), may be compared with the late Cretaceous age given by Colter and Havard (1981) for the south to north migration of oil at the time of the late Alpine tectonic events. A realistic genetic model for the brines would be to envisage considerable groundwater flow at intervals during the Alpine period with very little subsequent movement except for further exchange by diffusion between adjacent formations allowing salinities to increase.

The waters now found in Upper Palaeozoic aquifers are probably also all modified meteoric waters. As previously mentioned, the enrichment of Ca and depletion of Mg in Carboniferous brines are considered to be caused by dolomitisation and the slight depletion of potassium relative to sea water is probably associated with the diagenesis of clay minerals. The high total mineralisation is more likely to have originated by ultrafiltration by mudstones in the Carboniferous sequence rather than by halite dissolution, which appears to predominate in the Mesozoic basins.

GEOTHERMAL WATERS IN GRANITE

Apart from Bath, the only other region of Britain where thermal waters at more than 40°C were known, historically, was the tin mines of Cornwall. This potential geothermal development area has been explored and studied in detail and the geochemical studies (Burgess and others, 1982; Edmunds and others, 1984; Edmunds and others, 1985) have thrown light upon the origin of the thermal waters, especially their high salinity. It is important to note that the Carnmenellis Granite is the only granite in Britain in which thermal (and saline groundwaters) have been recorded. It is also the site of the Hot Dry Rock experiment, referred to earlier. Geochemical investigations of fluids produced during the testing of the doublet well system at the site have been carried out and are discussed below.

The thermal waters in the mines occur as springs at depths up to 800 m. The highest recorded temperature is 53°C although at present the warmest accessible spring is 41.5°C at the South Crofty Mine. The flow paths in the granite are summarised in Figure 6.5. The temperature, flow and the high salinity (19 000 mg/l) of the South Crofty spring have remained more or less constant for some 100 years. It is clear on the basis of stable isotope composition that the waters are of meteoric origin (Figure 6.3) and their salinity cannot have been derived from sea water. It is also clear that the mine waters which have different salinities represent a mix of an older saline water and a recent water which contains tritium and that the mixing has been induced by the mining. It was considered initially that the level of tritium (about 5 TU) might be anomalous, caused by the in-situ production of ^{3}H from the reaction ^{6}Li (n,α) ^{3}H but it has now been shown that this effect cannot be significant (Andrews and Kay, 1982).

The principal interest has been the origin of the salinity of the groundwater. Some of the principal chemical characteristics are demonstrated in Figure 6.6. The Na contents are notably depleted and the Ca enriched relative to equivalent sea water composition and this has been attributed to reaction between the water and plagioclase feldspar. The groundwaters also have very high chloride contents and are enriched in Li to a maximum of 125 mg/l. It has been shown that both Li and Cl are derived from the breakdown of biotite by a reaction such as:

$$K_2(Mg,Fe)_4(Fe,Al,Li)_2[Si_6Al_2O_{20}]OH_2(F,Cl)_2 + 3H_2O + 12H^+ \rightarrow Al_2Si_2O_5(OH)_4 + 4H_4SiO_4 + 2K^+ + 2(Mg^{2+} + Fe^{2+}) + 2(Fe^{3+} + Al^{3+} + Li^+) + F^- + Cl^-$$

Chlorite (or some other sink for Mg) is also a probable reaction product as well as, or in addition to, kaolinite since the Mg/Ca ratios in the waters are very low. The Li/Cl ratio in the groundwater is much lower than would be expected if it were controlled solely by biotite breakdown and it has been suggested that a considerable degree of chloride loss from interlayer exchange involving Cl and OH has taken place without loss of Li which is structurally bound in octahedral positions in the biotite lattice.

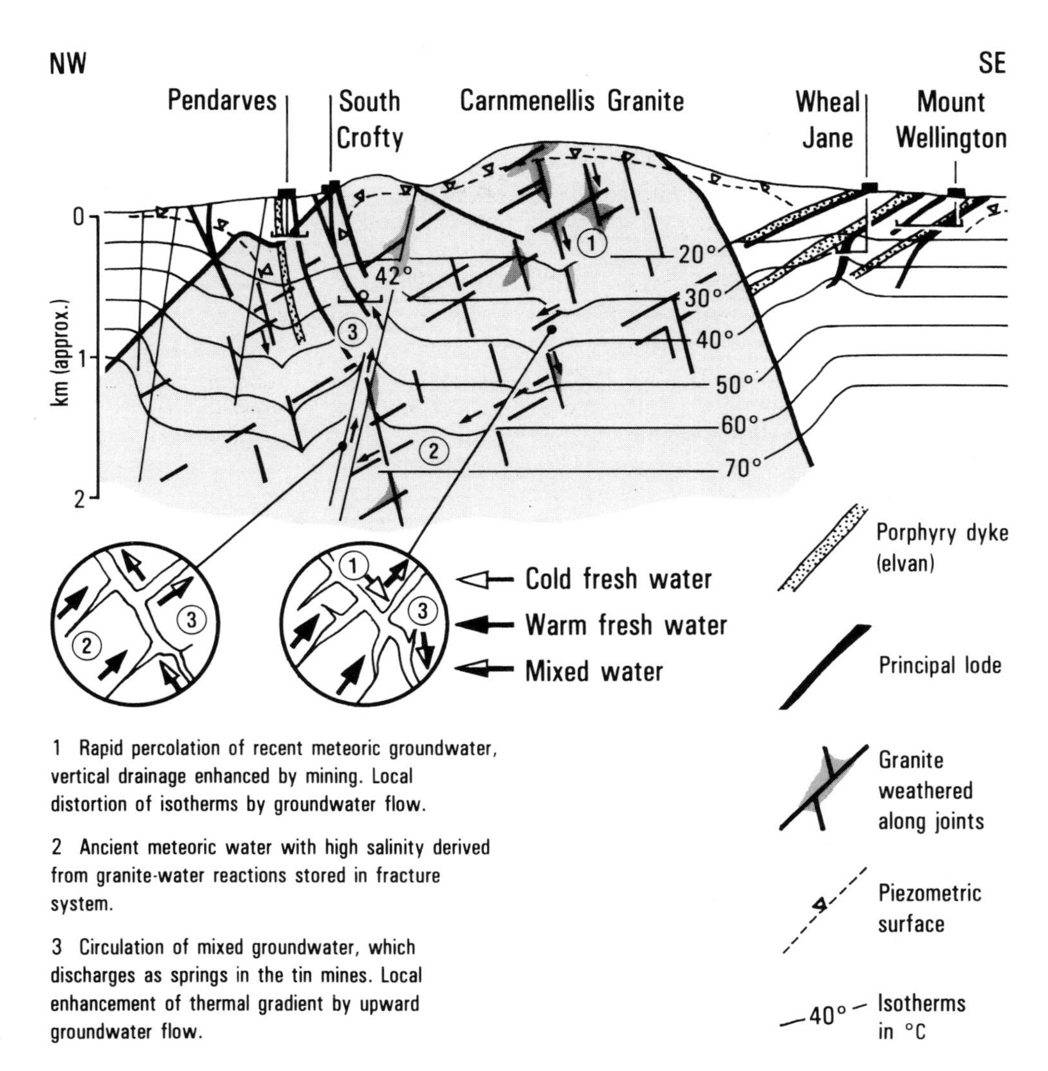

6.5 Conceptual model of groundwater circulation in the Carnmenellis Granite, Cornwall.

6.6 Na/Cl and Li/Cl plots of thermal and non-thermal mine-waters and shallow groundwaters in the Cornubian granite. Note the Cl/Na enrichment and Li/Cl enrichment relative to sea water. The Li/Cl ratio of fresh biotite and of solutions produced during biotite leaching experiments are shown, as well as the composition of fluids produced during Hot Dry Rock circulation tests in the granite in 1983.

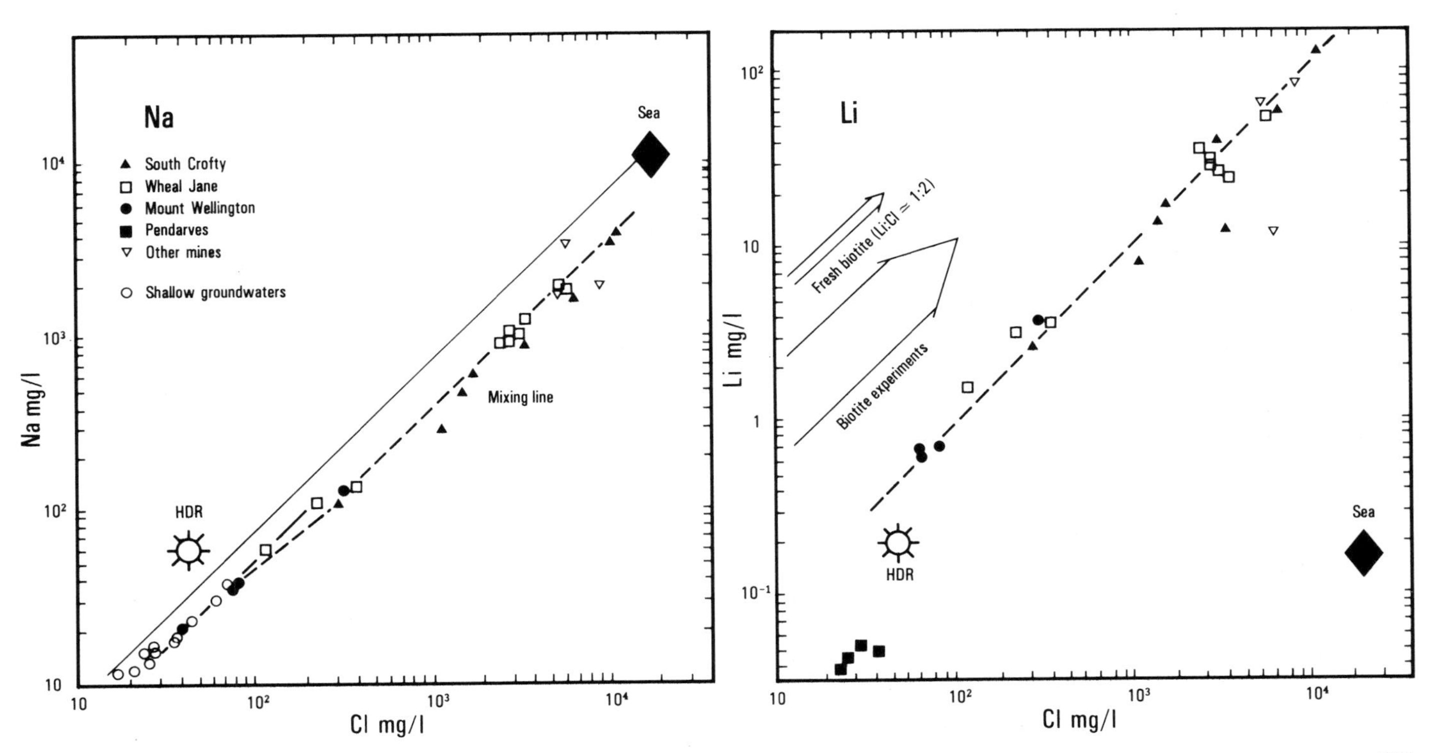

These hydrolysis reactions are considered to be the main reactions leading to salinity in the groundwaters and predominate over any other process such as release of salinity from fluid inclusions. Quartz and chalcedony are both saturated in the groundwaters and thus silica deposition rather than solution should be taking place, encouraging the growth of new fluid inclusions.

The radiogenic ^{4}He content, $^{40}Ar/^{36}Ar$ ratios and uranium series geochemistry suggest that the saline, thermal component has a likely residence time of at least 5×10^4 years and probably of the order of 10^6 years. The storage must be considerable and it is calculated that the volume of water discharged from the underground spring at South Crofty during the last 100 years only represents some 1% of the total saline water likely to be stored in the granite. The silica and Na/K geothermometers show that an equilibration temperature of about 54°C must have been reached during circulation and this corresponds to a depth of 1.1 to 1.2 km.

Monitoring of fluids produced during experiments at the Hot Dry Rock site in the Carnmenellis Granite shows that very similar chemical reactions occur to those in the natural systems, though at reduced salinities, demonstrating that the same water-rock reactions are taking place during very short time scales (Macartney, 1984; Edmunds and others, 1985). The proposed model for the genesis of saline water, therefore, links together or explains several processes—groundwater movement and mixing, convective heat transport, the distinctive chemistry of the water, water-rock interaction, secondary mineral formation and fluid inclusion formation and stability.

GEOCHEMICAL CONSIDERATIONS IN THE UTILISATION OF LOW ENTHALPY GEOTHERMAL BRINES

The main problems connected with exploitation of geothermal brines from the chemical viewpoint are the possibilities of corrosion, scaling and the need to dispose of the brines after use. All of these problems can be assessed from a geochemical standpoint, although once the brine has been delivered to the surface, engineering, metallurgical and various environmental problems must be considered. The problem associated with low enthalpy brines and other waters is essentially one of mineral and materials solubility and stability. Compared with high enthalpy development, where there is flashing of water to a steam phase, the problems are relatively straightforward and can be investigated using reliable chemical analyses and geochemical models.

Low enthalpy waters commonly have a low salinity. In the Paris Basin the salinity of the waters from theJurassic, that are being commercially developed, is usually less than 60 g/l. However, waters at more than 60°C in the Sherwood Sandstone of the Wessex Basin are very saline and resemble the brines found in deep Triassic aquifers in Denmark, West Germany and France, where the range is generally between 100 and 300 g/l.

Comprehensive and reliable chemical analyses have been obtained from only three sites during the present programme, Bath, Marchwood and Southampton (Western Esplanade), although good reconnaisance data have been collected for several others. The most important parameters required for corrosion/scaling prediction are major ion analyses, some minor elements (e.g. Sr, Mn, Ba, F, Si), and in situ measurements of temperature, pH, H_2S, HCO_3, E_H. The modelling of these data can be carried out using one of several computer programs. In this study, the program WATEQF (Plummer and others, 1976) has been used to compute the equilibrium distribution of aqueous ionic species and the states of reaction (departures from equilibrium) of up to 150 minerals. This program is adequate for waters of low ionic strength (e.g. as at Bath), but the methods used to calculate single ion activity coefficients may be inadequate for some species at high ionic strength and the results given for saline waters must be considered provisional.

The extent to which the geothermal waters are either in equilibrium with the minerals of the reservoir or are undersaturated with potential scaling products can be seen in Table 6.4. The saturation index (SI) for various minerals is computed as the log ratio of the measured values to the predicted concentrations of solute species. In the case of Bath, a value of zero ± 0.2 probably indicates saturation with the mineral; a more positive reading would indicate supersaturation and a more negative reading undersatur-

Table 6.4 Saturation indices for minerals in various formation waters calculated using the WATEQF program

	Bath Kings Spring	Marchwood Sherwood Sandstone	Southampton Sherwood Sandstone	Winterborne Kingston Sherwood Sandstone
Temperature	46.5	72.5	74.9	85.0
pH	6.75	6.75	6.0	6.9
Total mineralisation	2200	103 450	124 100	299 740
Eh	−57	−300	−200	—
Saturation index				
Albite	—	−0.04	0.39	—
Aragonite	−0.21	0.00	−0.71	0.23
Barite	0.57	−0.66	−0.88	−1.64
Calcite	0.05	0.27	−0.44	0.51
Celestine	−0.91	−1.14	−1.30	−2.34
Chalcedony	0.18	−0.09	−0.03	−0.14
Cristobalite	0.20	−0.12	−0.06	−0.19
Dolomite	−0.32	0.42	−0.98	1.05
Fluorite	0.12	−1.75	−2.05	—
$Fe(OH)_{AMORPH}$	−3.53	−6.04	−6.80	—
Fe S ppt	—	0.34	—	—
Gypsum	−0.12	−0.28	−0.39	−0.92
Halite	−5.96	−1.57	−1.37	−0.19
Pyrite	—	4.98	—	—
Quartz	0.59	0.22	−0.28	0.13
Rhodochrosite	−1.76	−2.09	−2.87	−2.64
Siderite	−0.70	−0.73	−1.57	—
$Silica_{AMORPH}$	−0.30	−0.59	−0.52	−0.63
Strontianite	0.76	0.48	−0.32	−0.16

ation. In the more saline waters the errors will be generally larger. In the case of Marchwood, the SI values for calcite, quartz, albite, gypsum and barite, which are all important minerals in the aquifer, are near to saturation. The SI for halite in Winterborne Kingston water is also close to zero indicating near-saturation.

The brine at Marchwood is supersaturated with iron sulphide minerals but, since sulphide is well below 1 mg/l (Table 6.3), sulphides will not cause a scaling problem. Silica minerals, quartz, chalcedony and cristobalite, are at or close to saturation in these waters but are undersaturated with respect to amorphous silica, indicating that silica is unlikely to precipitate from these waters. Fournier (1973) has shown that silica will remain in solution on cooling, even though saturated with quartz or chalcedony, until saturation with amorphous silica is exceeded.

During the Marchwood production test, coupons were suspended in the well. The only scaling products identified (in trace amounts) were aragonite, calcite and galena (the latter may have been derived from the grease used in the casing joints). The presence of carbonates may be significant in view of the apparent saturation in most brines of aragonite, calcite, dolomite and strontianite, although their solubilities increase with decreasing temperatures thereby lowering the possibility of carbonate precipitation as long as closed conditions (which avoids CO_2 loss) are maintained. Despite near-saturation with gypsum there should be no problem with these during any exploitation.

Corrosion problems are commonly associated with CO_2 and H_2S. Oxygen is not present in deep formation waters but it becomes an important corrosive agent in surface pipe-work or in re-injection waters if steps are not taken to reduce its activity.

Corrosion tests were carried out on a range of metals during the Marchwood test, both on coupons suspended downhole and on surface pipelines as well as corrosion meter measurements. The attack on a range of stainless steels, nickel alloys and titanium was negligible. Corrosion rates on a range of low alloy steels and copper base alloys were all less than 0.025 mm per year. No evidence of localised attack or pitting could be found. These low corrosion rates are consistent with the geochemical data, notably the absence of oxygen and a very low H_2S content; traces of H_2 gas are also present. The general conclusion must be that these brines, despite their very high salinities do not present a serious problem to materials likely to be used in low enthalpy engineering works, provided rigorous closed system conditions are maintained.

The main problems connected with the exploitation of brines, such as those found in the Wessex Basin, is likely to be their surface disposal. Apart from high salinities they contain high levels of iron (for example 4.0 mg/l), although levels of toxic metals and other substances are very low. Aeration, settling and dilution of the brines are probably the only treatment needed but these problems would be overcome entirely if a doublet well system were used, thereby allowing a closed circulation system with reinjection to the aquifer.

Chapter 7 Modelling of Low Enthalpy Geothermal Schemes

J. A. Barker

INTRODUCTION

The management of a geothermal aquifer scheme involves the choice between a large number of options. Even when the properties of the aquifer and the size of the heat load to be met have been determined, there are still several important decisions to be made; these include what pumping rate to use and, therefore, what size of pump to purchase; and, whether or not to drill further wells for either abstraction or injection of water. It is not possible to carry out experiments to determine the consequences of each of the possible decisions, and therefore it is necessary to model the system. Models also provide predictions of the long-term behaviour of a system and can reveal forms of response which would not be observed during short-term tests.

The purpose of this chapter is to present some simple formulae and graphs describing the hydraulic and thermal behaviour of low-temperature geothermal reservoirs under various development schemes. The results given are approximations based on relatively simple models under various simplifying assumptions, but they should provide an adequate basis for judging the viability of schemes when only limited data are available. Once a particular scheme is initiated, these results should be superseded by those based on more complex (analytical and numerical) models, using the more detailed geological and geophysical data that become available.

In general, these results will need to be considered in conjunction with some form of economic model in order to determine the viability of a scheme. However, in some cases it will be clear that technical constraints alone preclude the development of certain reservoirs.

BASIC PRINCIPLES

Any low-temperature geothermal scheme involves the abstraction of water at some temperature θ_a and volumetric flow rate Q. Energy can be extracted from this water at a rate P give by

$$P = \varrho_w c_w (\theta_a - \theta_{rej}) Q ,$$

where ϱ_w and c_w are the density and specific heat of the water, respectively, and θ_{rej} is the reject temperature of the heat extraction system.

It will be assumed that before development begins the reservoir has a stable temperature, θ_0, which will also be the initial abstraction temperature. Once the abstraction temperature falls below some value, θ_{min}, the scheme will be considered unworkable, probably because of a combination of economic factors and technical constraints associated with the heat extraction process. It is possible, however, to envisage a situation where θ_{min} is not a constant but depends to some extent on the flow rate Q.

The water is pumped from one or several wells and various arrangements will be considered later. This pumping will cause the water level in an abstraction well to fall by a distance (the drawdown) which will be represented by s_w. As pumping continues the drawdown generally increases unless water is replenished in the reservoir. There will be some natural replenishment (recharge), but in deep geothermal systems this is unlikely to have any significant effect during the lifetime of most development schemes. Therefore, if the drawdown is to be limited there must be some form of artificial recharge, such as reinjection of the abstracted water. It will be assumed that a scheme will not be viable once the drawdown exceeds some critical value, s_{max}. This maximum drawdown will often represent the deepest point at which a pump can reasonably be installed.

The above ideas are summarized in Figure 7.1. Any scheme starts at point A (temperature θ_0 and zero drawdown) and remains viable provided the temperature and drawdown remain within the region ABCD. In the absence of any recharge, a single well will operate along the line AB. If water is recharged at the same rate that it is extracted, then the drawdown will be limited by some value s_0 for a given pumping rate. Initially such a scheme will follow the line AE, but eventually the cooling effect of the injected water will reach the abstraction well and the scheme will then operate along the line EF. Should the pumping rate be

7.1 Various paths across the viable operating region of a geothermal scheme. Recharge is assumed to equal abstraction. The rate of movement along the various paths will be highly variable.

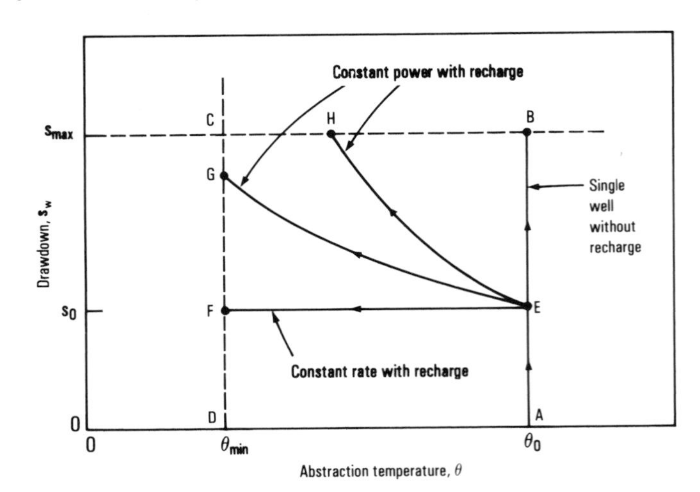

increased—perhaps to keep the power output constant—then the scheme would operate along lines such as EG or EH in Figure 7.1.

The rest of this chapter is concerned mainly with estimating the time that various development schemes take to traverse the viable region, ABCD, in Figure 7.1. Consideration will be given in turn to: a single well in the absence of recharge; a single abstraction well along with a single recharge well (a doublet system); and, finally and briefly, multiple well systems.

SINGLE-WELL SCHEMES

In the absence of any significant amount of natural recharge, a single-well scheme will produce geothermal water at a nearly constant temperature, θ_0, throughout its lifetime (line AB in Figure 7.1). Therefore the lifetime of the well is determined by the rate at which the drawdown increases with time. The main factors determining the drawdown behaviour of a well are: the intrinsic hydraulic characteristics of the reservoir, including the presence of any hydraulic barriers, the geometry of the well (its diameter, completion etc), and the presence of any other wells in the reservoir.

Infinite homogeneous aquifers

As a standard case, consider a well of radius r_w fully penetrating a confined, homogeneous geothermal aquifer of infinite extent. If the aquifer has a transmissivity T and storage coefficient S, then the drawdown in the well after pumping at rate Q for time t is approximated by:

$$s_w = \frac{Q}{4\pi T} W\left(\frac{Sr_w^2}{4Tt}\right), \qquad 7.1$$

where W is known as the Theis well function. This function is plotted in Figure 7.2 as (dimensionless) drawdown against the logarithm of (dimensionless) time; the graph is nearly linear since the approximation

$$s_w = \frac{Q}{4\pi T} \ln \frac{2.25Tt}{Sr_w^2} \qquad 7.2$$

is accurate to within about 6% for times greater than 2.5 $r_w^2\ S/T$, and therefore adequate for all practical purposes.

Equation 7.2 can be rearranged to give the time, t_{max} at which the drawdown exceeds s_{max} (point B in Figure 7.1):

$$t_{max} = \frac{Sr_w^2}{2.25T} \exp\left(\frac{4\pi T}{Q} s_{max}\right) \quad . \qquad 7.3$$

This formula should be used with considerable care since the predicted lifetime will be very sensitive to the parameters Q, T and s_{max}.

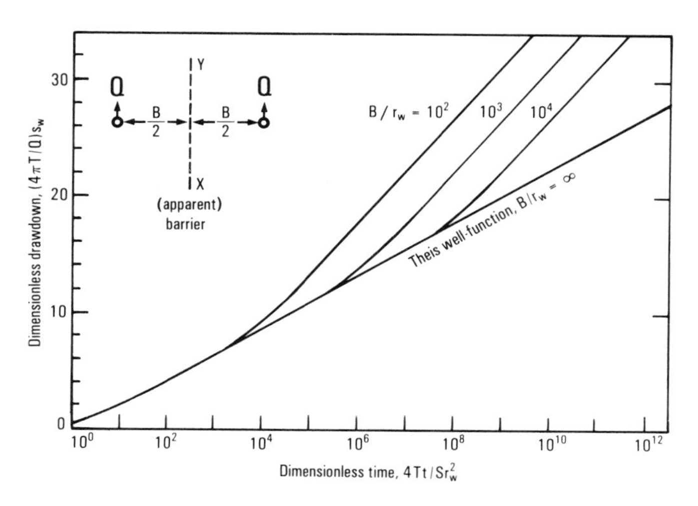

7.2 Drawdown variation with time in a well of radius r_w in the vicinity of an impermeable linear barrier at a distance $B/2$ or, equivalently, an identical well at distance B.

Hydraulic barriers and other wells

In practice, an aquifer cannot be regarded as being homogeneous and of infinite extent. In general it is necessary to take account of the presence of hydraulic barriers of various kinds, and of other wells abstracting from the same aquifer. Many different configurations can be envisaged but just two will be considered here in order to give some indication of the effects that can be expected. (For further cases see Section 7.3.3 in Allen and others, 1983).

The inset in Figure 7.2 can be taken to represent two different situations: either a single well at distance $B/2$ from an infinitely long impermeable barrier, XY; or two identical wells, separated by a distance B, pumping for the same period at the same rate. In the latter case the line XY is an apparent barrier since there will be no flow across it. The curves in Figure 7.2 show how the drawdown increases with time for various values of B/r_w. Initially, for a period of the order of $B^2S/4T$, the drawdown follows the curve corresponding to the Theis well-function. Then the drawdown increase accelerates because of the lack of flow across XY, and the drawdown tends to a line given by:

$$s_w = \frac{Q}{2\pi T} \ln \frac{2.25Tt}{Sr_wB} \quad . \qquad 7.4$$

Equation 7.4, which can be used for times greater than about 2.5 $B^2\ S/T$, can be rearranged to give an estimate of the time at which the drawdown reaches s_{max}:

$$t_{max} = \frac{Sr_wB}{2.25T} \exp\left(\frac{4\pi T}{Q} s_{max}\right) \quad . \qquad 7.5$$

The second case considered here is that of a single well which is encircled by a hydraulic barrier at distance R (inset in Figure 7.3). This could be taken as an approximation to a well near to the centre of any finite reservoir of area πR^2, or even to a large number of identical wells in the same aquifer where each draws water from an area of about πR^2. The curves in Figure 7.3 show that the drawdown behaves as if the reservoir were unbounded up until a time of about 2.5 $R^2\ S/T$, when the drawdown increases rapidly (note the

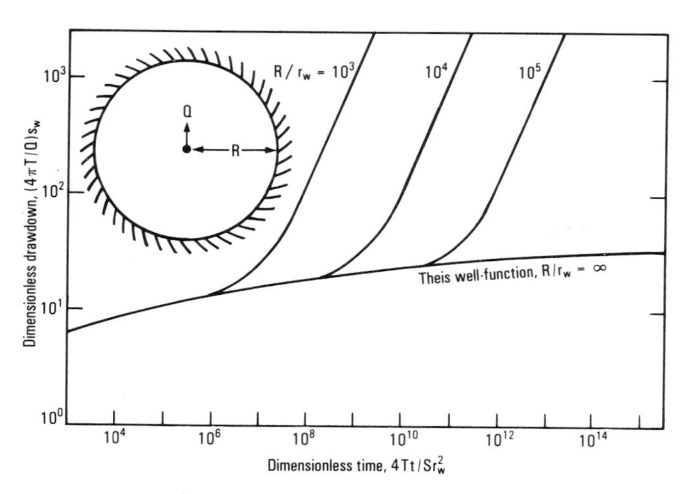

7.3 Drawdown for a well at the centre of a circular reservoir of radius *R*.

logarithmic drawdown scale) and tends to a line given by:

$$s_w = \frac{Qt}{\pi R^2 S} + \frac{Q}{2\pi T}\left(\ln\frac{R}{r_w} - \frac{3}{4}\right) \quad . \qquad 7.6$$

This equation can be used for times greater than 25 $R^2\ S/T$ and gives the time at which the drawdown reaches s_{max} as:

$$t_{max} = \frac{\pi R^2 S}{Q}\left[s_{max} - \frac{Q}{2\pi T}\left(\ln\frac{R}{r_w} - \frac{3}{4}\right)\right] \quad . \qquad 7.7$$

A generalisation of Equation 7.6 is given by:

$$s_w = \frac{Qt}{A_R S} + \frac{Q}{4\pi T}\ln\frac{2.25A_R}{C_A r_w^2} \ , \qquad 7.6a$$

where A_R is the reservoir area and values of the shape factor, C_A, can be found in Table C.1 of Earlougher (1977) for various reservoir shapes and well positions.

The very marked effect that boundaries can have on the lifetime of a geothermal well can be demonstrated by comparing results obtained from Equation 7.7, with those from Equation 7.3, for an unbounded reservoir. For example, using: $T = 10.5\ m^2/d$, $S = 5 \times 10^{-5}$, $r_w = 0.15m$, $Q = 30\ l/s$, and $s_{max} = 500\ m$, Equation 7.3 gives a lifetime of 15 years (this is very sensitive to the choice of T, Q and s_{max}). By comparison, using the same parameters for a bounded reservoir with an area (πR^2) of 100 km², Equation 7.7 gives a lifetime of only 0.6 years.

Improvement by artificial fracturing

It is clear from Equation 7.3 that the lifetime of a single well can be extended by increasing its radius, r_w. Although only a very limited range of borehole sizes can be drilled economically, the effective radius of a well can be increased considerably by fracturing. The technique of hydraulic fracturing is widely used in the petroleum industry and, at present, is applied to more than one third of all wells (Veatch, 1983). The fracturing process is complex, expensive and restricted in its application, and is far from being entirely predictable. For example, hydraulic fracturing is usually only worthwhile in reservoirs that have a low permeability—generally less than 50 mD. The yield is usually increased by a factor of not more than four. The existence of such practical restrictions should be borne in mind throughout the remainder of this chapter.

Consider a vertical fracture which fully penetrates the geothermal aquifer and extends a distance b from the borehole (that is, it has total length $2b$). The solution of the flow equations is greatly simplified by assuming that the fracture produces water at an equal rate across its whole surface. Normally this is the assumption made in analysing well-test data to determine the size of a fracture, and can therefore be justified on the basis of consistency. Given these assumptions, the drawdown in the well (that is at the centre of the fracture) after pumping for time t will be given by:

$$s_w = \frac{Q}{4\pi T}\left[(\pi\tau_f)^{\frac{1}{2}}\ \mathrm{erf}(\tau_f^{-\frac{1}{2}}) + W(\tau_f^{-1})\right] \ , \qquad 7.8$$

where:

$$\tau_f = \frac{4Tt}{Sb^2}, \qquad 7.9$$

erf is the error function, and W is the Theis well-function.

In Figure 7.4 the drawdown given by Equation 7.8 is compared with that for a well of radius b (that is the second term in Equation 7.8). After times greater than about 2.5 Sb^2/T, Equation 7.8 can be approximated by:

$$s_w = \frac{Q}{4\pi T}\left[2 + \ln\frac{2.25Tt}{Sb^2}\right] \ , \qquad 7.10$$

so a drawdown of s_{max} will be achieved at time:

$$t_{max} = \frac{Sb^2}{2.25T}\exp\left(\frac{4\pi T}{Q}s_{max} - 2\right) \qquad 7.11$$

7.4 Drawdown in a well with a fully-penetrating vertical fracture of length 2*b*. The lower curve shows the Theis well-function for a well of radius *b*.

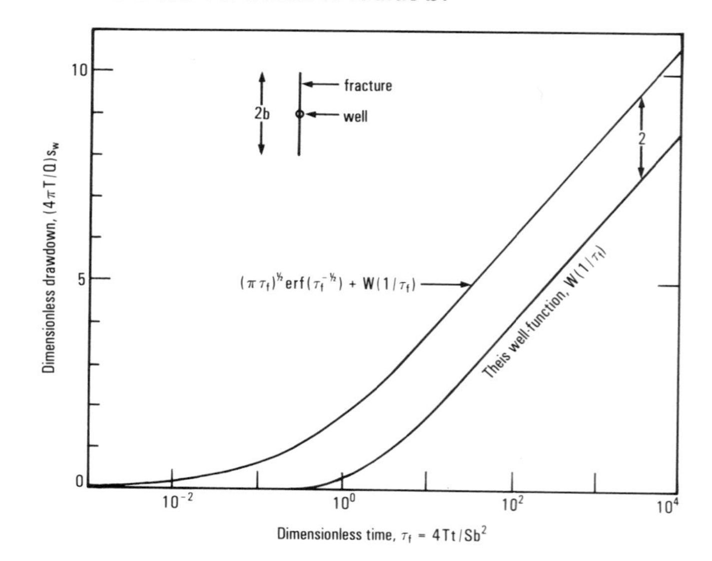

This represents a very significant increase in well lifetime, by a factor of about b^2/r_w^2, over that predicted from Equation 7.3. Unfortunately, such an improvement is unlikely to be realised in practice because of the presence of hydraulic barriers.

For example, consider a fractured well at the centre of a circular barrier of radius R ($R \gg b$). After a long period of pumping the drawdown in the well can be approximated by:

$$s_w = \frac{Qt}{\pi R^2 S} + \frac{Q}{2\pi T}\left[\ln\frac{R}{b} + 1\right] \quad . \qquad 7.12$$

Comparing this with Equation 7.6 for an unfractured well, it is seen that fracturing increased the lifetime by:

$$\Delta t_{max} \approx \frac{R^2 S}{2T}\left[\ln\frac{b}{r_w} - 1.75\right] \qquad 7.13$$

This leads, for example, to an increase of only 140 days by using:

R = 5 km, $S = 5 \times 10^{-5}$, T = 15 m²/d, b = 25 m and r_w = 0.15 m.

7.5 a) Section through a simple geothermal doublet showing the steady-state potentiometric surface.

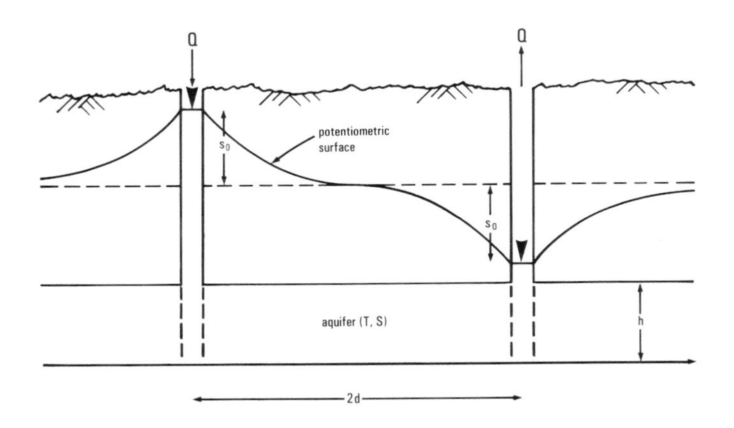

b) Flow net for the same doublet.

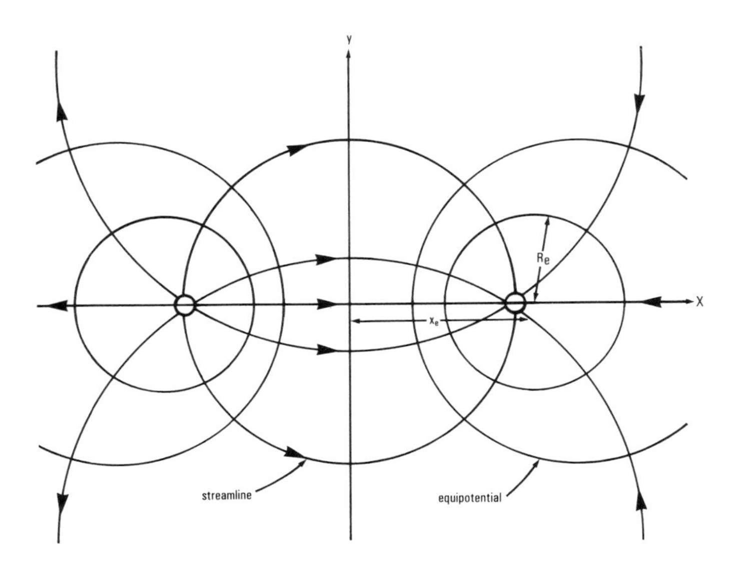

DOUBLET SCHEMES

A geothermal doublet consists of two similar wells; water is abstracted from one well and injected through the other, normally the same water at the same rate. The main advantage of such a system over a single well is that the drawdown in the abstraction well is limited for any given pumping rate. It also has the advantages of providing a method for disposing of the water, allowing more efficient development of an aquifer and, in unconsolidated formations, of preventing subsidence which can result from excessive pressure reduction in such formations. The main disadvantage of a doublet, apart from capital and operating costs, is that the abstraction temperature eventually begins to fall because of the cooling effect of the injected water.

Consider a doublet system in an aquifer with transmissivity T and storage coefficient S (Figure 7.5). If the wells are separated by a distance $2d$ then the drawdown will reach a near steady-state condition (point E in Figure 7.1) after pumping for a period of about Sd^2/T, which will always be very small in comparison to the lifetime of the system. The system will then operate at a constant temperature and drawdown up until the time—known as the breakthrough time—when the first effect of the colder injected water is felt at the abstraction well. The abstraction temperature then begins to fall and eventually tends to the injection temperature. Therefore for any doublet system essentially three things need to be estimated: the steady-state drawdown, the breakthrough time, and the rate of fall of the abstraction temperature after breakthrough.

In the following sections a detailed study of the behaviour of a simple doublet will be followed by a consideration of the possible advantage of fracturing.

Simple (unfractured) doublet

Hydraulic behaviour

A simple doublet system under steady-state flow conditions is depicted in Figure 7.5, where the effects of viscosity variations, due to temperature variations, have been ignored. Water flows along arcs of circles which have their centres equidistant from the two wells. The amount of flow across the y-axis (Figure 7.5b) between points $(0, -y_0)$ and $(0, y_0)$ is given by:

$$q_x(y_0) = \frac{2Q}{\pi}\tan^{-1}\frac{y_0}{d} \quad . \qquad 7.14$$

For example, 80% of the flow crosses the y-axis between points $(0, -3d)$ and $(0, 3d)$.

Equipotentials are circles, of radius R_e, centred on points $(x_e, 0)$ where $x_e^2 = R_e^2 + d^2$ and on which the drawdown is:

$$s(R_e) = \pm\frac{Q}{2\pi T}\ln\left(\frac{d + x_e}{R_e}\right) , \qquad 7.15$$

(the negative sign corresponds to points closer to the injection well than to the abstraction well). In particular, the drawdown in the abstraction well is given approximately by:

$$s_0 = \frac{Q}{2\pi T}\ln\frac{2d}{r_w} \quad . \qquad 7.16$$

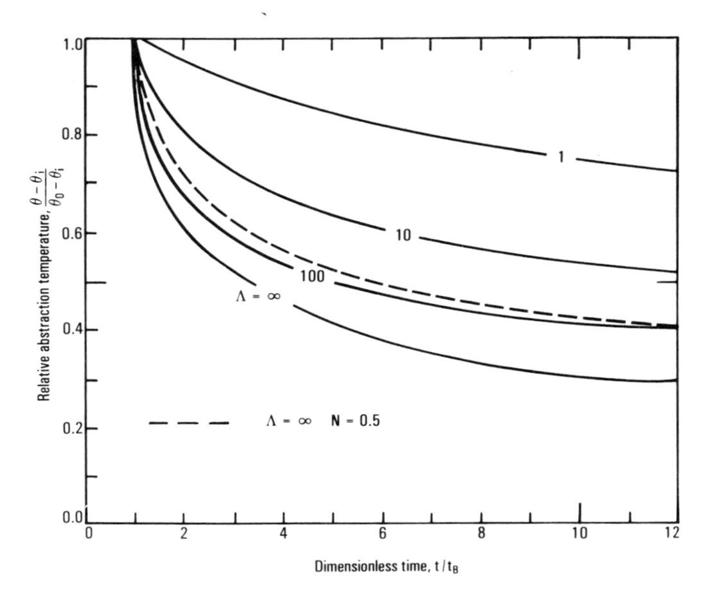

7.6 Variation of abstraction temperature with time for a simple doublet based on the theory of Gringarten and Sauty (1975). The dashed line shows the effect of viscosity increase with temperature reduction. Λ is defined by Equation 7.20.

In a typical Permo-Triassic aquifer the transmissivity may be 15 m²/d, so, for a doublet with a 1 km separation, pumping at 30 l/s with $r_w = 0.15$ m, the drawdown would stabilise at a value of 240 m.

Equation 7.15 can be rearranged to give the region outside which the drawdown (or head increase) does not exceed Δs:

$$R_e = d/\sinh\left(\frac{2\pi T}{Q}\Delta s\right) . \qquad 7.17$$

This represents two circles which are themselves contained within a rectangle with sides $2R_e$ and $2R_e + 2(R_e^2 + d^2)^{\frac{1}{2}}$. This result can be used to estimate how close to a hydraulic barrier, or other wells, a doublet can be placed before there is any significant interaction.

Thermal behaviour

The abstraction temperature for a geothermal doublet remains constant at the initial aquifer temperature, θ_0, up until the breakthrough time, t_B, and thereafter declines towards the injection temperature, θ_i. Gringarten and Sauty (1975) considered a steady-state flow model of a doublet which gave the breakthrough time as:

$$t_B = \frac{4\pi h d^2}{3\beta Q} , \qquad 7.18$$

where h is the aquifer thickness, $2d$ is the well separation and β is the ratio of the heat capacity of the water to that of the aquifer (both the water and the matrix), that is:

$$\beta = \frac{\varrho_w c_w}{\varrho_a c_a} . \qquad 7.19$$

For a sandstone with 20% porosity, Equation 7.18 becomes $t_B = 3hd^2/Q$. Therefore, for an aquifer of thickness 50 m, a doublet with a 1 km separation and pumping at 30 l/s would have a breakthrough time of about 40 years.

The abstraction temperature variations deduced by Gringarten and Sauty (1975) are shown in Figure 7.6, where (dimensionless) temperature is plotted against the pumping time (in units of t_B) for various values of a parameter Λ. This parameter provides a measure of the extent to which the injected water is heated by (vertical) conduction through the rocks above and below the aquifer, and is given by:

$$\Lambda = \frac{\alpha Q h}{4d^2} , \qquad 7.20$$

where

$$\alpha = \frac{\varrho_w c_w \varrho_a c_a}{\lambda_r \varrho_r c_r} . \qquad 7.21$$

the subscript r denoting the surrounding rock and λ_r its thermal conductivity. (Typically α will have a value in the range 1 to 5 × 10⁶ s/m²). When the effects of vertical conduction are small (giving large Λ values), the abstraction temperature falls rapidly after time t_B (Figure 7.6) so t_B represents the effective lifetime of the system. However, Λ values of the order of unity can often be expected, in which case the lifetime of the doublet can be many times the breakthrough time. (For example, $\Lambda = 1$ when: $\alpha = 4 \times 10^{-6}$ s/m², $Q = 20$ l/s, $h = 50$ m and $d = 1$ km).

Other factors affecting doublet behaviour

The above description of the behaviour of a geothermal doublet is grossly simplified; further details can be found in the following references: Gringarten and Sauty (1975); Gringarten (1979); Sauty and others (1980); Goblet (1980). Of the other factors affecting the behaviour of a doublet, stratification and viscosity variations are sufficiently important to warrant a brief discussion here.

All aquifers exhibit some form of heterogeneity. In particular different horizontal layers of an aquifer may have permeabilities which differ by several orders of magnitude. The effect of this can be that breakthrough occurs much sooner than predicted using an average permeability for the whole aquifer. This stratification will also cause the abstraction temperature to fall in a rather irregular manner. The details of this behaviour are complex since thermal conduction between the layers can be important. Goblet (1980) used a numerical model to gain some insight into the extent of these effects.

The injected water is colder and hence more viscous than the abstracted water. For example, if the abstraction temperature, θ_a, is about 75°C and the injection temperature, θ_i, is about 30°C; then the ratio of the viscosity of the abstracted to the injected water, N, will be about 0.5. This contrast will distort the flow pattern from that shown in Figure 7.5 and, more importantly, will cause the injection head to exceed the abstraction head (drawdown) by a factor of $1/N$. The effect on the abstraction temperature is discussed by Sauty and others (1980), and the dashed line in Figure 7.6 suggests the extent to which this effect might prolong the lifetime of a doublet.

Optimization of a doublet

The optimal design and operation of a doublet have been studied by Sauty and others (1980), and by Golabi and others (1981). In general, this is a complex task during which many economic as well as physical parameters must be considered. This topic will not be pursued other than to make the following observation. In Equations 7.16, 7.18 and 7.20, the parameters over which there is control are the well discharge, Q, and the well separation, d (and to an insignificant extent r_w). Ideally, one would like to be able to design for a given drawdown, pumping rate and lifetime but (given only two control parameters) once any two of these are fixed then the other is determined automatically. Fracturing of the wells introduces a further design parameter (the fracture size) which may allow the required control: the effects of fracturing are considered below.

Fractured-well doublets

As with a single borehole, the drawdown (and injection head) required to operate a geothermal doublet can be significantly reduced by hydraulic fracturing of the wells. White (1983) suggested that fractured-well doublets may allow the exploitation of geothermal reservoirs with permeabilities as low as a few millidarcys. These fractured systems have been studied by Andrews and others (1981) and many of the following results are based on their work.

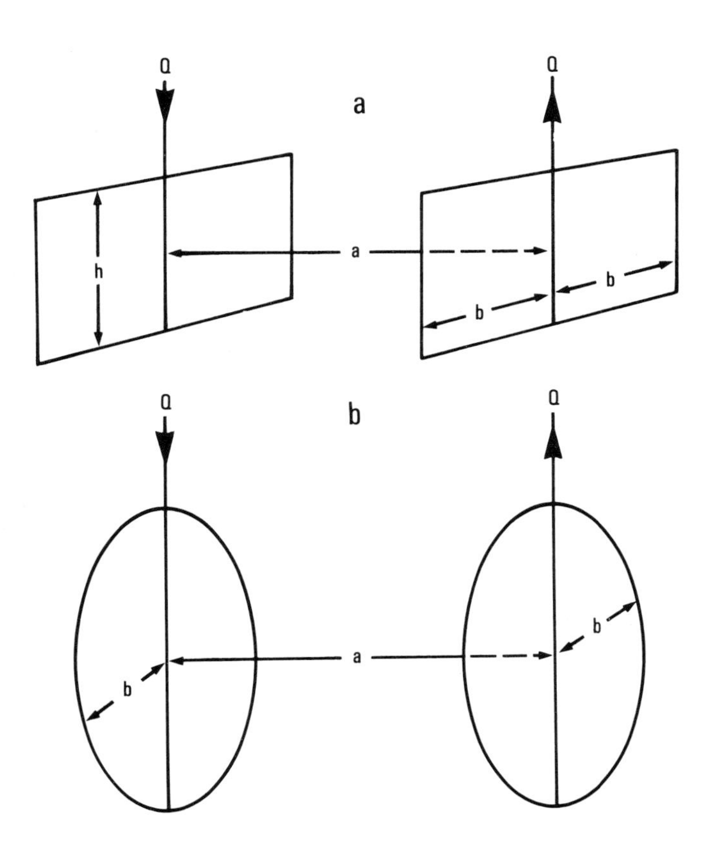

7.7 Parallel fracture arrangements studied by Andrews and others (1981); (a) two-dimensional flow between rectangular fractures which fully penetrate a confined aquifer, (b) three-dimensional flow between disc-shaped fractures.

Two-dimensional flow

Firstly, consider the case depicted in Figure 7.7a where parallel, vertical fractures are assumed to fully-penetrate a confined aquifer of thickness h; each fracture has an area $2bh$. The drawdown in the abstraction well is approximated by:

$$s_0 = \frac{Q}{2\pi T(0.25 + 0.83\, b/a)} \quad \text{for } 1 \leqslant a/b \leqslant 15 \qquad 7.22a$$

$$= \frac{Q}{2\pi T} \ln(1 + b/a) \text{ for } a/b \geqslant 15 \quad . \qquad 7.22b$$

The breakthrough time is given by:

$$t_B = \frac{\pi h a^2}{\beta Q}(0.25 + 0.83 b/a) \qquad 7.23$$

(cf. Equation 7.18), where β is given by Equation 7.19.

It is of interest to consider what fracture size would be required to limit the drawdown to a particular value, s_{max}, in an aquifer with a given thickness and permeability when the abstraction rate and breakthrough time are also fixed. Figure 7.8 shows some results based on Equations 7.22 and 7.23. The kinematic viscosity (required to convert from permeability to transmissivity) was assumed to be 0.46 mm^2/s, as for the brine found in the Western Esplanade Well in Southampton. The heat capacity ratio, β, was taken as 2.

The solid lines in Figure 7.8 give the required fracture size while the dashed lines indicate the corresponding geometry through values of a/b. For $a/b \leqslant 4$ the breakthrough time, t_B, essentially represents the lifetime of the doublet. A scheme with a/b less than two does not seem viable because of the possibility of fracturing between the wells.

When using Figure 7.8 (and Figure 7.9 below) it must be remembered that the range of fracture sizes that can be achieved in a given reservoir will be severely restricted by technical and economic considerations. As a very rough guide, given present-day technology, fracture production is difficult if the permeability is in excess of about 50 mD. Fracture heights up to about 100 m can be achieved (50 m would be more typical), and fracture half-lengths are usually limited to about 300 m. Much larger fractures are technically feasible but the cost is likely to be very high for half lengths of more than 300 m.

As an example, for a reservoir of thickness 50 m with a permeability of 10 mD, a viable scheme could be established under the conditions inset in Figure 7.8a by fracturing to a distance of about 180 m from wells with a separation of about 350 m (a/b = 3). However, if the permeability were only 5 mD the fractures would need to be uneconomically long (half-lengths of about 400 m) and dangerously close together (a/b would be less than about one).

Three-dimensional flow

The second case to be considered is that of parallel circular fractures (Figure 7.7b) of radius b separated by a distance a.

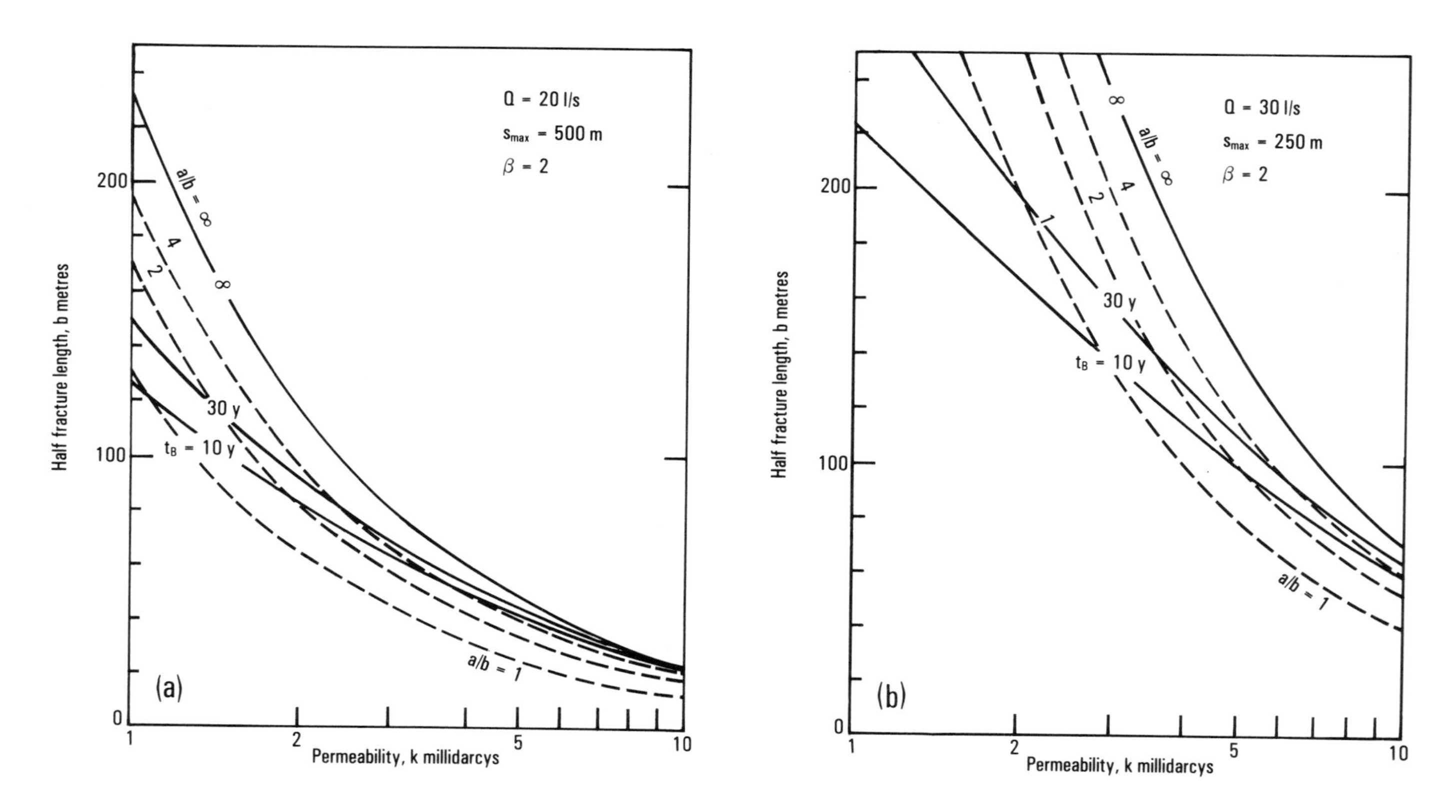

7.8 Minimum fracture lengths (Figure 7.7a) required to limit the drawdown to s_{max} for a breakthrough time t_B. Dashed lines represent contours of a/b.

Following Andrews and others (1981), it is assumed that the distance to any confining layer is large compared with b, so the aquifer can be considered to be of effectively infinite extent (this assumption is discussed below). The drawdown in the abstraction well is then approximated by:

$$s_0 = \frac{Q}{8Kb}\left(1 + \frac{\pi b}{4a}\right)^{-1}, \qquad 7.24$$

where K is the hydraulic conductivity. The breakthrough time can be approximated by:

$$t_B \approx \frac{a^2}{2\beta K s_w}\left[0.18\,\frac{a}{b} + \exp\left(-0.18\,\frac{a}{b}\right)\right] \qquad 7.25$$

These two formulae were used to construct Figure 7.9 which shows the minimum fracture radius required to limit

7.9 Minimum fracture radii (Figure 7.7b) required to limit the drawdown to s_{max} for various breakthrough times. Dashed lines represent contours of a/b.

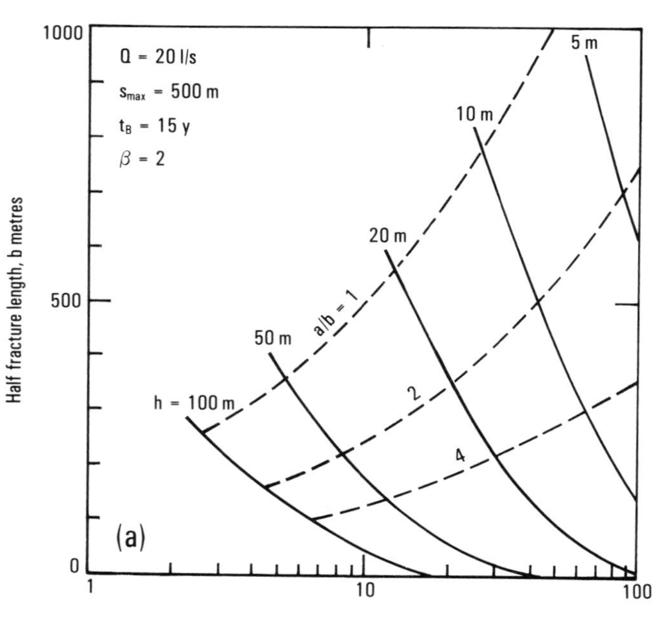

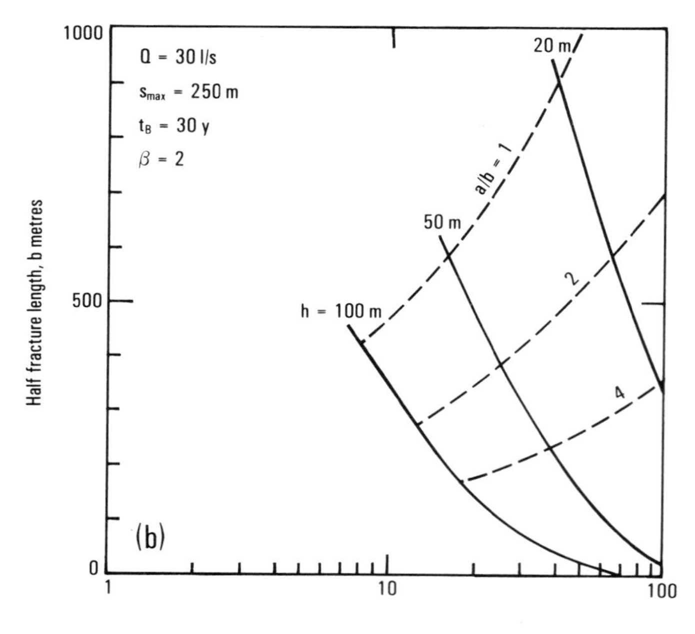

the drawdown to s_{max} for various breakthrough times. The conversion from hydraulic conductivity to permeability was based on a kinematic viscosity of 0.46 mm²/s.

The impression given by Figure 7.9 is that relatively small fractures (b less than about 200 m) can produce long lifetimes with moderate drawdowns. This impression is enhanced when it is noted that the breakthrough of the cold front will not be as sharp as in the previous case, and t_B is therefore an underestimate of the lifetime. However, the following considerations show that these results are, generally, very optimistic.

The geometry of the doublet is characterised by the value of a/b which is contoured in Figure 7.9 using dashed lines. The regions of Figures 7.9a and 7.9b corresponding to values of a/b less than about two may not correspond to viable schemes because of the danger of fracturing between the boreholes. When a/b is larger than about four, Equations 7.24 and 7.25 are not expected to be accurate because the assumption that the aquifer is of infinite extent becomes unreasonable: this assumption results in the prediction that about half of the flow (that is of Q) will be along paths which travel beyond a distance a from the axis of the fractures. This is not to say that viable schemes do not exist for $a/b \geqslant 4$, but rather that the curves in Figure 7.9 are grossly optimistic in this region.

The effects of stratification, which normally result in a very anisotropic bulk permeability, also give reason to question the applicability of Equations 7.24 and 7.25. In the extreme case of the vertical permeability being negligible in comparison with the horizontal permeability, the model depicted in Figure 7.7a replaces the three-dimensional flow model of Figure 7.7b.

MULTIPLE-WELL SCHEMES

It is likely to be desirable for certain geothermal reservoirs to be developed by several wells. Some of the formulae given above can be used to estimate how close wells can be placed before they have any significant interaction during their lifetimes. When wells do interfere it is important to consider their optimal positioning and operation.

Gringarten (1979) investigated a few simple arrangements of identical doublets and concluded that certain symmetric arrangements give the best hydraulic behaviour and longest breakthrough times; thermal behaviour after breakthrough was not considered. With abstraction and injection wells alternating on an infinite square grid (referred to as a *five-spot pattern*) the breakthrough time is 1.5 times that for a doublet with the same separation. However, such an arrangement is rather unrealistic since about 50 wells are required before more than half of them are surrounded by others.

Tsang and others (1977, 1980) have used a numerical model to predict both the hydraulic and thermal behaviour of several complex arrangements of abstraction and recharge wells. However, it is difficult to draw any general conclusions from their results and it appears that each case needs to be studied individually.

The presence of any hydraulic barriers within a reservoir could have a significant effect on the efficiency of any multiple-well scheme. This potential problem does not appear to have been investigated.

SUMMARY

The formulae and graphs that have been presented can be used to estimate the technical viability of various schemes for developing low-temperature geothermal aquifers.

The lifetime of a single-well scheme depends on the time it takes for the drawdown to become too large. This time will be strongly influenced by the presence of any hydraulic barriers but, under some conditions, can be extended by fracturing the well.

The lifetime of a doublet scheme is given by the time it takes for the cooling effects of the injected water to reduce the abstraction temperature to some minimum value. This time will be affected by vertical heat conduction from the surrounding, impermeable rock. Aquifer heterogeneity (particularly stratification) can have a significant effect on the thermal behaviour of a doublet but can only be investigated, theoretically, using complex numerical models.

Fractured-well doublets have been reported as having considerable potential for the development of low-permeability reservoirs (Andrews and others, 1981; White, 1983). However, for a confined aquifer this may require that the fractures should be hazardously close to one another. For three-dimensional flow between disc-shaped fractures, previously used formulae are found to be valid for only a limited range of geometries, and previous predictions of behaviour are generally very optimistic. Therefore, low-permeability reservoirs can be developed, but under restricted conditions which seem to require further study; particularly in relation to the limitations (both technical and economical) of the fracturing process.

Multiple-well schemes appear to require individual study using complex numerical models.

Chapter 8 Assessment of the Geothermal Resources

I. N. Gale and K. E. Rollin

DEFINITIONS AND TERMINOLOGY

Before a realistic assessment can be made of the potential for either low enthalpy or Hot Dry Rock development, it is necessary to estimate the total geothermal resource and its distribution. Resource assessments are undertaken for all types of mineral resources prior to the implementation of major, expensive exploration studies. They are periodically revised as more reliable data become available and as economic conditions change or technology improves. Such assessments are valuable for comparing the resources of different areas and also for comparing different types of resources. For this to be effective a consistent methodology and terminology must be used so that as far as possible reliable direct comparisons can be made.

Classifications of mineral resources make use of three basic terms — *resource base, resource* and *reserve*. The *resource base* refers to the total resource while *resource* relates to the proportion which may be produced under present or future technological or economic conditions and *reserve* is the quantity that has been proven and can be recovered economically at the present time. This general classification has been applied to geothermal resources by Muffler and Cataldi (1978) and Muffler (1979) and the terminology adopted here is essentially the same as that given by Muffler (1979) although there are some differences.

The total *Geothermal Resource Base* includes all the heat stored in both rock and fluid at temperatures above the mean annual surface temperature. However, only heat that is accessible by drilling wells can realistically be considered to be a resource base and therefore the *Accessible Resource Base (ARB)* is defined as the heat stored in rocks and fluids above the practical limit of economic drilling technology which is taken as 7 km. The Accessible Resource Base represents total heat in place within the defined limits but the adjective 'accessible' tends to give a false impression of availability for it does not relate to recoverable heat. It represents an estimate of the total thermal energy stored beneath an area to a depth of 7 km regardless of whether its existence is proven or unproven and regardless of whether it can be extracted economically.

The heat stored in rocks between the 100°C isotherm and a depth of 7 km is referred to as the *Hot Dry Rock Accessible Resource Base (HDR ARB)*. It represents the thermal energy stored in that volume of rock, again with the same qualifications.

The term *Geothermal Resource* (which is not to be confused with Geothermal Resource Base referred to above) is defined as that part of the Accessible Resource Base that could possibly be extracted economically at some specified time in the future. It may be regarded as the 'useful' component of the Accessible Resource Base. The Geothermal Resource for Hot Dry Rocks has not been assessed because the feasibility of the concept on an economic scale remains to be proved. However, in assessments of the potential of this form of energy, a conservative recovery value of 1% has been assumed (ETSU, 1982).

The Geothermal Resource for low enthalpy resources in permeable rocks has been assessed for the Mesozoic basins only. Application to low enthalpy reservoirs necessarily implies that fluids can be withdrawn at appropriate temperatures and at economic rates and, therefore, a minimum acceptable transmissivity of 5 darcy metres (D m) and a minimum temperature of 20°C are prescribed. The depth limit for the reservoirs covered in this report is the base of the reservoirs. This only exceeds 3 km in the Cheshire Basin where the maximum depth of the reservoir is in excess of 4 km.

The Geothermal Resource represents the total heat-in-place in a reservoir within the limits defined. It can be regarded as the low enthalpy accessible resource base for a particular reservoir for, again realistically, all the heat cannot be recovered. For this reason a further sub-category is the *Identified Resource*. This is the proportion of the Geothermal Resource that is more likely to be available for economic exploitation in the future. It is calculated by multiplying the Geothermal Resource by a factor referred to as the maximum recovery factor. The basis for this latter term is discussed in more detail later but it includes the temperature at which the fluid is rejected after use, the hydraulic properties of the reservoir and the method of disposing of the fluid after use. It provides a more realistic assessment of the maximum amount of energy that could be recovered.

Geothermal reserves are the proportion of the Geothermal Resource that could be recovered at costs competitive at the present time with other commercial energy sources. Definition of a reserve implies reasonable certainty in the assessment and proven reserves are based on direct evidence, obtained by drilling and testing, that a geothermal reservoir exists and can be developed.

Exploration in the UK has not been taken to the stage where reserves can be quantified. However, potential reservoirs are known and the most favourable areas for exploration drilling have been identified and are referred to as 'potential geothermal fields'. The criteria for defining these are that the temperature exceeds 40°C and the transmissivity exceeds 10 D m; areas where the temperature exceeds 60°C are also delineated. It must be emphasised that recognition of potential geothermal fields is not based on extensive drilling information but rather on the consensus of the known facts about the regional properties of the reservoir.

ACCESSIBLE RESOURCE BASE

Prediction of sub-surface temperatures

As defined previously, the Accessible Resource Base (ARB) represents the total thermal energy store in an area regardless of the energy demand or extraction costs. The main factors that determine the heat content are the temperature or enthalpy of the rock and the fluid it contains, and the volume of the rock; temperature, however, is the over-riding factor. A graph of temperatures measured in boreholes, corrected to equilibrium values, against depth shows that the least squares fit geothermal gradient is close to 26.4°C/km (see Plate). If this gradient is extrapolated to a depth of 7 km it implies a temperature of 197°C. Assuming a mean rock density of 2.70 g/cm^3 and a mean rock specific heat of 0.84 J/g °C, then an estimate of the heat store to a depth of 7 km in an area of one square kilometre can be derived from the equation for thermal capacity:

Heat store = mass × specific heat × mean temperature

= 14 × 10^5 tera joules (TJ)
= 3.9 × 10^2 tera watt hours (TW h).

This indicates the general principle of the calculation of the ARB and the approximate magnitude of the heat store. The range of rock densities and specific heats is limited, so variation in the magnitude of the ARB is caused largely by variation in the sub-surface temperatures.

There is a substantial variation in the observed mean geothermal gradient (from about 15 to 35°C/km) caused by local variations in heat flow and the thermal conductivity of the rocks. This affects the variation of temperature at depth, and to assess this variation sub-surface temperature profiles have been calculated for each 10 km National Grid square using appropriate heat flow and thermal conductivity data. In all some 2577 approximate temperature profiles were calculated for the UK land area. The temperature was calculated at 1 km intervals to a depth of 7 km and also at intermediate depths on the basis of the surface geology and the inferred deep geological structure in each grid square. Mean values of rock density, thermal conductivity and heat production were allocated to each of the major geological formations in the UK (Gale and others 1984d, appendix 1), and it was assumed that these mean physical properties could be applied nationally. For the purpose of estimating mean temperature differences, it was also assumed that the mean annual temperature in the UK is 10°C.

To calculate the temperature at any depth, it is necessary to know the heat flow and the thermal conductivity and the heat production of the rocks to that depth. Calculation of the sub-surface temperature at any depth is based on a solution to Poisson's equation for steady state conduction, assuming that the thermal conductivity and heat production are dependent only on depth (see Chapter 2). Various models have been used to describe the variation of heat production with depth, although for the uniform, linear, inverse and exponential models the variation in the calculated temperature at a depth of 7 km in granitic terrains is only about 5%. The exponential model has been shown to be valid under conditions of differential erosion of plutonic terrains (Lachenbruch, 1970), and it is consistent with thermodynamic theory (Rybach, 1976). Consequently, heat production in granite and metasedimentary terrains is frequently assumed to decrease exponentially with depth to at least a level D, which is a dimension of length defined by the observed surface heat flow-surface heat production relationship (see Chapter 2).

Over granitic terrains, assuming thermal conductivity is dependent on temperature, the solution for θ_z is given by Equations 2.5 and 2.10, which are repeated here for convenience:

$$\theta_z = a' \exp[(q_0 z - f(z))/(a'\lambda_0)] - b' , \qquad 8.1$$

$$\text{where } f(z) = A_0 D [z - D(1 - \exp(-z/D))] . \qquad 8.2$$

With regard to these equations two points are relevant. Firstly, observed heat flow, q_0, need not appear in the solution for θ_z since $q_0 = q^* + A_0 D$. Secondly, over some granites with high heat production, where the q_0-A_0 relationship has not been verified (that is where q_0 has not been measured and/or q^* is not equal to 27 mW/m^2 or D is not 16.6 km, as proposed by Richardson and Oxburgh (1978)), θ_z will be overestimated. In such cases a new parameter D' has been calculated such that A_0 fits the relationship $q_0 = 27 + A_0 D'$.

To calculate sub-surface temperatures in sedimentary rocks a uniform heat production with depth has been assumed so that the solution for θ_z is given by Equation 8.1 with

$$f(z) = \frac{A_0 z^2}{2} . \qquad 8.3$$

Equations 8.1 and 8.2 or 8.3 have been used to calculate sub-surface temperatures on a 10 km grid across the UK. Since granitic intrusions form only about 4% of the outcrop geology of the UK, the uniform heat production function used in Equation 8.3 has formed the basis of most of the sub-surface temperature calculations.

The predominant surface geology within each 10 km National Grid square was identified by a numerical code identical to that used on the 1:625 000 geological maps of the UK published by the British Geological Survey. This formation code takes values between 1 and 115, and generally increases as the formations become younger.

Values for thermal conductivity, density and heat production were assigned to each of the 115 rock formations (Gale and others 1984d, appendix 1). In general, the thermal conductivity tends to increase with age; conductivities of Mesozoic and Tertiary rocks are usually less than

Palaeozoic rocks. Palaeozoic and older rocks generally have conductivities in the range 2.5 to 3.5 W/m K, although an important exception is the Westphalian division of the Carboniferous. Because of the insulating property of coal, the Westphalian has been assigned a low mean thermal conductivity of 1.91 W/m K.

For the purpose of predicting sub-surface temperatures, three distinct groups of formations have been recognised in terms of thermal conductivity. These are the basement rocks (mostly pre-Carboniferous); the productive Coal Measures (Westphalian) and the Mesozoic and Tertiary sequence. If granitic rocks are also considered as a separate group, for they are significant because of their high heat production, the outcrop geology of the UK can be classified into one of four groups. Using the outcrop geology and these four groups, each grid square can be assigned to one of the five geological sequences or terrains indicated (I to V) in Figure 8.1. From this model temperatures can be calculated at depth as required (Gale and others, 1984d).

The observed and estimated heat flow values in the UK were interpolated by computer to give an estimated value for each 10-kilometre National Grid square. This series of heat flow values, together with the data-set of geological and physical properties, was used to calculate sub-surface temperature profiles for the 2577 grid squares covering the UK and produce maps of the sub-surface temperature distribution.

The calculated temperature at a depth of 7 km is shown in Figure 8.2. Over most of the country where the geology consists of sedimentary rocks, these estimates of sub-surface temperatures are based on generalised interpretations of the geological structure, as deduced from nearby deep boreholes, and generalised values for thermal conductivity. The estimates of sub-surface temperatures for the granite areas are considered to be more reliable since observed values of conductivity and heat production have been used in preference to the means given by Gale and others (1984d, appendix 1). Some detailed temperature profiles at individual sites where heat flow has been observed are given in Chapter 3.

In south-west England temperatures might reach 260°C within the granites, and in the Lake District and north Pennines values could exceed 240°C, again because of the presence of radioactive granites. However, in central-southern England and over much of Wales less than 140°C is more probable.

In Scotland there are relatively few heat flow observations. In the granites of the Eastern Highlands the observed values are all less than 80 mW/m^2, despite very high measurements of surface heat productions. These granites do not conform to the heat flow-heat production relationship of Equation 2.8 using the values published previously for q^* and D of 27 mW/m^2 and 16.6 km respectively (Richardson and Oxburgh, 1978) and would appear to lie in a separate heat flow province characterised by a smaller value of D (see Chapter 3).

Calculated temperatures at a depth of 7 km in the granites of Cairngorm, Mount Battock and Ballater are all below 150°C, because the parameter D' has been used in the calculation of θ_z (Equation 8.1) instead of D (which is 16.6 km). The heat flow-heat production relationship of the Eastern Highlands (Figure 3.12) indicates that the value for D' is close to 6 km if q^* is approximately 35 mW/m^2. If heat production in these granites is assumed to be uniform

8.1 Model of the crust in the UK used for the calculation of the Accessible Resource Base. Sub-surface temperatures were calculated at depth intervals of 1 km in addition to the depths shown on the diagram.

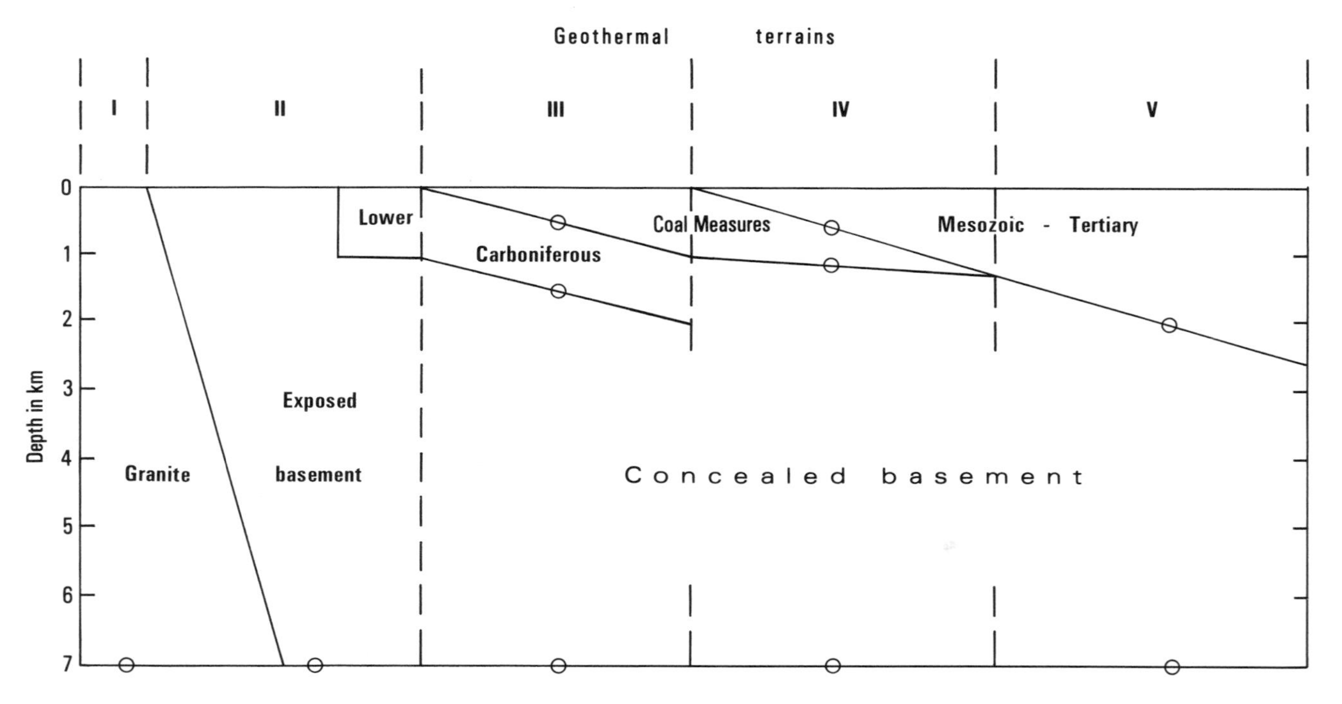

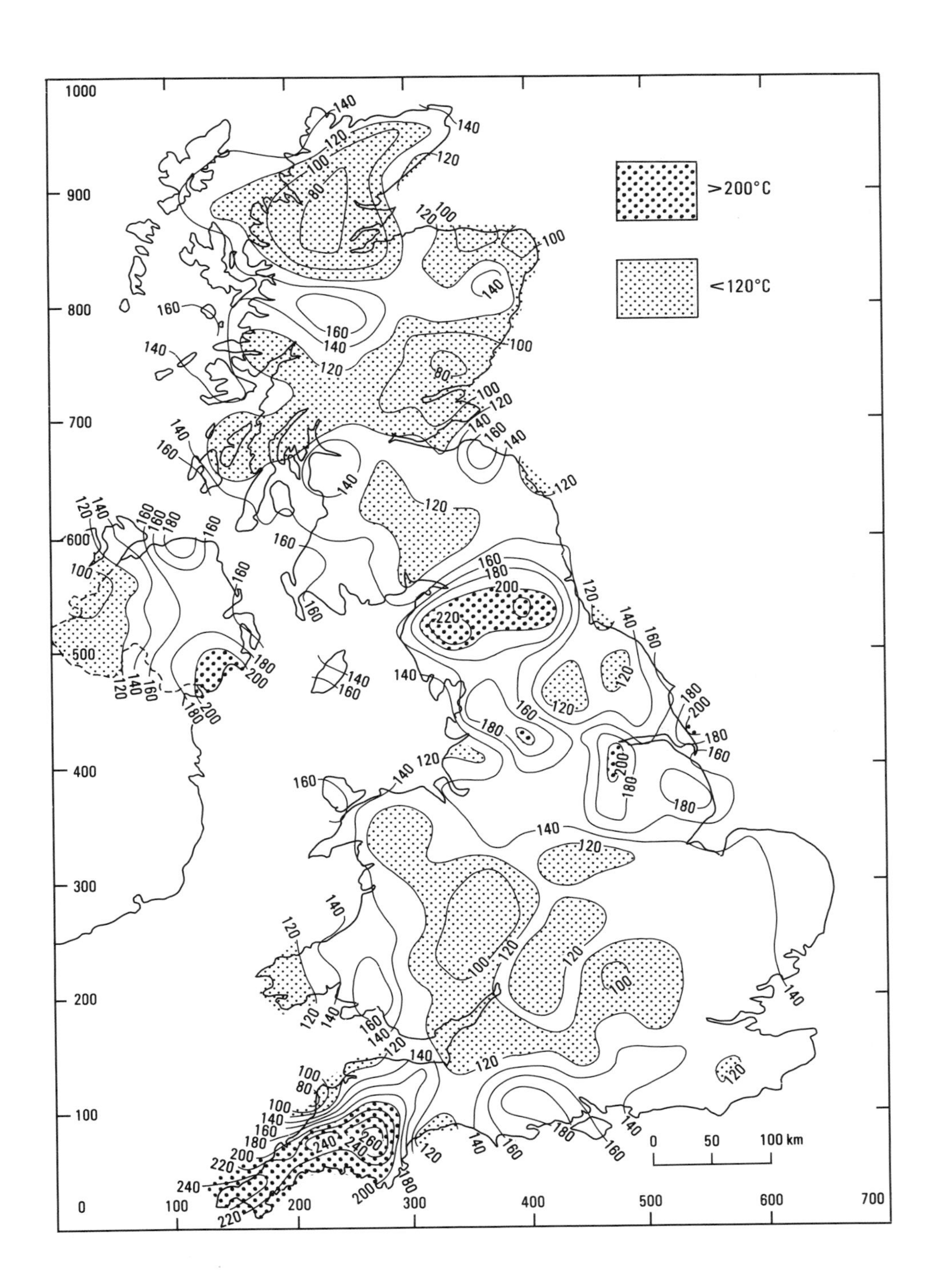

8.2 Estimated temperature at a depth of 7 km (in °C).

with depth (Equation 8.3) then the calculated temperatures at a depth of 7 km for the Cairngorm and Bennachie intrusions would be 120°C and 140°C respectively.

An apparently significant thermal anomaly in the Highlands occurs beneath the Great Glen and requires some explanation. It is caused by two high heat flow observations made in Loch Ness (Appendix 2) which have been manually filtered from the heat flow map shown in Figure 2.8 but which were included in the heat flow data set used for the calculation of subsurface temperatures and resources. Since heat flow measurements made in lake sediments are of dubious reliability, this positive temperature and resource anomaly, which appears on Figures 8.2, 8.3, 8.4 and 8.5, should be treated with caution.

Estimation of the Accessible Resource Base

Over the temperature range of interest, the specific heat of rocks tends to increase with increasing temperature (Kappelmeyer and Haenel, 1974). For each 10 km National Grid square the specific heat was calculated at each of three sub-surface temperatures using the relationship:

$$c_z = c_0(1 + 0.001\,35\,\theta_z) \,, \qquad 8.4$$

where c_0 is the specific heat at 0°C, assumed to be 0.75 J/g °C, and θ_z is the temperature at a depth z. Using the appropriate density values (Gale and others 1984d, appendix 1) to calculate the rock mass, the Accessible Resource Base to 7 km for each grid square was calculated.

The values have been smoothed and contoured to give the map shown in Figure 8.3.

Since only one specific heat was assumed for all rock formations and, since the variation of rock density is not large, the variation of the ARB shows a pattern similar to the temperature distribution. Values range from about 8×10^5 TJ/km^2 in parts of central England to over 20×10^5 TJ/km^2 in south-west England. In the Lake District and north Pennines the ARB exceeds 16×10^5 TJ/km^2 and this value is also attained in eastern England. In southern England, near Bournemouth, values exceed 16×10^5 TJ/km^2, the combined result of an above-average heat flow (about 70 mW/m^2) and a thick sequence of low-conductivity cover rocks producing elevated temperatures at depth.

The total ARB has been estimated to be about 2.75×10^{11} TJ (7.6×10^7 TWh) and the mean value 1.07×10^{18} J/km^2. The regional variation is summarised in Table 8.1 for 13 geographical areas shown in Figure 8.3. The geothermal significance of south-west England, where the mean heat store is about one and a half times the national mean, can be seen from the values in Table 8.1. However, east Yorkshire and Lincolnshire, northern England and Northern Ireland all have values above the national mean.

HOT DRY ROCK ACCESSIBLE RESOURCE BASE

It is generally assumed that energy stored in hot dry rocks at temperatures in excess of 200°C would be used to generate electricity. However, hot dry rocks at lower temperatures could be used for other energy conversions and because of

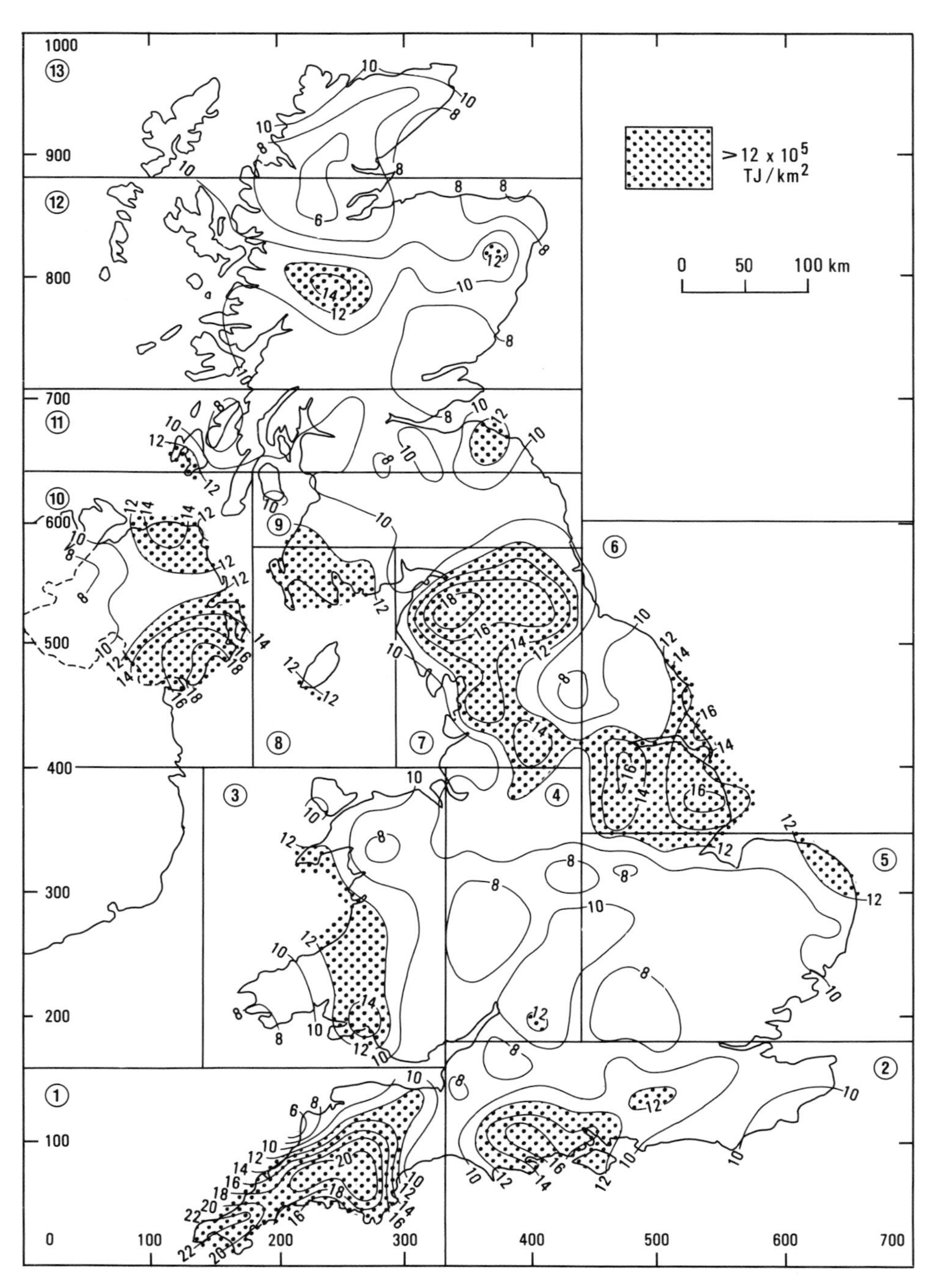

8.3 Accessible Resource Base (in 10^5 TJ/km^2). Figures encircled indicate regions in Table 8.1.

Table 8.1 Regional distribution of the Accessible Resource Base. The regions are shown in Figure 8.3

Region	Area (km^2)	Accessible Resource Base (joules $\times 10^{18}$)	(joules/km^2 $\times 10^{18}$)
1. South-west England	12700	19245	1.52
2. South-east England	25000	27149	1.09
3. Wales	20700	20660	1.00
4. The Midlands	24200	23655	0.98
5. East Anglia	32300	32221	1.00
6. East Yorkshire & Lincolnshire	15400	20170	1.31
7. Northern England	22300	30381	1.36
8. South-west Scotland & Isle of Man	3800	4378	1.15
9. Southern Uplands	12800	12313	0.96
10. Northern Ireland	15800	18501	1.17
11. Midland Valley of Scotland	14600	14086	0.96
12. Grampians	41400	37191	0.90
13. Northern Highlands	16700	15178	0.91
Totals	257700	275126	1.07

The regions are shown in Figure 8.3.

this a minimum temperature of 100°C has been used to define the HDR ARB. Moreover, it should not be overlooked that temperatures as low as 60°C could be used directly for space heating if favourable conditions for applying the HDR concept occurred at shallower depths.

The HDR ARB is the heat stored between the depth of the 100°C isotherm and 7 km, and it has been calculated by extending the methods used to estimate the ARB. Firstly, the depth to the 100°C isotherm was interpolated for each grid square using the calculated sub-surface temperatures (Figure 8.4). Depths greater than 7 km have not been contoured. Sub-surface temperatures in excess of 100°C have been observed at only four sites in the UK so that the contours shown in Figure 8.4 are based almost entirely on calculated values. In some areas the calculated temperature at a depth of 7 km is less than 100°C so these areas have no HDR resource.

Using the depths to the 100, 150 and 200°C isotherms the heat store in each grid square was evaluated for the three temperature zones: above 100°C, above 150°C and above 200°C (Table 8.2). The first of these heat stores is the HDR ARB, and is shown in Figure 8.5. Values range from zero where the temperature at a depth of 7 km is less than 100°C to about 10^{18} J/km^2 over parts of the granite batholith in Cornwall. The main areas where heat is available suitable for future HDR development are south-west England, the north Pennines and also east Yorkshire, Lincolnshire and Wessex. In these areas it would be necessary to drill to depths of 3 or 4 km to attain more than 100°C. The size of the HDR ARB is inversely proportional to the drilling depth necessary to exploit any resources above 100°C; the larger the HDR resource the shallower the drilling depth.

Table 8.2 Regional distribution of the Hot Dry Rock Accessible Resource Base. The regions are shown in Figure 8.5

Region	Area	Hot Dry Rock Accessible Resource Base					
		>100°C		>150°C		>200°C	
	(km^2)	(joules $\times 10^{18}$)	(joules/ $km^2 \times 10^{16}$)	(joules $\times 10^{18}$)	(joules/ $km^2 \times 10^{16}$)	(joules $\times 10^{18}$)	(joules/ $km^2 \times 10^{16}$)
1. South-west England	12700	5279	45.1	2200.0	17.3	564.0	4.44
2. South-east England	25000	2945	11.8	247.0	1.0	3.9	0.02
3. Wales	20700	1898	9.2	66.0	0.3	0.0	0.00
4. The Midlands	24200	2161	8.9	35.9	0.2	0.0	0.00
5. East Anglia	32300	2580	8.0	34.4	0.1	1.1	0.00
6. East Yorkshire and Lincolnshire	15400	3775	24.5	627.0	4.1	35.9	0.23
7. Northern England	22300	6871	30.8	1572.0	7.1	111.0	0.50
8. South-west Scotland & Isle of Man	3800	629	16.6	21.4	0.6	0.0	0.00
9. Southern Uplands of Scotland	12800	889	7.0	8.4	0.1	0.0	0.00
10. Northern Ireland	15800	3200	20.3	514.0	3.3	15.3	0.10
11. Midland Valley	14600	1170	8.0	44.3	0.3	0.0	0.00
12. The Grampians	41400	2797	6.8	105.0	0.3	0.4	0.00
13. Northern Highlands	16700	1160	7.0	0.1	0.0	0.0	0.00
Totals	257700	35804	13.9	5476.0	2.1	732.0	0.30

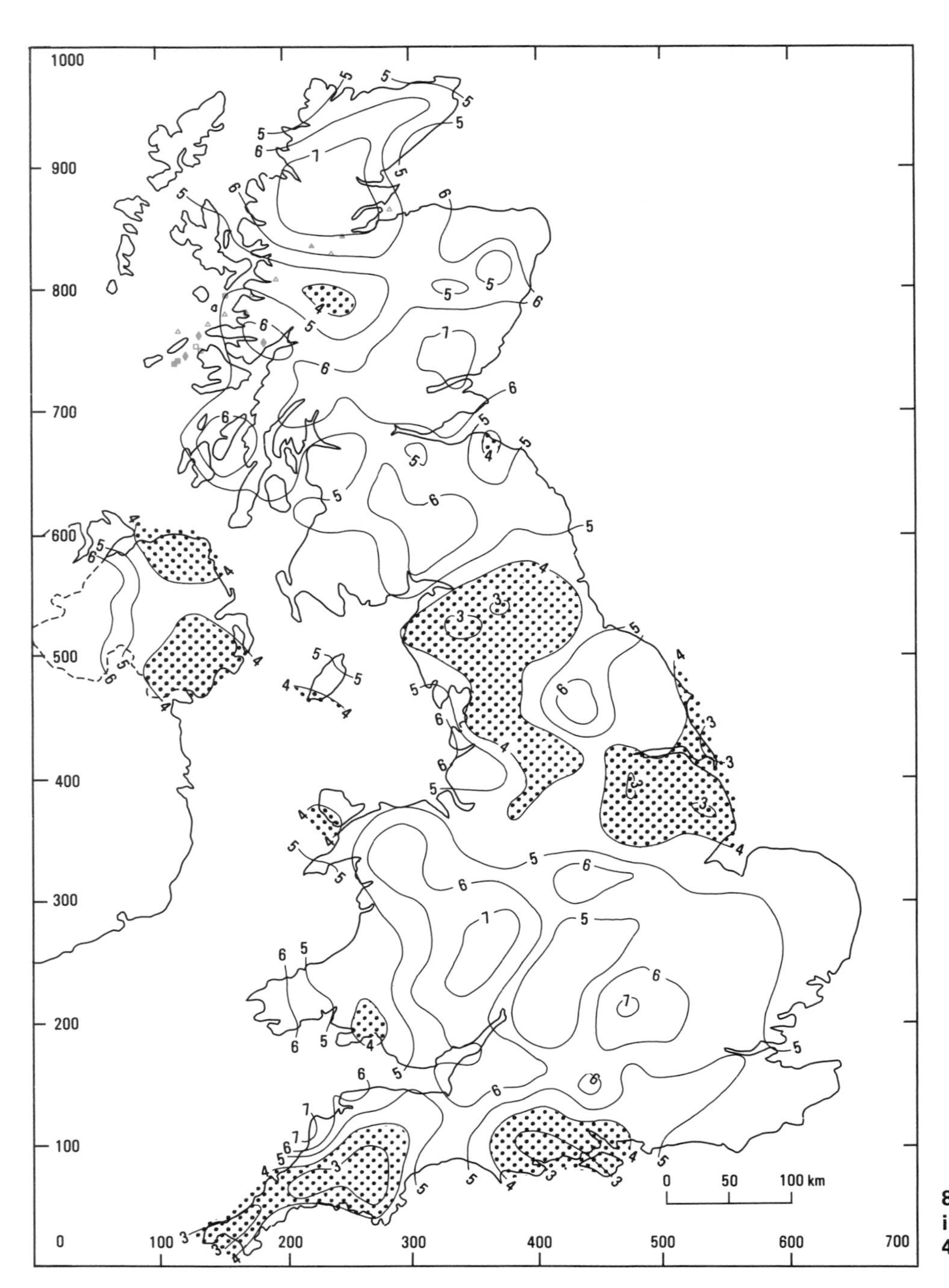

8.4 Depth to the 100°C isotherm in the UK (in km). Depths less than 4 km are shown by stipple.

The total HDR ARB has been calculated to be 3.58×10^{10} TJ, or about one-tenth of the total ARB. Mean values for the 13 geographical regions are given in Table 8.2, together with mean values and totals for the different temperature subdivisions of the resource.

LOW ENTHALPY RESOURCES OF DEEP AQUIFERS

Method of calculating resources

To make realistic assessments of the geothermal resources of deep aquifers, practical values for transmissivity and sub-surface temperature have to be used to define the limits of the resource. The values adopted are a transmissivity of at least 5 D m and a mid-point temperature, for the productive zone in the aquifer, of 20°C or greater.

The Geothermal Resource (H_0) may be calculated from:

$$H_0 = [\phi \varrho_w c_w + (1 - \phi)\, \varrho_m c_m] V(\theta_m - \theta_g) \text{ joules} \qquad 8.5$$

where ϕ = porosity
ϱ_w and ϱ_m are densities of water and matrix respectively in g/m³
c_w and c_m are specific heats of water and matrix respectively in J/g °C
V = volume of aquifer in m³
θ_m = mean reservoir temperature in °C
θ_g = mean annual ground temperature in °C (assumed to be 10°C).

A sensitivity analysis of errors involved in estimating the

various parameters indicated that, in the United Kingdom, the equation may be reduced, without undue loss of accuracy, to:

$$H_0 = 2.6 \times 10^6 V(\theta_m - 10) \text{ joules} \quad 8.6$$

The Geothermal Resources represent the total heat contained in aquifers. The proportion of the Geothermal Resources more likely to be available for development is the Identified Resources which are calculated by multiplying the Geothermal Resources by a maximum recovery factor, defined as (Lavigne, 1978):

$$\left[\theta_m - \theta_{rej})/(\theta_m - \theta_g)\right]F \quad 8.7$$

where θ_{rej} is the reject temperature of the heat extraction system and F is a factor related to the hydraulic properties of the aquifer and to the method of abstracting the hot water. In the Dogger aquifer in France, where water is reinjected into the aquifer after use, a value of 0.33 has been found to be applicable for F as, when a doublet system is used, more heat is swept from the aquifer by the reinjected water. In the absence of specific studies in the UK, a value of 0.25 has been used for a hypothetical doublet system but if abstraction is from a single well (as for example where disposal of brine into the sea at a coastal site would be possible) then the sweeping effect would not take place and a value of 0.1 has been assumed. Both values (0.1 and 0.25) have been used with reject temperatures of 10°C and 30°C to calculate the Identified Resources of an aquifer. The two reject temperatures relate to systems incorporating heat pumps and to those using the heat directly.

In order to identify areas of greatest geothermal potential a predicted well productivity has been calculated where

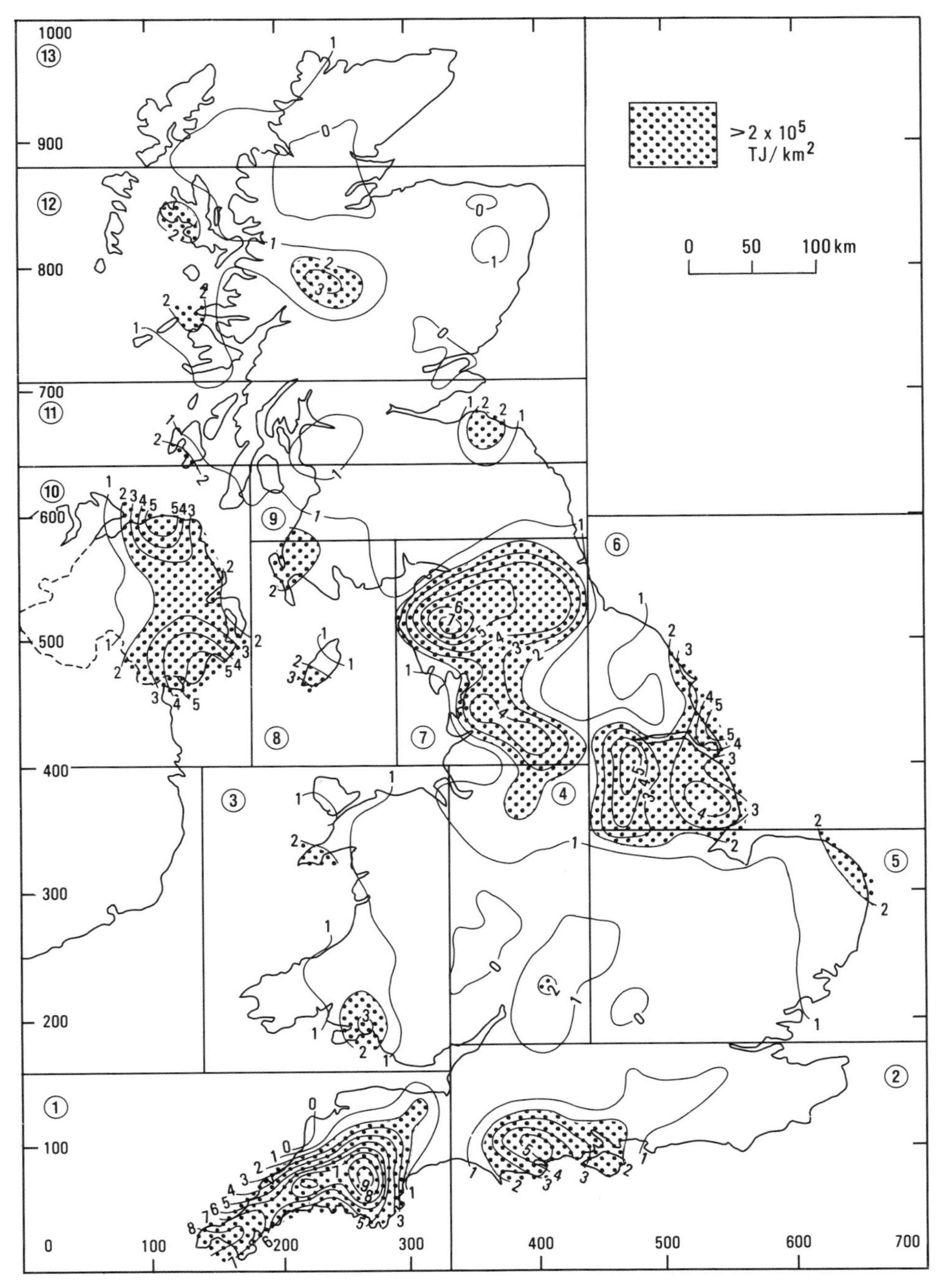

8.5 Hot Dry Rock Accessible Resource Base showing regions referred to in Table 8.2 (in 10^5 TJ/km^2).

sufficient data are available. It is expressed as Unit Power and defined as the quantity of thermal power that can be obtained from a well per unit loss of head, per unit drop in the temperature of the produced water. Unit power was calculated assuming efficient installation of a well, 310 mm in diameter, pumped under steady state conditions, from:

$$\text{Unit Power} = 7.07 \times 10^{-5}(kh\varrho_w c_w/\mu) \text{ MW/bar °C} \quad 8.8$$

where k is permeability in darcys
h is the aquifer thickness in m
ϱ_w is the density of the fluid g/cm^3
c_w is the specific heat of the fluid in J/g °C
μ is the dynamic viscosity in centipoises.

The distribution of Geothermal Resources in an aquifer can also be expressed as the quantity of power per unit area, in megawatt years per square kilometre (MW a/km^2), calculated from:

$$H_0/(3.15 \times 10^{13} \times \text{area}) \quad 8.9$$

where the area is in square kilometres. This is equivalent to

$$8.25 \times 10^{-2}\ h(\theta_m - 10°\text{C}) \text{ MW a/km}^2. \quad 8.10$$

Both of these methods have been used, where sufficient data exist, to illustrate the distribution of the resources of the aquifers by estimating the various parameters for each square in a grid pattern of suitable size. Where data are too limited for this, a mean value for the aquifer or temperature zone has been calculated.

Insufficient data exist to estimate the geothermal reserves of aquifers but, as previously mentioned, areas of greatest geothermal potential have been identified and are referred to as 'potential geothermal fields'. They have been defined where the temperature exceeds 40°C and the transmissivity exceeds 10 D m; but the definition of such areas is not always based on extensive drilling information.

Using the methods and resource limits outlined above, the low enthalpy geothermal resources of the Permo-Triassic and Lower Cretaceous sandstones have been calculated for each basin in the United Kingdom. It should be appreciated that it would be impracticable to develop all the Identified Resources as this would require many wells distributed over an entire aquifer, where heat loads would not necessarily exist. Realistically the energy that could be used is smaller than the estimates given.

GEOTHERMAL RESOURCES OF THE PERMO-TRIASSIC SANDSTONES

East Yorkshire and Lincolnshire Basin

The only aquifers with significant geothermal potential in this basin are the Triassic Sherwood Sandstone Group and the Basal Permian Sands (Gale and others, 1983). Existing data on the permeability of these aquifers indicate that, at depths where temperatures are suitable for development, average values for permeability of 250 and 150 mD may be applied to the Sherwood Sandstone Group and the Basal Permian Sands respectively. Because of the great thickness of the Sherwood Sandstone Group, the extent of the geothermal resource is limited by the 20°C isotherm but, as the Basal Permian Sands are relatively thin and very variable in thickness, the resource in this aquifer is limited to where the water-bearing thickness exceeds 33 m, that is in north-east and east Lincolnshire.

The heat flow is above the mean value for the UK in Nottinghamshire and Lincolnshire and this, together with the effect of the thermal insulation of the sediments overlying the Sherwood Sandstone Group, results in thermal gradients of 26 to 30°C/km. The estimated temperatures at

Table 8.3a Geothermal Resources of the East Yorkshire and Lincolnshire Basin

Temperature zone (°C)	Area (km^2)	Thickness (km)	Mean temperature (°C)	Geothermal Resources (J × 10^{18})	Geothermal Resources (MWa/km^2)
Triassic Sherwood Sandstone Group					
20–40	9570	0.24	30	120	400
40–60	2900	0.38	45	99	1084
Basal Permian Sands					
20–40 (Norfolk)	510	0.06	35	2.0	125
40–60 (Lincolnshire)	1060	0.05	53	5.4	162
>60 (Lincolnshire)	210	0.05	61	1.3	200

Total Geothermal Resources: Triassic = 220 × 10^{18} joules
Permian = 8.7 × 10^{18} joules

Table 8.3b Identified Resources of the East Yorkshire and Lincolnshire Basin

Temperature zone (°C)	Identified Resources (joules × 10^{18}) Reject temperature: 10°C F=0.1	Reject temperature: 10°C F=0.25	Reject temperature: 30°C F=0.1	Reject temperature: 30°C F=0.25
Triassic Sherwood Sandstone Group				
20–40	12.0	30.0	–	–
40–60	9.9	24.8	4.2	10.6
Totals	21.9	54.8	4.2	10.6
Basal Permian Sands				
20–40 (Norfolk)	0.2	0.5	0.04	0.1
40–60 (Lincolnshire)	0.54	1.4	0.29	0.72
>60 (Lincolnshire)	0.13	0.33	0.08	0.2
Totals	0.87	2.23	0.41	1.02
Total Identified Resources	22.8	57.0	4.6	11.6

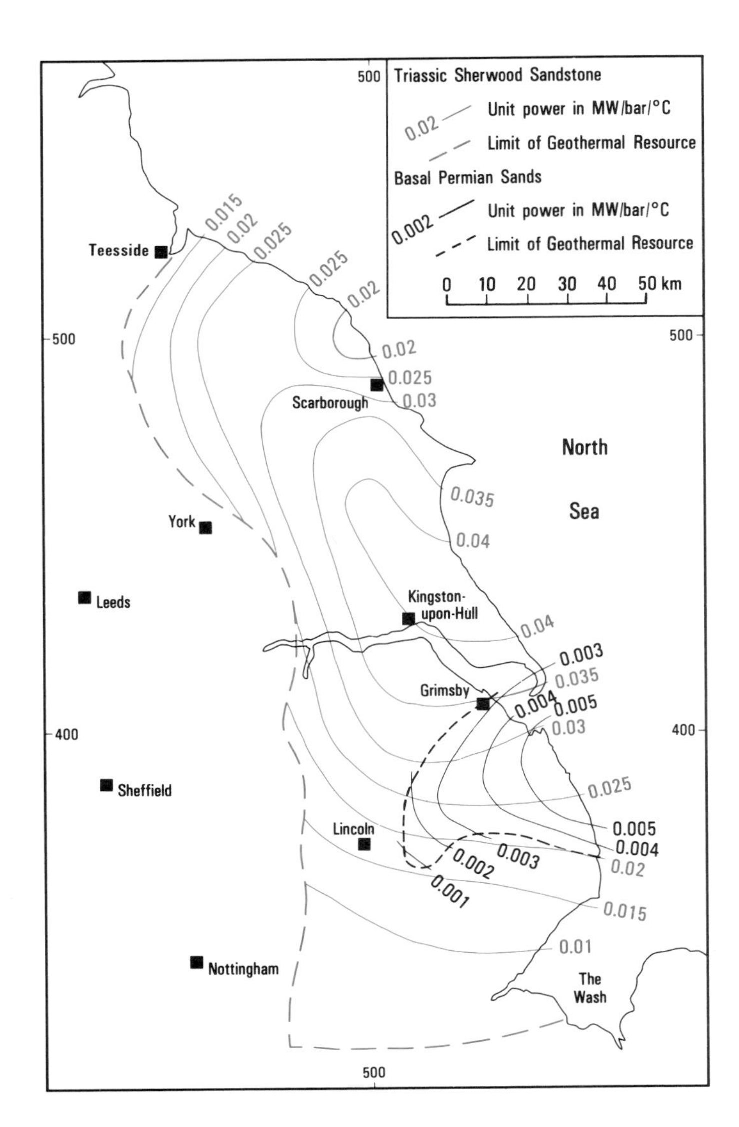

8.6 Unit power of the Sherwood Sandstone and the Basal Permian Sands in the East Yorkshire and Lincolnshire Basin.

the mid-point of the Sherwood Sandstone Group and at the base of the Lower Permian are shown in Figures 4.6 and 4.9 respectively. Both sets of isotherms increase towards the east, the Sherwood Sandstone Group attaining 50°C and the Basal Permian Sands 70°C in Humberside. A breakdown of the calculated Geothermal and Identified resources is given in Tables 8.3a and b which also give the main distribution of the resources for the various temperature zones. In the Sherwood Sandstone Group the Geothermal Resources exceed 140 MW a/km^2 along the east Yorkshire coast from Flamborough Head to Holderness, reflecting the location of the greatest thickness of sandstone and the highest temperatures. Despite the higher temperature found in the underlying Basal Permian Sands the Geothermal Resources only exceed 200 MW a/km^2 along the Lincolnshire coast.

In order to estimate the probable thermal yield of a well at a particular site the Unit Power has been calculated for both aquifers (Figure 8.6). In the Sherwood Sandstone Group it increases towards the coast from 10^{-2} to 4×10^{-2} MW/bar °C and in the Basal Permian Sands from 10^{-3} to 5×10^{-3} MW/bar °C.

The 'potential geothermal fields' in the East Yorkshire and Lincolnshire Basin are shown in Figure 8.7. Although the Sherwood Sandstone Group is the only aquifer that meets the defined criteria (that is, temperatures greater than 40°C and transmissivity greater than 10 D m) the area where the Lower Permian aquifer exceeds 40°C and has a transmissivity in excess of 5 D m is also indicated, together with the major towns in the area which would be potential demand centres for geothermal energy.

Four towns, Scarborough, Bridlington, Kingston-upon-Hull and Grimsby, lie within, or partially within, the area underlain by aquifers that are classified as representing potential geothermal fields. Estimated aquifer parameters for each of these locations and the calculated thermal yield expected under different abstraction conditions are given in Table 8.4. All four are situated so that brine could be discharged into the North Sea or the Humber Estuary and so only single well completion has been considered, although reinjection would be preferable for long-term development. For a pumping rate of 30 l/s from the Sherwood Sandstone Group the thermal yield would range from 1.3 to 2.5 MW. These yields would require quite modest water level drawdowns and if they were increased to only 100 m the thermal yields would rise to between 2.2 and 3.5 MW (Table 8.4).

Although the Lower Permian aquifer has an estimated mean temperature some 15°C higher than the Sherwood Sandstone, its geothermal potential is smaller because of its lower permeability and thickness. However, if both aquifers were pumped at the same rate the thermal yield

8.7 Potential geothermal fields in the East Yorkshire and Lincolnshire Basin.

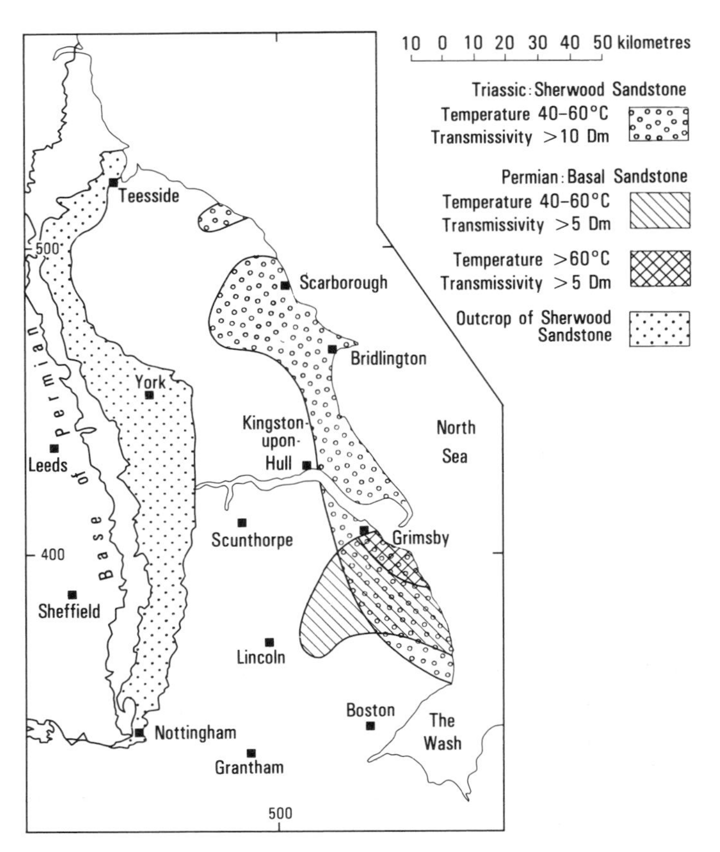

Location	Aquifer*	Thickness† (m)	Mean temperature (°C)	Thermal yield‡(MW) If yield is 30 l/s	If draw down is 100 m (11 bars)
Scarborough	SS	200	45	1.9	2.6
Bridlington	SS	300	43	1.6	3.4
Kingston-upon-Hull	SS	250	40	1.3	2.2
Grimsby	SS	200	50	2.5	3.5
	BPS	35	65	4.0	0.6

* Aquifer: SS Triassic Sherwood Sandstone
BPS Basal Permian Sands

† Thickness is estimated total water-bearing thickness

‡ Assuming rejection temperature of 30°C

Table 8.4 Estimated thermal yields of aquifers at population centres in East Yorkshire and Humberside

from the Lower Permian would be approximately twice that of the Sherwood Sandstone although the resultant drawdown would be ten times greater in the Lower Permian, thus increasing the pumping costs.

Since the assessment for Grimsby was prepared (Table 8.4) a geothermal exploration well has been drilled in Cleethorpes. The Sherwood Sandstone proved to be 400 m thick, and the total thickness of the main water-bearing horizons is about 250 m. The transmissivity, derived from a gas-lift test, was 60 Dm although analysis of core samples and interpretation of geophysical logs suggests it could be as high as 200 D m. A short production test by gas lifting with nitrogen yielded 20 l/s for a drawdown of less than 7 m. This implies a long-term potential of some 30 l/s for a pumping water level of 70 m below the surface after production for about 20 years, assuming development with a single well. The temperature of the water at the level of the aquifer during the production test was 53°C.

The potential of the deeper Basal Permian Sands was disappointing. The permeable horizons were only 8.5 m thick and the total transmissivity was less than 2 D m although the temperature of the reservoir was 64°C. The sequence is possibly unrepresentative of east Lincolnshire as a whole for it comprised only 26 m of fluviatile sandstones and conglomerates.

Because of the lower temperature of the Sherwood Sandstone, development would require the conjunctive use of heat pumps and larger volumes of water would have to be pumped for a given thermal yield. The higher thermal yields derived from a drawdown of 100 m, given in Table 8.4, would require abstraction rates of about 60 l/s, which is probably the maximum feasible pumping capacity for a typical deep well. Because there are so few measured reliable data from the Sherwood Sandstone and the Lower Permian aquifers in the deeper parts of the East Yorkshire and Lincolnshire Basin, it must be stressed that all the above calculations and estimates are based on estimated parameters. Estimates of temperature, aquifer geometry and water chemistry are regarded as reasonable and any errors are unlikely to cause great differences in the resultant thermal yield and drawdown. However, there are few permeability data, and because of the presence of cement, clay content and variation in grain size and lithofacies, values for permeability can vary considerably regionally and even locally; the estimated values could be an order of magnitude in error with a proportional effect, either increase or decrease, on the calculated thermal yield, discharge and drawdown.

Wessex Basin

In the Wessex Basin the major geothermal resource is in the Sherwood Sandstone Group which attains thicknesses of up to 250 m (Allen and Holloway, 1984). However, only part of the formation is water-bearing and the aquifer is actually represented by thin discrete sandstone units with porosities up to 25% and permeabilities up to 5 D. The total transmissivity exceeds 10 D m over appreciable areas (Figure 4.14). The temperatures at the centre of the Sherwood Sandstone Group range from 10°C at outcrop to over 80°C in the deepest parts of the basin (Figure 4.16). The average temperature gradient is 30°C/km, which is significantly above the average for the UK. The heat flow is greater than 60 mW/m^2 and attains 70 mW/m^2 in localised areas.

A summary of the Geothermal Resources and the Identified Resources is given in Tables 8.5a and b. The total Geothermal Resources amount to 26 $\times$ 10^{18} joules, and the Identified Resources for the 20 to 40°C, 40 to 60°C and over 60°C temperature zones are 0.8 $\times$ 10^{18}, 2.8 $\times$ 10^{18} and 2.9 $\times$ 10^{18} joules respectively, if a doublet system is used and if a reject temperature of 10°C can be attained by using heat pumps.

Table 8.5a Geothermal Resources of the Sherwood Sandstone Group in the Wessex Basin

Temperature zone (°C)	Area (km)^2	Mean thickness (m)	Mean temperature (°C)	Geothermal Resources (J $\times$ 10^{18})	(MWa/km^2)
20–40	990	58	31	3.1	97
40–60	1590	68	50	11.3	225
>60	1580	49	67	11.6	233

Total Geothermal Resources = 26 $\times$ 10^{18} joules

Table 8.5b Identified Resources of the Sherwood Sandstone Group in the Wessex Basin

Temperature zone (°C)	Identified Resources (J $\times$ 10^{18}) Reject temperature: 10°C F = 0.1	F = 0.25	Reject temperature: 30°C F = 0.1	F = 0.25
20–40	0.31	0.78	–	–
40–60	1.13	2.83	0.57	1.41
>60	1.16	2.90	0.74	1.85
Totals	2.60	6.51	1.31	3.26

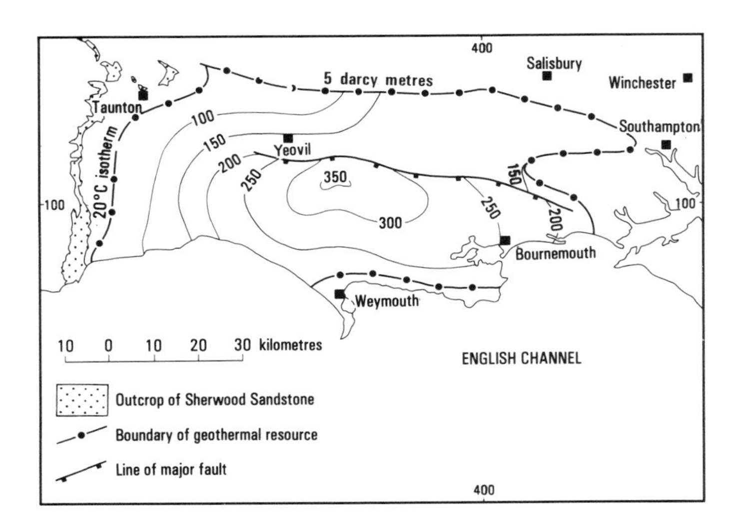

8.8 Geothermal Resources of the Sherwood Sandstone Group in the Wessex Basin (in MW a/km^2).

Well location	Discharge (l/s)	Aquifer tempera-ture (°C)	Thermal yield (MW) Reject temperature 30°C	10°C
East Bourne-mouth	13	65	1.8	2.8
Dorchester	25	55	2.4	4.3
Bridport	39	45	2.4	5.4
Lyme Regis	41	35	0.9	4.1

Table 8.6 Estimated thermal yield at selected sites in the Wessex Basin when a drawdown of 100 m is imposed

The resource distribution across the Wessex Basin is shown in Figure 8.8. The maximum resource is to the north of Dorchester, where the temperature at the centre of the Sherwood Sandstone Group is about 60°C. Values decrease to the west, mainly because the temperature decreases as the sandstone becomes shallower, and to the east, north and south because the aquifer thins in these directions.

A Unit Power distribution map for the Wessex Basin (Figure 8.9) has been produced from transmissivity (Figure 4.14), temperature (Figure 4.16) and salinity data (Figure 4.15). The highest values are in the extreme west of the basin, and the lowest in the northern and eastern parts, a variation that is similar to that for transmissivity, showing the interdependence of the two. Thermal yields have been estimated for four sites in the basin to demonstrate the regional variation (Table 8.6). The information in the table does not enable the relative merits of the different sites to be fully compared because it does not consider factors such as the different costs of drilling at each site or the variation in operating costs, such as pumping costs and the costs of brine disposal or reinjection. However, it does indicate that significant thermal power outputs could be obtained from wells over a large part of the Dorset Basin for a drawdown of 100 m and greater thermal yields than those given in the table could be produced if larger drawdowns were imposed.

The potential geothermal field in the Wessex Basin is shown in Figure 8.10 to extend from west of Bridport to east of Bournemouth and as far north as Sherborne. As the transmissivity decreases towards the east while the temperature increases, the optimum thermal yield from a site depends on the correct balance of drawdown, depth to the aquifer and the temperature at which the water is rejected after use. The Sherwood Sandstone contains brines with salinities up to 300 g/l and this would result in low piezometric levels, as much as 200 m below surface, which would significantly increase pumping costs. In addition the apparently intercalated nature of the water-bearing horizons make it unlikely that individual thin, high permeability, water-bearing zones would extend for more than a few kilometres without meeting some form of hydraulic boundary. In these circumstances a reinjection well would be necessary to maintain reservoir pressure. In areas where the aquifer is thick and transmissivity is high, cross-connection of such discrete zones may be sufficient to maintain hydraulic continuity throughout the aquifer.

8.9 Unit power of the Sherwood Sandstone Group in the Wessex Basin (in MW/bar °C).

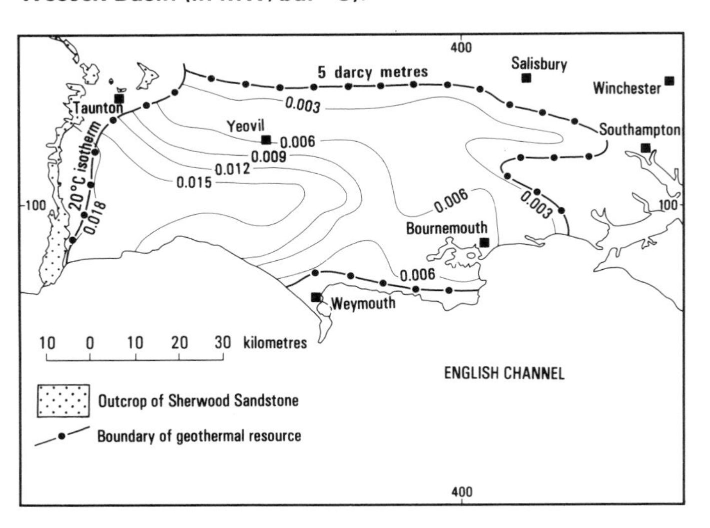

8.10 Potential geothermal fields in the Wessex Basin.

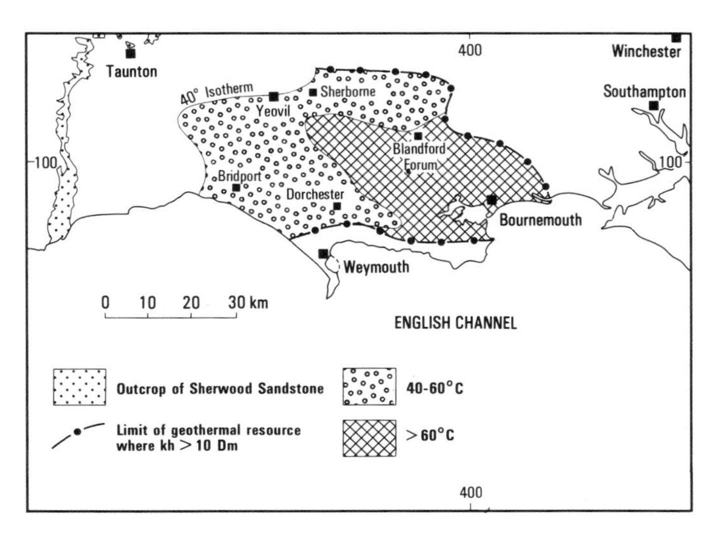

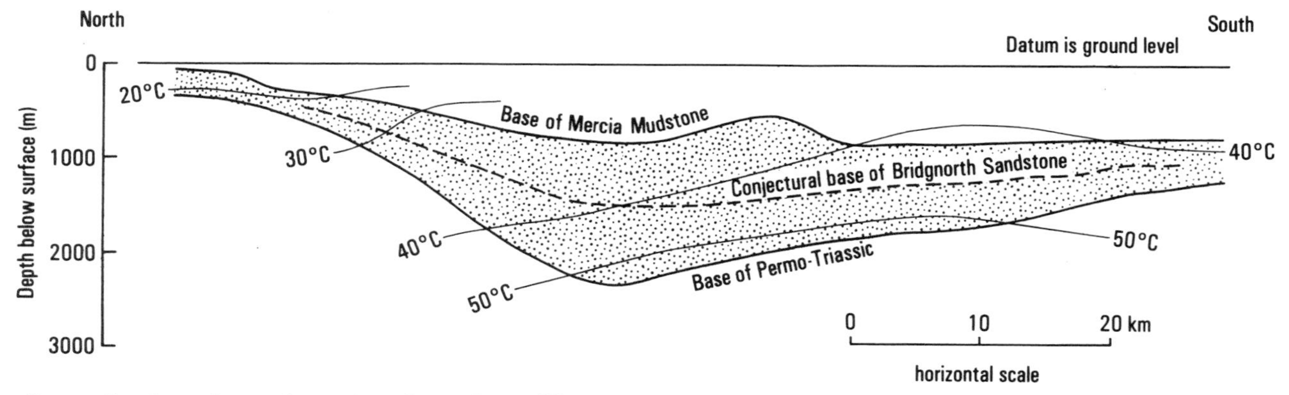

8.11 Generalised north-south section through the Worcester Basin.

Worcester Basin

The Lower Permian Bridgnorth Sandstone and the Triassic Bromsgrove and Wildmoor sandstone formations, which are part of the Sherwood Sandstone Group, are the aquifers with geothermal potential in the Worcester Basin. The Permo-Triassic sandstones attain thicknesses in excess of 2000 m and over much of the basin exceed 1000 m (Smith and Burgess, 1984). Mean permeability is probably about 100 mD, indicating a transmissivity of up to 100 D m for the Bridgnorth Sandstone and about 80 D m for the Triassic sandstones.

The layer of thermal insulating mudstones overlying the Triassic sandstones is relatively thin and this influences the temperature at the top of the sandstones, the range being 20 to rather more than 40°C (Figure 4.21). Furthermore, the temperature gradient in the sandstones is relatively low, generally of the order of 18°C/km. So, despite the great thickness of the sandstones, it is unlikely that 60°C is attained anywhere in the basin (Figure 4.22 and Figure 8.11). Temperatures at the top of the sandstones exceed 40°C in a small area to the north-east of Cirencester but more than 30°C is found over an extensive part of the basin (Figure 4.21 and 4.22).

The Geothermal Resources and the Identified Resources for the Permo-Triassic sandstones in the basin are given in Tables 8.7a and b. The total Geothermal Resources amount to 85 $\times$ 10^{18} joules; the Identified Resources for temperature ranges of 20 to 40°C and 40 to 60°C are 18 $\times$ 10^{18} and 3 $\times$ 10^{18} joules respectively, if a doublet system is employed and the use of heat pumps allows a reject temperature of about 10°C.

In order to develop low enthalpy geothermal resources for direct use a minimum temperature of 60°C is required and, as this is unlikely to be achieved anywhere in the basin, a heat pump would probably have to be incorporated into any development system. The geological formations are relatively permeable and large quantities of groundwater are probably available in the range 30 to 50°C, from which substantial thermal yields could be maintained; Table 8.8 indicates the Unit Power possible from different sandstones in the basin.

The largest thermal yields may be expected from the Bridgnorth Sandstone near the northern gravity minimum at Worcester and from the Wildmoor and Bromsgrove sandstones in the south of the basin. Because of the increasing thickness of the Mercia Mudstone Group and Jurassic

Table 8.7a Geothermal Resources of the Worcester Basin

Temperature zone (°C)	Area (km)^2	Thickness (km)	Mean temperature (°C)	Geothermal Resources (J $\times 10^{18}$)	Geothermal Resources (MWa/km^2)
20–40	1350	1.09	30	72.7	1710
40–60	500	0.25	48	12.0	763

Total Geothermal Resources = 84.7 $\times 10^{18}$ joules

Table 8.7b Identified Resources of the Worcester Basin

Temperature zone (°C)	Mean temperature (°C)	Identified Resources (J $\times 10^{18}$): Reject temperature 10°C, F = 0.1	Reject temperature 10°C, F = 0.25	Reject temperature 30°C, F = 0.1	Reject temperature 30°C, F = 0.25
20–40	30	7.3	18.2	–	–
40–60	48	1.2	3.0	0.6	1.4
Totals		8.5	21.2	0.6	1.4

Table 8.8 Unit power at selected sites in the Worcester Basin

Location	Formation	Depth (m)	Unit Power (MW/bar°C $\times 10^{-3}$)	Expected temperature (°C)
Worcester	Bromsgrove/ Wildmoor	1200	30	25
Worcester	Bridgnorth	2300	50	40
Evesham	Bromsgrove/ Wildmoor	1300	35	32
Evesham	Bridgnorth	1770	15	42
Cheltenham	Permo-Triassic sandstones	1280	30	33
Cirencester	Permo-Triassic sandstones	1500	30	42

strata towards the south, temperatures in the Wildmoor and Bromsgrove sandstones in the south (about 45°C) are actually higher than in the statigraphically lower Bridgnorth Sandstone in the north of the basin (about 40°C). The lower temperature of the Bridgnorth Sandstone near Worcester is offset by its greater thickness and consequently higher transmissivity. The Wildmoor and Bromsgrove sandstones in the south of the basin, say west of Stow-in-the-Wold, could be exploited by a borehole about 1700 m deep, whereas to exploit the Bridgnorth Sandstone near Worcester a 2300 m borehole would be necessary.

Clearly, the Worcester Basin offers a geothermal resource only if heat pumps become economical. The high transmissivity should ensure adequate yields from boreholes in the temperature range 30 to 45°C over most of the area, with higher temperatures available locally. However, most of the area is rural in character, although Worcester, Cheltenham and Cirencester represent potential heat loads.

Cheshire and West Lancashire basins

The potential geothermal aquifers in the Cheshire and West Lancashire basins are the Triassic Helsby, Wilmslow and Kinnerton sandstones and the Permian Collyhurst Sandstone (Gale and others, 1984a). The Manchester Marl acts as an impermeable barrier between the Triassic and Permian sandstones in the West Lancashire Basin and in the northern and central parts of the Cheshire Basin. In the southern part of the Cheshire Basin, where the Manchester Marl does not occur, two aquifers can be recognised. The upper is in the Sherwood Sandstone and the lower is mainly Collyhurst Sandstone but includes some Triassic sandstones and for this reason is referred to as the Permo-Triassic sandstones aquifer.

In the Cheshire Basin, because of the great thickness of the sandstone units and the uncertainty of where, within a unit, the main water-bearing intervals occur, temperatures 500 m below the top of the Tarporley Siltstone and 250 m above the base of the Permo-Triassic sandstones were chosen as most representative of average temperatures of the Sherwood Sandstone Group and the Permo-Triassic sandstones respectively for the purpose of estimating the geothermal resources. The temperature of the Sherwood Sandstone Group exceeds 40°C over an area of approximately 550 km^2 and in the deeper Permo-Triassic aquifer exceeds 60°C over an area of approximately 320 km^2. The cumulative thickness of water-bearing sandstones in the Sherwood Sandstone Group and the Permo-Triassic sandstones is believed to be 300 and 200 m respectively. In the West Lancashire Basin maximum temperatures are unlikely to attain 40°C although an area of approximately 450 km^2 may be underlain by aquifers with mean temperatures of 20 to 40°C.

The estimated Geothermal and Identified Resources of the Sherwood Sandstone and the Permo-Triassic sandstones in the Cheshire and West Lancashire basins are shown in Tables 8.9a and b. Development of the resources in the West Lancashire Basin and of the major part of the resource of the Sherwood Sandstone in the Cheshire Basin would require the use of heat pumps. The deeper Permo-Triassic sandstones in the Cheshire Basin contain water at more than 60°C over some 320 km^2 but they occur at depths of 3000 m because of the thick sequence of sandstones with low thermal gradients. The higher temperatures occur below rural areas except for Crewe and Northwich. Below

Temperature zone (°C)	Area (km^2)	Thickness (km)	Mean temperature (°C)	Geothermal Resources (J × 10^{18})	Geothermal Resources (MWa/km^2)
CHESHIRE BASIN					
Triassic Sherwood Sandstone Group					
20–40	820	0.3	30	12.8	495
40–60	540	0.3	50	16.9	990
Permo-Triassic sandstones (undifferentiated)					
20–40	640	0.2	30	6.7	330
40–60	900	0.2	50	18.7	660
>60	320	0.2	65	9.2	908
Total Geothermal Resources: of Cheshire Basin:				Triassic = 30 × 10^{18} joules	Permian = 35 × 10^{18} joules
WEST LANCASHIRE BASIN					
Permo-Triassic sandstones (undifferentiated)					
20–40	450	0.1	28	2.1	150
Total Geothermal Resources of Cheshire and West Lancashire basins:		67 × 10^{18} joules			

Table 8.9a Geothermal Resources of the Cheshire and West Lancashire basins

Temperature zone (°C)	Identified Resources (J × 10^{18}) Reject temperature 10°C F = 0.1	Reject temperature 10°C F = 0.25	Reject temperature 30°C F = 0.1	Reject temperature 30°C F = 0.25
CHESHIRE BASIN				
Triassic Sherwood Sandstone Group				
20–40	1.3	3.2	–	–
40–60	1.7	4.2	0.9	2.1
Totals	3.0	7.4	0.9	2.1
Permo-Triassic sandstones (undifferentiated)				
20–40	0.7	1.7	–	–
40–60	1.9	4.7	0.9	2.3
>60	0.9	2.3	0.6	1.5
Totals	3.5	8.7	1.5	3.8
WEST LANCASHIRE BASIN				
Permo-Triassic sandstones (undifferentiated)				
20–40	0.2	0.5	–	–
Total Identified Resources	6.7	16.6	2.4	5.9

Table 8.9b Identified Resources of the Cheshire and West Lancashire basins

Crewe the temperature of the Sherwood Sandstone Group exceeds 40°C and the Permian Collyhurst Sandstone exceeds 60°C. Both aquifers are likely to have transmissivities of at least 10 D m and, if this is so, the thermal yield would be 2.3 and 4.0 MW from the Sherwood Sandstone and Collyhurst Sandstone respectively for production rates of 30 l/s, assuming a rejection temperature of 30°C.

As the areas of greatest geothermal potential in the Cheshire Basin are all at inland sites a second, reinjection, borehole would almost certainly be required to dispose of the produced fluid, which is likely to be a brine. Although this would increase the capital cost of the project it would maintain the pressure in the aquifer and allow a greater proportion of heat to be 'swept' from the aquifer. A reinjection borehole could also be advantageous if, as is likely, the extent of the aquifer is limited by faults or changes in facies.

Carlisle Basin

The Permo-Triassic sandstones in the Carlisle Basin attain a maximum thickness of approximately 1700 m to the east of Silloth. This is greater than the estimated depths of the adjacent Dumfries and Lochmaben basins and of the Vale of Eden which are unlikely to exceed 1000 m. Insulating cover provided by the Stanwix Shale, the Mercia Mudstone Group and Jurassic shales only occurs in the Carlisle Basin but despite this effect temperatures are only likely to exceed 40°C at the base of the Permo-Triassic in a small area (Chapter 4 and Gale and others, 1984b).

The Geothermal and Identified Resources for both the Triassic and Permian sandstones in the Carlisle Basin (Table 8.10a and b) are insignificant in comparison with other basins in the UK. Exploitation of the limited resource would require the use of heat pumps but the hottest aquifer underlies a largely rural area. If a borehole at Silloth were to be developed by pumping the Permian Penrith Sandstone at 30 l/s, the mean temperature would be about 33°C and the thermal yield 2.7 MW, assuming the water were rejected at 10°C from the heat pump.

Basins in Northern Ireland

Information about the nature of the Permo-Triassic sandstones in Northern Ireland, where they occur at appreciable depths below overlying confining beds, is very limited. The sandstones have been studied in greatest detail at four sites where deep boreholes have been drilled — Port More, Ballymacilroy, Newmill and Larne No.2 (Figure 4.35). Examination of data available for these sites leads to the general conclusion that the permeability is very much reduced below depths of 1500 to 1600 m, although exceptions do occur, as for example in the Ballymacilroy Borehole where the Permian includes permeable zones at depths of 1900 m (Chapter 4 and Bennett, 1983).

The implication is that the Sherwood Sandstone Group is the main formation with geothermal potential and the Geothermal Resources have only been estimated for this formation. This is a conservative approach because the Permian sandstones have yielded brines at temperatures of more than 60°C but in view of the uncertainties concerning their thickness and detailed nature, a resource estimate is not justified at this stage. To a large extent this view is influenced by the recent failure of the Permian at Larne to produce any significant yield because of the poor aquifer properties throughout the thick Lower Permian Sandstone sequence of 440 m. This condition may be atypical of the area as a whole (Downing and others, 1982b).

The Sherwood Sandstone Group attains thicknesses in excess of 500 m, but the producing thickness is much less, being probably of the order of 200 to 250 m, and tends to occur at or near the top of the formation. Therefore, although the positions of the producing zones are distributed over a greater thickness of the sandstones than 250 m, for the purpose of making an initial assessment of the resources they are assumed to be entirely at the top of the formation and a conservative thickness of 200 m is taken as the total aquifer thickness. The average permeability is considered to be between 50 and 100 mD, giving a transmissivity of 10 to 20 D m.

An average temperature gradient in Northern Ireland is about 30°C/km and the temperature at the top of the Sherwood Sandstone Group is shown in Figure 4.36. It is apparent that the temperature in the sandstone exceeds 20°C over most of the area where the sandstones are confined by younger rocks and over appreciable areas exceeds 40°C, but more than 60°C is only attained north of Lough Neagh.

The Geothermal and Identified resources are summarised in Tables 8.11a and b. Most of the resources are away from the coast and in any development scheme a second reinjection well would be mandatory to dispose of the

Table 8.10a Geothermal resources of the Carlisle Basin

Temperature zone (°C)	Area (km^2)	Thickness (m)	Mean temperature (°C)	Geothermal Resources ($J \times 10^{18}$)	Geothermal Resources (MWa/km^2)
Triassic Sherwood Sandstone (Kirklinton Sandstone)					
20–40	100	0.1	25	0.5	120
Permian Penrith Sandstone					
20–40	200	0.1	27	0.9	140
>40	Resource too uncertain and limited to warrant calculation				

Total Geothermal Resources: 1.4×10^{18} joules

Table 8.10b Identified Resources of the Carlisle Basin

Temperature zone (°C)	Mean temperature (°C)	Identified Resources ($J \times 10^{18}$), Reject temperature: 10°C, F = 0.1	F = 0.25
Triassic Sherwood Sandstone (Kirklinton Sandstone)			
20–40	25	0.04	0.1
Permian Penrith Sandstone			
20–40	27	0.09	0.22
Total Identified Resources		0.13	0.32

Temperature zone (°C)	Area (km^2)	Mean thickness (km)	Mean tempera- ture (°C)	Geothermal Resources ($J \times 10^{18}$)	(MWa/ km^2)
20–40	2355	0.2	30	24.5	330
20–40	1291	0.2	50	26.9	660
>60	327	0.2	60	8.5	825
Total Geothermal Resources:				60×10^{18} joules	

Table 8.11a Geothermal Resources of the Sherwood Sandstone Group in the Northern Ireland Basins

Temperature zone (°C)	Identified Resources ($J \times 10^{18}$) Reject temperature 10°C F = 0.1	F = 0.25	Reject temperature 30°C F = 0.1	F = 0.25
20–40	2.55	6.1	–	–
40–60	2.70	6.7	1.3	3.4
>60	0.95	2.1	0.5	1.3
Totals	6.20	14.9	1.8	4.7

Table 8.11b Identified Resources of the Sherwood Sandstone Group in the Northern Ireland Basins

geothermal brine after the heat had been extracted. Providing heat pumps could be incorporated in a heating system, useful thermal yields of the order of 1 to 4 MW could probably be obtained from the Sherwood Sandstone Group. But most of the area within which such yields could be anticipated is rural in character. The larger conurbations are all located outside the area underlain by potential geothermal fields, as defined by the 40°C isotherm. Towns such as Larne, Antrim, Ballymena and Coleraine, which lie near or in the area of interest, are relatively small but have some significant heat loads which could probably be met from a geothermal source. In the vicinity of Antrim town at the north-east corner of Lough Neagh, the underlying Sherwood Sandstone aquifer probably contains brine at a temperature of 60°C or more.

GEOTHERMAL RESOURCES OF THE LOWER CRETACEOUS SANDSTONES

The Lower Greensand may contain a geothermal resource at temperatures of between 20°C and at least 30°C along the south coast between Portsmouth and Worthing. Small areas could also exist near Salisbury and on the Isle of Wight. In these areas the aquifer is over 400 m deep and is about 20 to 50 m thick (Figure 4.17). The transmissivity could exceed 5 Dm and on that basis over an area of some 550 km^2 the Geothermal Resources would amount to about 0.6×10^{18} joules. This would be reduced to 0.05×10^{18} joules when a maximum recovery factor is applied to obtain the Identified Resources, assuming development by single wells and a rejection temperature of 10°C; by using doublets it would be increased to 0.16×10^{18} joules. To put these figures in perspective, the resource available from a site at Portsmouth may be considered. The thickness there could amount to 25 m and if the permeability is 250 mD, a yield of 25 l/s should be obtained for a drawdown of no more than 20 m. This would give a thermal yield of at least 1.5 MW if the water was rejected after use at 10°C.

Clearly the resource is small compared with the Permo-Triassic aquifers and would require the use of heat pumps to exploit it. Potential heat loads exist along the south coast but a coastal location may not be of paramount importance as the water could be relatively fresh and might even be suitable for use as a potable supply after the heat had been extracted (Chapter 4 and Gale and others, 1984d).

Chapter 9 Engineering and Economic Aspects of Low Enthalpy Development

R. A. Downing

The engineering and economic aspects of low enthalpy geothermal schemes are outside the scope of this volume, but some understanding of what is involved is necessary to appreciate the possible application of the geological details and the resource assessments given in earlier chapters.

The initial stage in a geothermal investigation is a regional survey which takes the form of the basin studies described in Chapters 4 and 5. The objective of these is to identify potential geothermal reservoirs with suitable aquifer properties and temperatures by means of geological, geophysical, hydrogeological and geochemical studies, particularly by collating and interpreting existing information.

The next stage is to make in-depth assessments of the geothermal potential in selected areas where heat loads exist. This involves more detailed geological studies and possibly seismic surveys leading to the identification of a potential development site which may then be investigated by drilling a preliminary exploration well. In the UK three exploration wells have been drilled to investigate the Permo-Triassic sandstones as possible geothermal reservoirs. The first was at Marchwood, near Southampton, in the Wessex Basin, the second was at Larne in Northern Ireland and the third at Cleethorpes on the south Humberside coast. In addition a development well was drilled by the Department of Energy on behalf of Southampton City Council in the centre of Southampton. This well may be used to heat civic offices and a swimming pool in the city, with financial support from the EEC.

Exploration wells differ from development wells because the purpose of the former is to identify reservoirs and evaluate their potential whereas a development well is drilled into a reservoir that has already been defined and the drilling programme is more reliable from the outset. However, it should be borne in mind that the most important aspect of any investigation programme is to determine the extent of the reservoir and the position of any barriers or boundaries limiting flow to the well, such as faults or facies changes. This can only be reliably achieved when at least two and ideally three or four wells have been drilled and tested over an extended time period. Geothermal energy development differs from oil development because even the limited oil produced during an extended evaluation test can be used, whereas geothermal development requires significant engineering works before the energy can be used. It also differs because a potential oilfield will be investigated by drilling an exploration well which, if successful, will be followed by appraisal wells and ultimately development wells. A successful field will be exploited by a considerable number of wells. The same approach would be adopted when developing a major steam field. But the direct use of low enthalpy resources is normally to meet a local demand for energy and will usually only require two wells — a production and a reinjection well to form a doublet system — and this makes an assessment of the total resource from well tests less reliable.

The primary objective of the exploration wells drilled at Marchwood, Larne and Cleethorpes was to determine the nature and thickness of any geothermal reservoirs, their properties and the temperature of any water they contained. This objective may be amplified under the following headings to indicate the type of information required:

Depth of aquifer
Thickness of aquifer
Lithology of aquifer
Identification of water-bearing zones
Aquifer properties — vertical distribution of porosity and permeability
Static reservoir pressure
Fluid properties — temperature
— geochemical properties, in particular salinity and corrosion and scaling potential
— specific heat
Yield of the aquifer from production tests.

A geothermal well in a sedimentary basin is drilled with conventional rotary drilling equipment applying the technology pioneered by the oil industry, although some differences in approach are necessary, particularly with regard to the method of completing the well.

A typical geothermal well in the UK, aimed at developing water at 60°C or more, will be about 2 to 2.5 km deep. A pump chamber is created by drilling at a diameter of $17\frac{1}{2}$ inches and running $13\frac{3}{8}$ inches casing to a depth of between 500 and 800 m, depending upon the anticipated pressure reduction in the reservoir when the well is produced. During the life of the well there will be a progressive reduction in reservoir pressure and this has to be allowed for when deciding upon the depth of the pump chamber. The reservoir itself is usually drilled at a diameter of $12\frac{1}{4}$ inches, though it may be under-reamed to increase the diameter. The well will be completed by leaving it as an open-hole if the reservoir is firm, or by inserting a wire-wrapped stainless steel well-screen or, if very long intervals are required for production, by using a slotted liner of appropriate slot-size. Should the reservoir be friable and unconsolidated, a gravel-pack completion may be necessary, although installation of significant lengths of gravel pack at depths of 2000 m can be difficult. The rock sequence between the pump chamber and the reservoir is usually cased with a liner $9\frac{5}{8}$ inches in diameter.

Wells are usually tested in two stages. A preliminary test is made by air-lifting or gas-lifting with nitrogen. Nitrogen has been used in the tests on UK geothermal wells for safety reasons since the Permo-Triassic sandstones are potential hydrocarbon reservoirs, but also because reservoir pressure reductions were expected to be large on the basis of interpretations of cores, geophysical logs and, at Marchwood, drill-stem tests. With a nitrogen system gas injection rates are higher than are possible with readily available air compressors and nitrogen prevents oxidation reactions in the well and eliminates corrosion of downhole pipes and equipment. The purpose of the gas-lift test is to develop the well, provide an assessment of the yield, derive the transmissivity of the aquifer and the temperature of the water and to define a specification for a suitable pumping unit.

If successful, a gas-lift test is followed by a more extensive test lasting about 30 days and using a downhole pump. The purpose of this test is to derive well characteristics and aquifer properties, identify any flow barriers near the well, measure a stable reservoir temperature and make an initial assessment of the size of the geothermal resource; in other words to provide information from which a decision can be made about whether or not commercial development is possible. Nevertheless, at this stage the actual size of the resource, and hence the rate at which it can be optimally developed, remains uncertain. This represents the risk that has to be taken at this time. The true size of a reservoir only becomes apparent once production has started and continued for some years.

The testing sequence adopted for the two wells at Southampton, together with the results, have been described in detail by Burgess and others, 1981; Price and Allen, 1982; Downing and others, 1982a; and Allen and others, 1983; and summarised by Downing and others, 1984.

Geothermal wells differ from oil wells in the important respect that in general flow rates are much higher. A successful geothermal well produces about 100 m^3/h and this may be yielded from several or even many permeable horizons over a thick reservoir interval. For example, the reservoir thickness may be 150 m with about 50% contributing significantly to the yield from a series of discrete permeable horizons. Highly permeable reservoirs are necessary to provide yields of 100 m^3/h for acceptable reductions of reservoir pressure but contributions are also necessary from the less permeable horizons. Permeable formations are particularly susceptible to damage by invasion of the drilling fluids and the selection of a fluid that does not damage the reservoir is important when drilling the reservoir itself. Polymer muds are commonly used that include potassium chloride to inhibit the swelling of clays, which can also lead to the migration of fine particles and constriction of pore spaces in the formation.

Geothermal schemes are capital-intensive and most of the capital is required in the initial stages. However, geothermal development of a major reservoir on a regional scale has the advantage of all groundwater schemes that it can proceed in stages, with the gradual input of capital as a scheme is extended, thereby avoiding capital expenditure for works that would not be fully used for an appreciable time although this does not apply, of course, in the case of a small local scheme. Geothermal schemes also have the advantage that they can provide energy on a local scale, for example for district heating or for horticultural purposes.

The drilling component of a geothermal scheme represents a major proportion of the total capital cost. Because geothermal fluids are invariably saline, a second well is required at inland sites to dispose of the brine after use, although at coastal sites discharge to the sea or a saline estuary may make this unnecessary. However, even at a coastal site reinjection has the advantage of maintaining reservoir pressure, thereby extending the life of the well and it is, of course, essential if a reservoir is to be developed by more than one well, or if the extent of the reservoir is restricted by faults or facies changes to a closed, relatively small, reservoir block. As most of the heat is stored in the rock rather than the water it contains, reinjection has a further advantage that as the colder injected water flows towards the production well it extracts heat from the rock and thus recovers a higher percentage of the available resource. A very stratified reservoir that contains bands of high permeability is not ideal for reinjection as this encourages early breakthrough of the injection water to the production well. The extent of the heterogeneity of the reservoir is reflected in the efficiency with which the injection water 'sweeps' heat from the rock.

When operating injection wells care has to be taken to avoid a reduction in the injection rate because of chemical precipitation or physical clogging in the aquifer or on well screens. The injection water must be compatible with the native groundwater and efficient filtering facilities are necessary to remove solids from the water; in general particles greater than 5 microns should be removed, and ideally particles greater than 1 micron; suspended solids should be less than 1 mg/l. The water should be de-oxygenated before injection to minimise corrosion and prevent precipitation of oxidation compounds, particularly iron compounds. The water should be chlorinated before injection to avoid bacterial growths in the well and the gas content should be controlled to prevent air entrainment and avoid gases coming out of solution in the aquifer, which would reduce the permeability. Any tendency to form scale from precipitation of calcium carbonate or iron compounds should be avoided by controlling the pH. Care must also be taken to avoid scaling, caused by pressure or temperature reductions, or corrosion in the pipeline connecting the two wells and in any chemical treatment system that may be necessary. Where practicable the production and injection system should form a closed circuit to prevent oxygenation and to restrict other physico-chemical changes prior to reinjection.

Ideally a reservoir should have a transmissivity of at least 10 D m. This requires, for example, an average thickness for the water-bearing intervals of 100 m if the average permeability is 100 mD. In practice a suitable reservoir is likely to include permeable horizons with permeabilities of perhaps 1 or 2 darcys or even more. To obtain a yield of 100 m^3/h from a reservoir with a transmissivity of 10 D m would require pressure reductions of the order of 225 and 260 metres after two and twenty years respectively if development was by a single well and reservoir pressure was not maintained by reinjection. Abstraction rates of this magnitude require large downhole pumps which may be either electrically or hydraulically driven. During the investigations in the Southampton area, where the transmissivity of the reservoir is only 4 D m, the Marchwood Well was tested with an electrical submersible pump. This was a 248 kW, 27-stage unit about 30 m long, capable of meeting the expected production rate of 22 to 37 l/s against a head of 500 m. A prototype, very compact, hydraulically driven, 175 kW downhole pumping unit was used on an experimental basis to test the Southampton Well at a rate of 20 l/s.

Unlike fossil fuels, geothermal energy must be used and converted to a convenient energy form at or within, say, one or two kilometres of the drilling site. This is necessary to conserve the energy by maintaining the temperature. With regard to exploitation it also means that the availability of a heat load is more likely to be the limiting factor than is the size of the resource (ETSU, 1982). Heat exchangers are used to transfer the heat to a suitable water circulation system. High performance, plate-type titanium or possibly stainless steel heat exchangers are required. Heat pumps are necessary to extend the use of geothermal fluids to other than the initial application and also to exploit fluids at temperatures less than 60°C. In some situations the availability of insulated, overnight hot-water storage tanks can be economic.

Basically a heat pump extracts heat from a lower temperature source and upgrades it to a higher temperature. This is done by circulating a refrigerant, such as a halocarbon, which is evaporated by contact with the lower temperature water source. The resulting gas is compressed, raising the pressure and hence the temperature. The heat is absorbed by water (in, for example, a heating system) flowing through a condenser which changes the refrigerant to a liquid before returning it through a throttle valve to the evaporator for the cycle to be repeated.

The sequence of a well, with a submersible pump, connected by pipeline to a heating station (containing a heat exchanger and possibly heat pumps) and a reinjection well, represents the engineering stages required at an inland site to make the energy of a hot brine available through a secondary circulation system to, say, a district heating network. If an injection well is required a water treatment facility and pump to reinject the brine under pressure would also be necessary.

For the efficient application of such a geothermal system, the maximum use of the heat must be attained. To achieve this the well is usually connected to a load having a peak demand of about 2 to 2.5 times the maximum output of the well. In practice, this allows the well to operate on a load factor of more than 60% and to supply about 70% of the annual energy needs of a typical space-heating load. The geothermal energy is generally supplied at a constant rate and temperature, and peak loads are met from an auxiliary or back-up heating unit based on coal or gas.

The running costs of a geothermal scheme, additional to interest charges on capital, are represented by pumping costs (that is electricity charges), depreciation charges to cover overhaul of the pump and its replacement, maintenance costs, and the cost of conventional fuels for the back-up boiler and to drive heat-pump compressors. If an injection well is included, water treatment costs are required to ensure that the injected water is compatible with the reservoir water and does not block the well-face prematurely, and pumping costs are necessary to inject the brine into the reservoir.

Golebi and others (1981) have addressed the problem of optimising the extraction of geothermal energy from a doublet comprising an abstraction and reinjection well and relating the various interrelated factors involved. The economics of geothermal district heating schemes have been reviewed by Harrison and others (1983) using data for low enthalpy developments derived mainly from experience in the Paris Basin. An analysis of these French schemes indicates that well drilling costs amount to 40% to 50% of the total cost when doublets are drilled, but 26% if there is only a single well, and that the cost of surface works amounts to 30% to 50%, with pipeline costs representing 10% to 25% of the total. The point is made that economic viability is generally only possible if the scheme provides an average heating output of about 3 MW.

The heat recovered from a geothermal scheme depends upon the flow rate, the heat capacity of the fluid and the temperature drop as the heat is extracted by heat exchangers and heat pumps. The parameters that have a strong influence on unit costs are (Harrison and others, 1983):

geothermal gradient
depth and hence temperature of the aquifer
transmissivity and hence permeability of the aquifer
well losses
production rate.

The key factors are the temperature of the reservoir and the permeability. The geothermal gradient controls the depth necessary to obtain the required temperature. This, together with the depth to the aquifer, determines the depth of the well and hence drilling costs which increase non-linearly with depth. The temperature of the water in the reservoir is the main factor determining the amount of useful heat and therefore the size of the resource.

The transmissivity is reflected in the reservoir pressure reduction for a particular flow rate and as such affects pumping costs and electricity consumption. The well losses have a similar effect by increasing the pressure reduction for a given flow rate. Well losses can be significant in deep wells and this emphasises the importance of avoiding, as far as possible, formation damage from the invasion of drilling fluids. Unit costs are also a function of production rate for a given transmissivity. The static fluid pressure in the reservoir, the thickness of the reservoir and its permeability (from which the transmissivity is calculated) determine the amount of water that can be produced from a well. The transmissivity controls the rate at which water can be abstracted, as the reduction in reservoir pressure is a limiting factor, and the static fluid pressure controls the depth to the pumping water level; both influence the life of the well. In this context the salinity of the water is important. Many reservoir waters are brines with high salinities and hence high densities. This results in deep rest water levels which in turn affect pumping water levels and hence have an impact on pumping costs.

For the successful promotion of a geothermal scheme it

must be economically attractive, that is, cheaper than alternative schemes, and it must be attractive relative to other investments. The factors that influence economic viability are the discount rate, the real cost of fossil fuels and the performance of the well. Geothermal schemes require large capital investments and there is a low financial return on investment over a long period. There is also a significant risk involved in drilling wells to 2 km or more to obtain water at over 60°C, for the well may not yield an adequate flow initially or over the planned life of the scheme. The risk is much less if aquifers at shallower depths containing water at about 40° to 50°C are the target because the aquifer properties are more likely to be favourable.

The attraction of geothermal energy as an energy source depends very much on the rate of increase in real terms (that is in addition to increases caused by inflation) of the cost of alternative fuels.

The economic feasibility of developing geothermal energy has been examined in some detail by the Southampton City Council with respect to the well at Western Esplanade, in the centre of the city. In the short-term this well, which was initially intended to supply energy for a major new development including offices and shops, is capable of yielding over 20 l/s at a temperature of 74°C. However, extensive pumping tests indicated that the reservoir is heterogeneous and is probably limited to an area of 200 km^2 by boundaries or barriers that will ultimately prevent or reduce flow to the well (Allen and others, 1983). In these circumstances a flow rate of 20 l/s or more could not be sustained in the long-term because the drawdown of the water level in the well would be too great. Had that not been the case the yield would have provided some 3 MW of thermal energy which would have supported a peak load of 10 MW and provided 87% of the total energy demand of the Western Esplanade heat load; the remainder would have been met by a back-up coal-fired boiler. In this scheme only one well would have been used, the brine being discharged after use to the adjacent Southampton Water, a saline estuary. This scheme was expected to become profitable after 8 to 9 years and would have provided energy at a cheaper rate than alternative fuel sources. The real return on the investment would have been about 6%.

If it had been necessary to reinject the brine into the aquifer after use, as for example at an inland site, a flow rate of about 40 l/s would have been necessary to cover the additional costs of a second well and water treatment before reinjection. The aquifer would require a transmissivity of at least 10 D m and ideally nearer 20 D m to give this yield.

Because the reservoir below Southampton is restricted in size, the long-term average yield of the well would have to be restricted to about 10 l/s to give a useful life of 15 to 20 years. As already mentioned, Southampton City Council may use the well to heat civic offices and a swimming pool. If so, about 8 l/s would be abstracted in the summer and 11.5 l/s during the winter heating season. This scheme could be the first practical use of geothermal energy in the UK.

Chapter 10 Review of the Geothermal Potential of the UK

R. A. Downing and D. A. Gray

The overriding factor in any consideration of the development of geothermal energy in the UK is that geologically speaking the country is very stable being part of the continental foreland of Europe. The high heat flows associated with less stable volcanic or orogenic regions are not found. Instead the heat flow is relatively low, averaging somewhat less than 60 mW/m^2. Nevertheless, in deep sedimentary basins temperatures of about 60 to 70°C are found at depths of 2 km and approaching 100°C at 3 km, temperatures that are adequate for many domestic and industrial applications. In the radiothermal granites of Cornwall and northern England heat flows and temperature gradients are above the average values for the UK.

In these circumstances the prospects of developing geothermal resources depend upon either the presence of permeable rocks in deep sedimentary basins or the successful development in economic terms of the Hot Dry Rock concept. In the case of developing hot groundwater resources in permeable rocks, the most favourable reservoirs are the sandstones of the Permo-Triassic although in some areas the sandstones of the Lower Cretaceous and aquifers in the Upper Palaeozoic may also have potential. For the application of the Hot Dry Rock concept, however, the granite regions are favoured where the regional heat flow is augmented by the heat produced from the decay of radioactive elements.

The basic factor that determines the geothermal potential of the UK is the terrestrial heat flow. Above average values are found in south-west and northern England, the east Midlands of England, the Hampshire Basin, the

10.1a Depths at which 200°C is attained in the UK

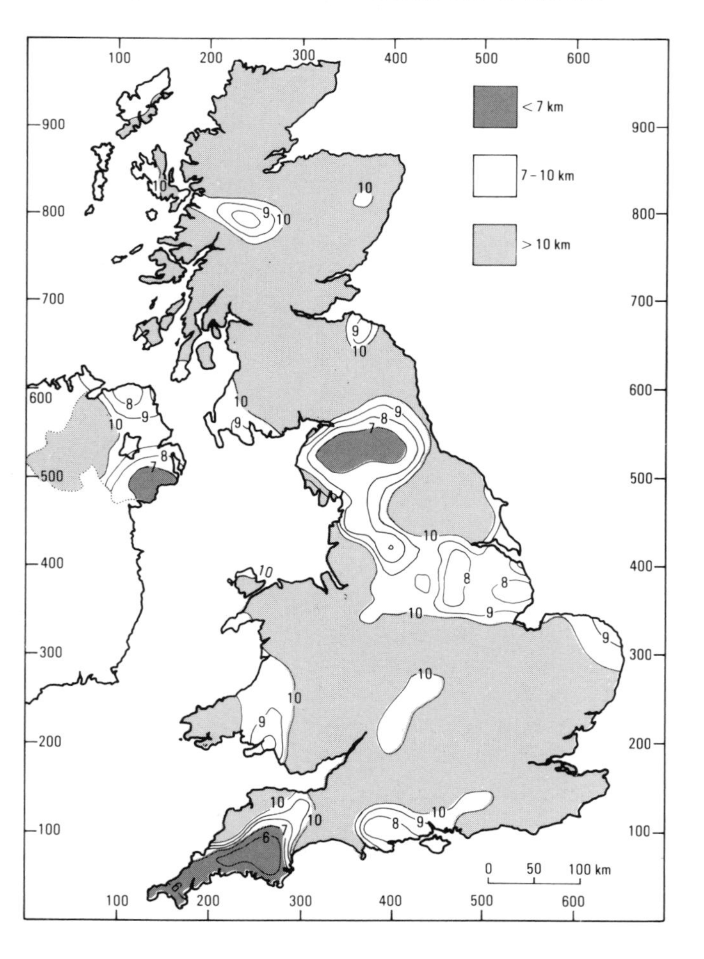

10.1b Depths at which 150°C is attained in the UK

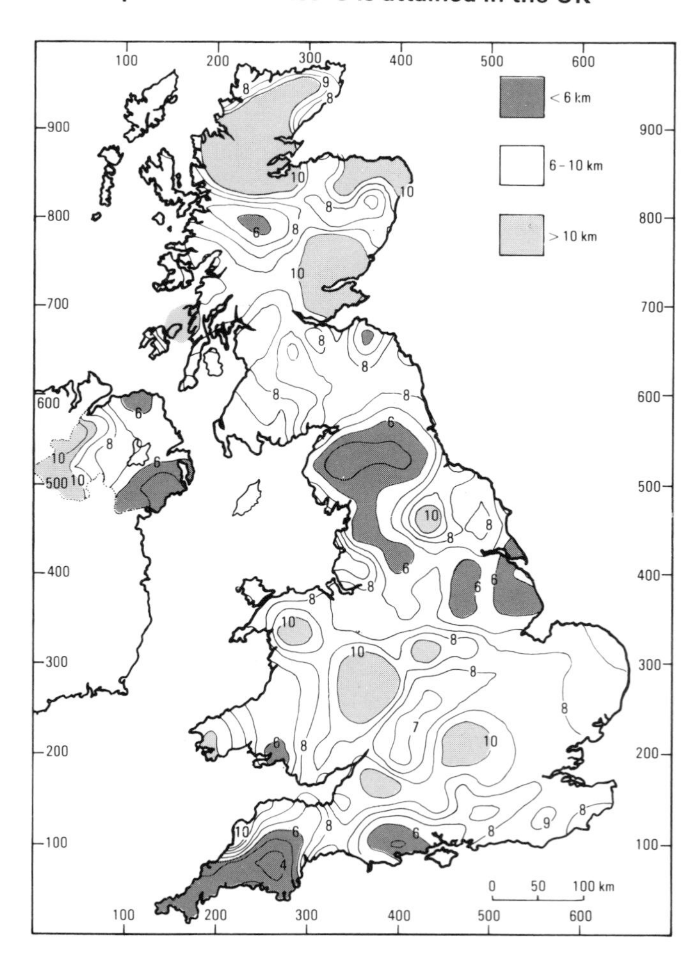

Eastern Highlands of Scotland, part of the Midland Valley of Scotland and in Northern Ireland, because of either above average concentrations of radioactive elements in the crust or regional groundwater flow. These are the regions favourable for geothermal investigation (Figure 2.8). The geothermal energy Accessible Resource Base to a depth of 7 km amounts to 2.75×10^{23} joules with a mean value of 1.07×10^{18} J/km^2; the distribution of this heat store reflects the variation of the heat flow.

HOT DRY ROCK RESOURCE

The highest heat flow values in the UK are found over the Cornubian batholith of south-west England and on this basis this is the most favourable area for HDR development. It attains this position because the heat flow from the lower crust and mantle is enhanced by heat production from radioactive elements in the granite which extends to mid-crustal depths. Predicted temperature profiles suggest 200°C might be reached at a depth of about 5.5 km (Chapter 3 and Figure 10.1a).

Other radiothermal granites occur in northern England and Scotland. Of these the concealed Weardale Granite appears, on the basis of limited current data, to be the most favourable site and a temperature of 200°C might be reached at a depth of about 6 km (Figure 10.1a). The HDR potential of the Lake District is less than that for the Weardale Granite but sub-surface temperatures are well above those found in basement rocks over most of the UK. The thermal anomaly over the Weardale-Lake District region is not so well known as that in south-west England and further work would be necessary to assess its full potential.

The granites of the Eastern Highlands are disappointing for, although they have high heat productions, the surface heat flow is relatively low and they do not present above average potential for HDR development. This is because the heat producing elements are concentrated near the surface and decrease quite rapidly with depth. A high surface heat production does not, therefore, necessarily mean a good HDR target and in this respect the Eastern Highlands granites contrast with the Cornubian granite.

In areas where granites do not occur and which do not benefit from enhanced surface heat flow from the presence of radioactive elements, it is necessary to drill to greater depths to attain equivalent temperatures but, if this is acceptable, the potential for HDR applications becomes more widespread, assuming that the technology can be applied to rocks other than granites. The most favourable situations are where sediments with relatively low thermal conductivity overlie Lower Palaeozoic and Precambrian basement as in Lincolnshire and the Wessex Basin. In these situations temperatures of 150°C might be attained at depths of less than 6 km in some areas (Figure 10.1b).

Temperatures of more than 200°C are required before electricity can be generated directly with steam obtained from geothermal energy. However, temperatures of less than 150°C can be used for power generation by involving a heat exchanger in which enthalpy from the geothermal fluid is transferred to a secondary working fluid which then expands through a turbo-generator. Compounds suitable for such binary-fluid cycles include lower molecular weight hydrocarbons, such as isobutane and propane, fluorocarbons and ammonia (see for example Tester, 1982). Waters at these temperatures could also be used for direct heating purposes.

The maps in Figures 10.1 a and b show the depths at which 150°C and 200°C may be anticipated. Heat stored at temperatures of more than 200°C, and which lies within the reach of current, economic, drilling technology, occur mainly in south-west and northern England where they are associated with the radiothermal granites. Over most of central and southern England and Scotland, 200°C would only be attained at depths of more than 10 km as the geothermal gradient is relatively low, generally less than 20°C/km. The influence of the radiothermal granites is again evident in the depth distribution of the 150°C isotherm but the existence of prospects in other areas is also apparent, particularly where basement rocks are buried beneath low conductivity sediments. If 100°C is an acceptable temperature for direct heating applications then potential development areas are more extensive. The shaded areas in Figure 8.4 show where 100°C would be attained at depths of less than 4 km, which is well within current drilling capacity.

The Hot Dry Rock Accessible Resource Base (the heat stored at depths of less than 7 km and at temperatures greater than 100°C) varies from zero in the Midlands of England to about 10^{18} J/km^2 over parts of the granite batholith in Cornwall. Values of more than 0.4×10^{18} J/km^2 occur in northern England and in other more local areas. The total HDR ARB at temperatures over 100°C amounts to some 36000×10^{18} joules. On a regional basis, highest values per square kilometre are 45×10^{16} joules in south-west England, 31×10^{16} joules in northern England and 24×10^{16} joules in east Yorkshire and Lincolnshire. As already mentioned, the heat stored at temperatures greater than 200°C and at depths of less than 7 km occur principally in south-west and northern England which together contain over 80% of the total.

The estimate for the HDR ARB is very large and represents many times the annual demand for energy in the UK. However, it is only of theoretical significance as it begs the question of recoverability and cost and only a small proportion could ever be developed. It is, however, significant as a base-line for the assessment of the geothermal resources, indicating the regional distribution of the resources and drawing attention to the scale of the resource which would become available if the formidable sub-surface engineering problems associated with developing the HDR resources can be overcome. It has been suggested (ETSU, 1982) that a heat recovery factor of 1% may be appropriate in the case of HDR resources, in which case the resources would amount to 360×10^{18} joules.

LOW ENTHALPY AQUIFERS

Groundwater in deep sedimentary basins can be a source of energy when it occurs at depths where suitable temperatures obtain. The aquifer should be confined by overlying sediments of low thermal conductivity which enhances the temperature of the water it contains. Ideally the base of the confining sequence should be at a depth where the temperature approaches that required. This eliminates the risk of cold water moving downwards into deeper hot water producing zones. The flow of water required from a successful deep well is large, of the order of 30 litres per second (2500 m^3/d), and the permeabilities of geothermal aquifers must be high for the depths at which they lie. Nevertheless, typical permeabilities of aquifers at depths suitable for geothermal development are an order of magnitude less than at outcrop, and large reductions of reservoir pressure are generally necessary to obtain appropriate yields.

The geothermal potential of the Mesozoic and Upper Palaeozoic rocks are described separately in Chapters 4 and 5 respectively. It would, however, be possible to develop the resources in both sequences in a single well. But the prospects of adequate additional yields from the Upper Palaeozoic, except in limited regions, are dependent upon fissures being intersected by the wells. This is a more speculative proposal than drilling for resources in the Permo-Triassic sandstones. Accordingly, the economics of deepening a well into the Upper Palaeozoic, when resources are available in overlying formations, are not generally attractive. In some circumstances it may prove advantageous to deepen a well from the Permo-Triassic into, for example, the Upper Carboniferous to take advantage of sandstone horizons, but this would be the exception rather than the rule and each site would have to be judged on its merits.

In Mesozoic basins the sandstones at the base of the Permo-Triassic lie at depths that are great enough for the water they contain to be at temperatures useful for geothermal development (Tables 10.1 and 10.2). In several of the basins they also have hydraulic properties that are favourable for the development of the hot water.

The Lower Permian sandstones in East Lincolnshire are at depths of over 2000 m and are between 30 and 60 m thick. The mean permeability is probably about 150 mD giving a transmissivity ranging from 5 to rather less than 10 D m. The temperature in the sandstones exceeds 60°C near the Lincolnshire coast and suitable heat loads exist in and around Grimsby. The Identified Resources for temperatures greater than 60°C are 0.2 × 10^{18} joules assuming rejection at 30°C and development by doublets (Table 10.3). Larger resources are available in the 40 to 60°C temperature range and almost 10^{18} joules are available at temperatures of more than 40°C assuming rejection at

Basin	Aquifer	Geothermal Resources (J x 10^{18})		
		20–40°C	40–60°C	>60°C
East Yorkshire and	SS	120.0	99.0	—
Lincolnshire	BPS	2.0	5.4	1.3
Wessex	SS	3.1	11.3	11.6
Worcester	P-T	72.7	12.0	—
Cheshire	SS	12.8	16.9	—
	P-T	6.7	18.7	9.2
West Lancashire	P-T	2.1	—	—
Carlisle	SS	0.5	—	—
	PS	0.9	—	—
Northern Ireland	SS	24.5	26.9	8.5
Totals		245.3	190.2	30.6

SS Sherwood Sandstone
BPS Basal Permian Sands
P-T Permo-Triassic sandstones (undifferentiated)
PS Penrith Sandstone

Table 10.1 Summary of the Geothermal Resources of the Permo-Triassic sandstones

Table 10.2 Summary of the Identified Resources of the Permo-Triassic sandstones

Basin	Aquifer	Identified Resources (J × 10^{18})			
		Reject temperature: 10°C		Reject temperature: 30°C	
		0.1*	0.25*	0.1*	0.25*
East Yorkshire and	SS	21.9	54.8	4.2	10.6
Lincolnshire	BPS	0.87	2.23	0.41	1.02
Wessex	SS	2.6	6.5	1.3	3.3
Worcester	P-T	8.5	21.2	0.6	1.4
Cheshire	SS	3.0	7.4	0.9	2.1
	P-T	3.5	8.7	1.5	3.8
West Lancashire	P-T	0.2	0.5	—	—
Carlisle	P-T	0.13	0.32	—	—
Northern Ireland	SS	6.1	14.9	1.8	4.7
Totals		46.8	116.5	10.7	26.9

* Reducing factor F (see Chapter 8)
SS Sherwood Sandstone
BPS Basal Permian Sands
P-T Permo-Triassic sandstones (undifferentiated)

30°C (Tables 10.3 and 10.4).

The Sherwood Sandstone attains a maximum thickness of over 500 m in east Yorkshire. In east Lincolnshire it thickens to the north from 100 m at the Wash to about 400 m at the Humber. The mean permeability is about 250 mD so that thicknesses in excess of 40 m have transmissivities of over 10 D m. As a consequence over very extensive areas the properties of the sandstone are favourable for development. The aquifer occurs at relatively shallow depths of between 800 and 1200 m and the maximum temperature for the mid-point of the aquifer just exceeds 50°C to the north and south of the mouth of the Humber, although over 40°C is exceeded over extensive areas. The Identified Resources amount to about 10 × 10^{18} joules for water at more than 40°C if development were with doublets and the reject temperature were 30°C. If heat pumps were used the Identified Resources would increase to 25 × 10^{18} joules (Table 10.4).

In east Yorkshire and Lincolnshire the resources are mainly in the Sherwood Sandstone although the temperature is lower than in the Lower Permian because of the shallower depth. Larger volumes of water would have to be pumped for a given thermal yield but the aquifer properties are favourable for high well yields.

The Wessex Basin is a very favourable area for geothermal development from the Sherwood Sandstone with the thickness and transmissivity gradually increasing in a westerly direction from Southampton. The Identified Resources for temperatures greater than 40°C amount to 3.26 × 10^{18} joules, assuming rejection at 30°C and the use of doublets (Tables 10.3 and 10.4) and over an appreciable area where the transmissivity is over 10 D m the temperature is greater than 60°C. In the extreme west, near the outcrop where groundwater temperatures are 20 to 40°C, the transmissivity exceeds 30 D m over extensive areas. To the east of a line lying somewhat to the west of Bridport and Yeovil, the temperature is greater than 40°C and over 70°C is attained some 20 km to the north-west of Bournemouth, where the greatest geothermal resource per unit area is concentrated, amounting to over 300 MW a/km^2.

Table 10.3 Summary of the Identified Resources at temperatures greater than 60°C

Basin	Aquifer	Identified Resources (J × 10^{18})			
		Reject temperature: 10°C		Reject temperature: 30°C	
		0.1*	0.25*	0.1*	0.25*
East Yorkshire and Lincolnshire	BPS	0.13	0.33	0.08	0.2
Wessex	SS	1.16	2.9	0.74	1.85
Cheshire	P-T	0.9	2.3	0.6	1.5
Northern Ireland	SS	0.9	2.1	0.5	1.3
Totals		3.09	7.63	1.92	4.85

BPS Basal Permian Sands
SS Sherwood Sandstone
P-T Permo-Triassic sandstones (undifferentiated)
* Reducing factor F (see Chapter 8)

Table 10.4 Summary of the Identified Resources at temperatures between 40 and 60°C

Basin	Aquifer	Identified Resources (J × 10^{18})			
		Reject temperature: 10°C		Reject temperature: 30°C	
		0.1*	0.25*	0.1*	0.25*
East Yorkshire and	SS	9.9	24.8	4.2	10.6
Lincolnshire	BPS	0.54	1.4	0.29	0.72
Wessex	SS	1.13	2.82	0.57	1.41
Worcester	P-T	1.2	3.0	0.6	1.4
Cheshire	P-T	3.6	8.9	1.8	4.4
Northern Ireland	SS	2.7	6.71	1.3	3.4
Totals		19.07	47.63	8.76	21.93

BPS Basal Permian Sands
SS Sherwood Sandstone
P-T Permo-Triassic sandstones (undifferentiated)
* Reducing factor F (see Chapter 8)

The thick sequence of Permo-Triassic sandstones in the Worcester Basin includes two aquifers, the Lower Permian Bridgnorth Sandstone and the Bromsgrove and Wildmoor sandstones of the Sherwood Sandstone Group. The layer of insulating mudstones above the Sherwood Sandstone Group is relatively thin and this reduces the temperature at the top of the sandstones. As the temperature gradient in the sandstones is low, generally of the order of 18°C/km, the groundwater in the deepest parts of the basin does not attain 60°C. The mean temperature in the Permo-Triassic sandstones as a whole exceeds 40°C over an appreciable area around and to the north of Circencester. In the deepest part of the basin, north-west and south-east of Evesham, groundwater attains temperatures of more than 50°C.

Despite the lower maximum temperatures in the Worcester Basin groundwater at more than 40°C represents Identified Resources of some 1.4×10^{18} joules assuming rejection at 30°C and the use of doublets (Table 10.4). The total Identified Resources for water above 20°C, using heat pumps (rejecting at 10°C), and doublets, amounts to over 20×10^{18} joules (Table 10.2). The greatest potential is in the north of the basin where boreholes 2000 to 2500 m deep would be necessary. In the south, the resources could be developed from depths of about 1700 m. The Permo-Triassic sandstones in the Worcester Basin represent a significant resource albeit in the lower temperature range of 20 to 60°C, but the region is worthy of consideration as the transmissivity is expected to be high, possibly between 50 and 100 D m in many areas.

In the Cheshire Basin the temperature gradients are low because of the low heat flow and the relatively high thermal conductivity of the Permo-Triassic sandstones. Nevertheless, because the sandstones attain a total thickness of more than 2000 m and occur at depths of over 4000 m, maximum temperatures at their base exceed 70°C. Few data exist about conditions at depth in the basin but the porosities of the sandstones are about 15 to 20% in the few deep boreholes that have been drilled and on this basis permeabilities would be expected to range from 10 to 100 mD. The temperature in the sandstones exceeds 40°C over some 550 km^2 and is greater than 60°C over 320 km^2; in these areas the transmissivity is believed to exceed 10 D m. The Identified Resources in the range 40 to 60°C and greater than 60°C could amount to 4.4×10^{18} joules (Table 10.4) and 1.5×10^{18} joules (Table 10.3) respectively, assuming development by doublets and rejection at 30°C. However, the only towns in the basin beneath which temperatures exceed 40°C are Crewe and Northwich. Because of the low temperature gradient, deep wells would be required; for example, at Crewe the base of the sandstones is at about 3.4 km but a well drilled to this depth could yield 4 MW from water at 65°C.

Limited data from deep drilling supplemented by interpretation of seismic surveys indicates that in the West Lancashire Basin the maximum depth of the Permo-Triassic sandstones is likely to be about 1800 m. With the geothermal gradient obtaining, 40°C is unlikely to be attained, although over some 450 km^2 temperatures between 20 and 40°C occur at depths of between 450 and 750 m.

As in the West Lancashire Basin, the Permo-Triassic sandstones in the Carlisle Basin are at relatively shallow depths, only attaining 1700 m in the deepest area. The temperature exceeds 40°C at the base of the sandstones over only a small area east of Silloth, but it is in the range 20 to 40°C over the deepest parts of the basin west of Carlisle. Although the geothermal resources of the Carlisle and West Lancashire basins are not large when compared with other basins in the UK, the heat store they contain could provide useful contributions to local energy demands if used in conjunction with heat pumps.

The Sherwood Sandstone in the deep sedimentary basins in Northern Ireland represents the main potential for geothermal energy in the Province. The permeability of the sandstones tends to be lower in Northern Ireland than elsewhere in the UK and the limited information about conditions at depth leads to the conclusion that generally the permeability is much reduced below depths of 1500 to 1600 m. The producing horizons tend to be 200 to 250 m thick near the top of the formation and the transmissivity of these horizons is likely to exceed 10 D m. Over extensive areas the temperature is greater than 40°C, and 60°C is exceeded over some 300 km^2 below and to the north of Lough Neagh. For temperatures greater than 60°C the Identified Resources amount to 1.3×10^{18} joules, assuming the use of doublets and rejection at 30°C (Table 10.3). Nearly three times as much is available in the 40 to 60°C range (Table 10.4).

The underlying Lower Permian sandstones contain brines at more than 60°C but their distribution and variation in thickness are unknown. Views about the prospects of developing these sandstones are currently coloured by the poor aquifer properties throughout the thick sequence encountered in the Larne No 2 geothermal exploration well but this could be atypical of the area as a whole (Downing and others, 1982b). Nevertheless, at this stage, in view of the uncertainty surrounding the distribution of the Lower Permian sandstones, the resource potential remains speculative.

Although the greatest potential for geothermal development in the Mesozoic basins is in the Permo-Triassic sandstones, the Lower Cretaceous sandstones along the south coast between Portsmouth and Worthing are believed to contain water at between 20°C and at least 30°C. The sandstones are only 20 to 50 m thick but nevertheless wells could yield say 25 l/s for relatively small water-level drawdowns. The Geothermal Resources could amount to 0.6×10^{18} joules. Information about the reservoir is limited and exploration would be necessary to confirm the assumptions made in making the resource estimates. It is possible that the water in these sandstones is fresh, in which case it might be possible to use it as a potable supply after the heat had been extracted. This factor together with the fact that the sandstones occur at relatively shallow depths, could mean that they represent an economic resource if heat pumps were used.

The Upper Palaeozoic sequences contain large thicknesses of potentially water-bearing rocks, and, because of their great thickness and depth, temperatures in excess of 60°C, and locally above 100°C, are found in many places within readily drillable depths. Gale and others (1984c) have calculated the total heat stored in the Carboniferous in the East Midlands of England to be 17.5×10^{20} joules. This is equivalent to 6.10×10^4 million tonnes of coal (mtce)

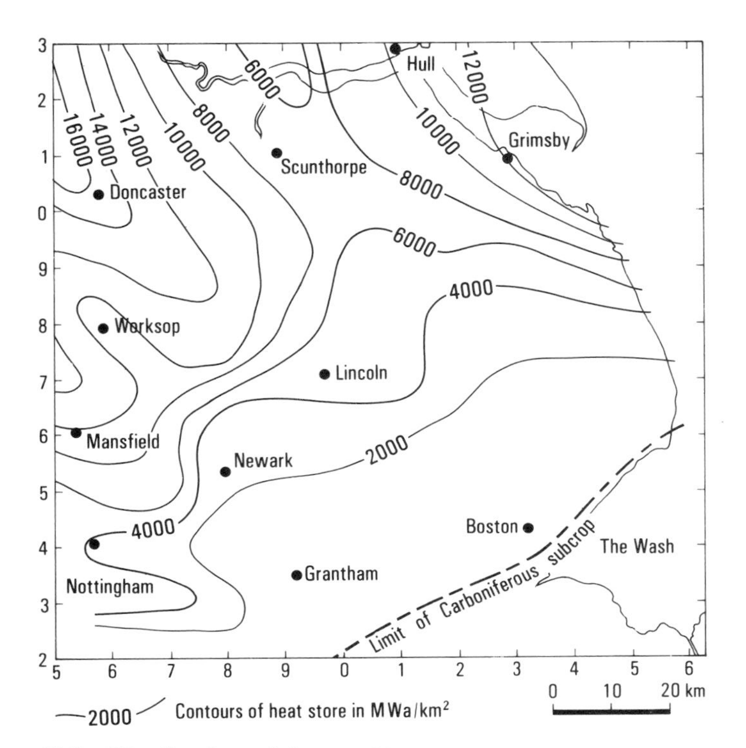

10.2 Distribution of the total heat store in the Carboniferous rocks of the East Midlands.

with a mean distribution of 2.0×10^{17} joules/km^2 (6340 MWa/km^2) which is equivalent to 6.6 mtce/km^2. The distribution of the heat store is related to the thickness and depth of the Carboniferous and is shown in Figure 10.2, expressed as MWa/km^2. Similar calculations have not been made for the rest of the country, but this one example illustrates the great potential of geothermal energy, even if only a small proportion of the total heat in place could be utilised. Unfortunately, these rocks, unlike the Mesozoic sequences, have been buried to great depths and subjected to the earth movements associated with the Hercynian orogeny. As a consequence they are typically hard, compact and indurated, and water flows through them mainly in fissures and fractures. Upper Palaeozoic rocks with the high permeability necessary for geothermal development have been encountered only locally. The average matrix permeability is generally less than 10 mD and commonly less than 1 mD, but where fractures exist the permeability is increased many times. For successful development of their water resources, wells must penetrate fracture systems that are extensive enough to drain adequate flows of water from the rock matrices. Such fracture systems are commonly associated with major structural disturbances and drilling to intersect them at depth is a speculative venture.

Generally, transmissivities are too low for economic development but this does not apply everywhere. In the Midland Valley of Scotland, the Upper Old Red Sandstone, in particular the Knox Pulpit Formation, has properties more akin to the Permo-Triassic sandstones of England. These favourable properties could persist at depth though this remains unproven. Similarly, the Fell Sandstone could have geothermal potential below Tyneside although again the nature of the rock at depths, where temperatures would be suitable, has not been proven.

Although, with these exceptions, the prospects for geothermal development in the Upper Palaeozoic appear discouraging, the fact remains that the only hydrothermal systems known in the UK are in these rocks. The Bath-Bristol system is the best known but similar systems occur in Derbyshire and South Wales. Anomalous heat flow patterns suggest others exist at depth below younger formations.

In these deep circulation systems, where heat is transferred from appreciable depths to the surface by water flowing along fractures or faults, the flow of water is controlled by the least permeable part of the conduit. This would be a controlling factor in any development rather than the total hot water stored at depth.

The hydrothermal system issuing in Bath and Bristol is probably typical of other systems in Upper Palaeozoic rocks. Groundwater flow in the Bath-Bristol system is predominantly through fissures in the Carboniferous Limestone which developed along pre-existing planes of weakness such as faults and bedding planes. The water originates as rain on the outcrops of the Carboniferous Limestone in the Mendip Hills, descending to depths of 2700 to 4300 m, possibly into the Old Red Sandstone, where it is heated to temperatures of 60 to 80°C before rising along southerly dipping fault planes to the surface in the Bath-Bristol area (Burgess and others, 1980). The evidence indicates that open fracture systems exist in the Carboniferous Limestone around Bath to depths of about 4 km and possibly somewhat less in the Bristol area.

The Carboniferous Limestone could store thermal water throughout the region between Bath and Bristol. While it is impractical to consider increasing the present level of geothermal development within the vicinity of Bath, in view of the possibility of interfering with the hydraulic regime of the thermal springs which already constitute a valued national asset, it is at least theoretically possible that a well elsewhere into the Carboniferous Limestone, for example in the Bristol area, could produce useful quantities of thermal water. However, the fact that the reservoir would be essentially a fracture system in a massive limestone has important consequences. Because of the marked heterogeneity of secondary permeability systems, the location of open, interconnected fissures at any particular site cannot be predicted. Thus, the siting of a successful well would either be by chance or would require the identification and location of fractures at depths to an accuracy not currently possible.

The question remains as to whether the permeability of Upper Palaeozoic rocks can be increased sufficiently by hydraulic fracturing to produce economic flows of water. If a rock formation has thickness of 300 m and average permeability of 10 mD, the yield of water would be about 1000 m^3/d (11 l/s) for a drawdown of 500 m. Hydraulic fracturing would generally be expected to increase the yield by a factor of no more than two or three, that is to say, to some 2500 m^3/d which would be an adequate supply at 60°C. Generally, however, Upper Palaeozoic rocks have a lower permeability than that assumed in the above example and an average value is more likely to be no more than 1 mD, which is too low to make hydraulic fracturing worthwhile.

Andrews and others (1981) have discussed the feasibility

of developing geothermal reservoirs by drilling two wells and creating parallel independent fractures from each well (Figure 7.7). One well would be a production well and the other for reinjection. Under certain limited conditions, rocks with matrix permeabilities of between 3 and 10 mD (Figure 7.8) could give economic yields but an initial matrix permeability of about 1 mD would be too low and generally the approach does not appear to be applicable to Upper Palaeozoic rocks.

In general the Dinantian limestones probably provide the best chances for large scale low enthalpy geothermal developments from Devonian and Carboniferous rocks. However, if this potential is to be realised many more geological and hydrogeological data are needed and more refined exploration methods necessary to identify suitable drilling sites. Apart from the Midland Valley of Scotland and south Northumberland, the sandstones have little potential. Indeed, taking the Devonian and Carboniferous rocks as a whole, the best chance of utilising the large amount of heat they store may well prove to be the Hot Dry Rock method. If this technique should prove feasible then a deep borehole could be drilled with the twin options of either developing natural water if available or creating a Hot Dry Rock fracture system if not.

Because of the uncertain knowledge about the detailed nature of the Upper Palaeozoic rocks at depth their geothermal resources have not been estimated at this stage. Nevertheless, it would be unwise and incorrect to dismiss them as having no potential for, although a significant resource has not been specifically identified, the possibility remains that resources exist in several regions.

SUMMARY OF LOW ENTHALPY RESOURCES

The total Geothermal Resources of the Permo-Triassic sandstones amount to 465 × 10^{18} joules with 30 × 10^{18}, or 6%, at temperatures of more than 60°C, mainly in Wessex, Cheshire and Northern Ireland (Table 10.1). About 41% is in the 40 to 60°C range and 53% between 20 and 40°C.

The Identified Resources are given in Table 10.2. Assuming development by doublets and a rejection temperature of 30°C, they amount to 27 × 10^{18} joules; if heat pumps were used, allowing rejection at 10°C, the total is increased to 116 × 10^{18} joules.

The Identified Resources for the highest grade of heat, that is at more than 60°C, are summarised in Table 10.3. Some 5 x 10^{18} joules are available assuming the use of doublets and rejection at 30°C, increasing to 7.6 x 10^{18} joules if heat pumps were employed. However, comparison of Tables 10.3 and 10.4 reveals that much more heat is stored in the 40 to 60°C range. It amounts to 47 × 10^{18} joules, assuming development with doublets and the use of heat pumps, allowing rejection at 10°C, and over 50% is in the East Yorkshire and Lincolnshire Basin. The Identified Resources, for particular development conditions, are also expressed in terms of tons of coal equivalent in Table 10.5.

As mentioned previously, particularly favourable areas for exploration occur where the transmissivity exceeds 10 Dm and the temperature exceeds 40°C; these areas have been referred to as 'potential geothermal fields' and they are shown in Figure 10.3. Appreciable areas occur where the temperature is in the 40 to 60°C range in the Sherwood Sandstone in East Yorkshire, East Lincolnshire, Wessex (mainly in Dorset) and Northern Ireland, and in the Permo-Triassic sandstones in Cheshire, as well as in Gloucestershire near Cirencester. A temperature of 60°C is exceeded in the Sherwood Sandstone in Wessex and Northern Ireland and in the Permo-Triassic sandstones of Cheshire. Of these Wessex is probably the most favourable exploration area for resources at more than 60°C. However, the largest store of low enthalpy thermal energy is in the Sherwood Sandstone of East Yorkshire and Lincolnshire where some 25 × 10^{18} joules (equivalent to 920 million tons of coal) occur in the temperature range 40 to 50°C assuming development by doublets and the use of heat pumps.

It should not be overlooked that the estimates of the Geothermal and Identified resources given in this volume relate to heat stored in aquifers over very extensive areas and only part of the heat could be realistically used. Development would be likely to start at individual sites where heat loads existed and, if successful, expand into small well-fields in relatively restricted areas. In this manner the resource estimates would be confirmed and estimates of the reserves could be made.

The feasibility of developing the hot water resources in the Permo-Triassic has been examined by drilling exploration wells in Southampton, in Larne, in Northern Ireland, and in Cleethorpes, and by carrying out drill-stem tests in hydrocarbon exploration boreholes. These have shown that the high water yields required can be obtained from discrete permeable horizons in the sandstones but, as expected, the

Table 10.5 Identified Resources of the Permo-Triassic sandstones expressed in units equivalent to million tonnes of coal (Mtce)

Basin	Aquifer	Temperature °C		
		20–40*	40–60*	>60†
East Yorkshire and Lincolnshire Basin	SS	1115	922	–
	BPS	19	52	7
Wessex	SS	29	105	69
Worcester	P-T	676	112	–
Cheshire	P-T	182	331	56
West Lancashire	P-T	19	–	–
Carlisle	P-T	12	–	–
Northern Ireland	SS	226	249	48
Totals		3278	1771	180

Estimates assume development with doublets (i.e. two wells — one for abstraction and the other for reinjection).

* Use of heat pumps assumed and hence a reject temperature of 10°C.

† Assumes heat pumps would not be used and hence rejection at 30°C.

BPS Basal Permian Sands

SS Sherwood Sandstone

P-T Permo-Triassic sandstones (undifferentiated)

Total annual coal consumption in the UK is about 100 million tonnes.

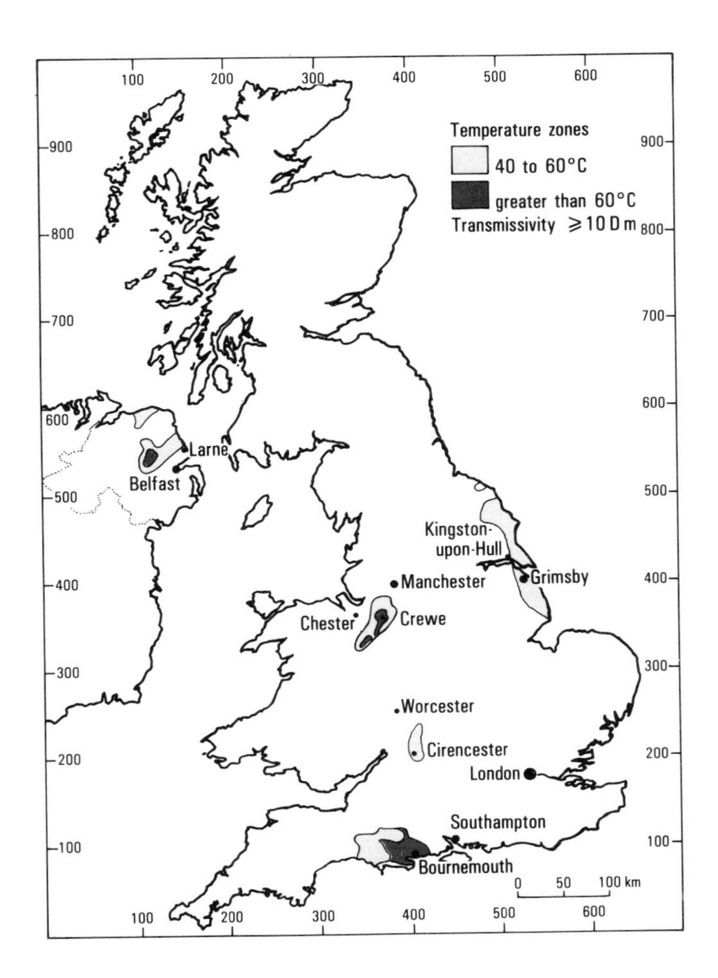

10.3 Potential low enthalpy geothermal fields in the UK as defined by a transmissivity of more than 10 D m and a temperature of more than 40°C.

Table 10.6 Results of investigations of the Permo-Triassic sandstones in hydrocarbon exploration boreholes

Borehole	Thickness of sandstone (m)	Transmissivity (Dm)	Temperature (°C)	Basis of data
Winterborne Kingston (Wessex Basin)	159	11	85	DST + cores
Ballymacilroy (Northern Ireland)	255	18	62	DST + cores
Kempsey (Worcester Basin)	1390	c. 150	32	DST + cores
Oil Bore 1*	1225	c. 100	36–54	Cores
Oil Bore 2	142	c. 20	62	Cores
Oil Bore 3	177	c. 10	74	DST + cores
Oil Bore 4	169	22	62	DST
Oil Bore 5	c. 40	0.5	73	DST
Oil Bore 6	200	10	75	DST

* Data relate to confidential oil exploration boreholes.

unpredictable occurrence of cement in the pores of the rock can significantly reduce the permeability. At Larne the Lower Permian sandstones are over 400 m thick but are cemented over virtually their entire thickness thereby reducing the yield to negligible proportions. In the two wells drilled at Southampton, the Sherwood Sandstone is relatively thin and the producing zones were only some 6 and 16 m thick. Nevertheless, these wells yielded over 30 l/s albeit for pressure reductions of as much as 3.7 MN/m^2 over the test period. But these results indicated that where the sandstones are thicker similar yields can be anticipated for smaller pressure reductions and this view has been supported by drill-stem tests and core analyses of selected horizons in hydrocarbon wells that penetrate greater thicknesses of sandstone (Table 10.6). At Cleethorpes, for example, where the Sherwood Sandstone is both thick and permeable, a short production test yielded 20 l/s for a drawdown of less than 7 metres.

The extended tests of the two wells at Southampton revealed that they are probably in a sandstone 'block' of some 200 km^2 that is bounded by impermeable boundaries—either faults or facies changes. This situation is probably not uncommon leading to the conclusion that even at coastal sites a second well is necessary to maintain the reservoir pressure and allow a greater proportion of the heat store to be abstracted.

THE RISK FACTOR

A problem of significant importance in any geothermal resource exploration is the risk factor, that is the possibility of drilling unsuccessful wells. In the oil industry the worldwide success ratio is about 1 in 15 but oil is a scarcer resource than hot water, although a larger rate of flow of hot water over a much longer time span is necessary to make a geothermal well successful as compared with an oil well at an on-shore site. Nevertheless, the risk of failure in drilling a geothermal well is not so great as for an oil well, given comparable conditions, except during the early exploration phase in unproven basins. Looking for a geothermal field is concerned not so much with finding a trap of limited size which contains the resource (as in oil) but with finding suitable reservoir properties because water at depths of up to 3 km is more or less ubiquitous though appropriate reservoir properties are not. Aquifers of regional significance extend throughout the larger depositional basins, but appropriate reservoir properties may occur only in limited areas. If a potential reservoir has not been proved before drilling then the major risks are clearly that it may not exist at all or that it may be too thin or that the porosity and permeability may be too low. If the reservoir has been proved then the risks are reduced and decrease further as the knowledge of the formation increases. The major risk is then resolved as to whether the aquifer properties will be suitable and in particular whether the presence of extensive cement will reduce the permeability to uneconomic values. Once the reservoir has been identified and found to have suitable properties, the next risk level is the size of the reservoir. If it is too small the resource will have an inadequate life-time. This could be overcome by maintaining the reservoir pressure by re-injection and thereby using more of the heat stored in the rocks near the wells.

As knowledge of an area increases then the consequence of drilling is more predictable and development becomes more logical and routine. This is the situation in the Paris Basin, where commercial schemes have been operating for over 10 years, but in the UK all the wells drilled to date have been for exploration purposes. This applies even to the well in the centre of Southampton, for more than one well is necessary to evaluate a reservoir when its thickness is marginal as was the case at Southampton.

Whether or not a well is deemed successful depends upon economic factors and these are intimately bound up with the cost of alternative fuels, at the time of development. The variables that particularly influence costs are:

Exploration costs: geophysical surveys, drilling and testing

Development costs: pumping unit, pipe-lines and surface plant

Operating costs: maintenance, pumping costs.

The capital investment required for exploration and development costs can be high but operating costs are generally low.

The success of a project depends almost entirely on the permeability of the reservoir and the temperature. A reservoir with a transmissivity as low as 5 D m can be economic providing the reservoir is extensive and development is with single wells, but 5 D m probably represents the lowest practical limit for development. Ideally a reservoir of at least 10 D m and more probably 20 D m, is necessary to reduce the water-level drawdown in the well under production conditions to favourable values. Currently, 20 D m is considered to be necessary for economic development in the UK but this could be reduced by a change in the price of alternative fuels. Twenty darcy metres requires a reservoir thickness of 200 m for an average permeability of 100 mD at depths of about 2 to 2.5 km. As the Permo-Triassic sandstones commonly contain thin horizons of much higher permeability, the total thickness necessary is likely to be less than this but these figures indicate the order of magnitude required.

Geothermal wells have to yield large volumes of water for extended periods, of the order of 20 to 25 years. The size of the reservoir is clearly of paramount importance but it is usually uncertain when development begins with a single well. Reservoirs, especially those that are thin or consist of a series of thin high permeability layers, may be discontinuous because of faults or lateral changes in the nature of the rocks. In these circumstances reinjection wells are advisable to maintain the reservoir pressure. As discussed previously (Chapter 9), this would be necessary at inland sites to dispose of the fluids after use as they are invariably brines.

An example of a favourable development situation would be a sandstone reservoir with a transmissivity of 20 D m developed with a doublet system, the wells being separated by 1000 m at the reservoir depth. Abstraction at a rate of 100 m^3/h would result in a stable drawdown of about 100 m, the stability being maintained by reinjection. If the water-yielding zone was 50 m thick the breakthrough time of the injected water at the production well would be of the order of 80 years.

CONCLUSIONS

The geothermal resources of the United Kingdom represent a significant energy source which can be quantified in general terms. A problem that remains, however, is the step from the resources estimate to a realistic assessment of the reserves. Identification of reserves, as distinct from resources, depends upon the extent of exploration in the several basins where resources exist. This is in turn a function of the technology available to exploit the resources and the economics of developing it compared with the cost of alternative energy sources.

The Hot Dry Rock Accessible Resource Base at more than 100°C amounts to 36 000 $\times 10^{18}$ joules and, while all this heat could never be used, recovery of 1% amounts to 360 $\times 10^{18}$ joules. If the engineering problems associated with the Hot Dry Rock concept can be overcome these values are sufficiently high to indicate that a very significant energy source is available and widely distributed, especially if the minimum temperature of 100°C is exploitable for direct heating purposes.

The low enthalpy resources are mainly in the Permo-Triassic sandstones and the Identified Resources at more than 40°C amount to 55 $\times 10^{18}$ joules if doublets are used and heat pumps allow rejection at 10°C; the Identified Resources under the same operating conditions but at more than 60°C are about 7 $\times 10^{18}$ joules. More heat is stored in the 40 to 60°C range and it is in reservoirs that are at shallower, more accessible depths. These resources are also in reservoirs with more favourable properties but exploitation at temperatures of less than 60°C with rejection at 10°C is dependent on further development of heat pumps including an improvement in their Coefficient of Performance to above the current (1984) value of about three.

One of the main results of the current programme has been the identification of areas particularly favourable for low enthalpy exploration — areas that represent potential geothermal fields. If data become available from exploration and development in these areas then the present assessment of the resources can be improved — for the performance of a reservoir only becomes apparent after production starts.

Geothermal energy could make a contribution to the energy demand of the UK if the economics are right but any contribution is likely to be as a local energy supplement, at least initially. A doublet developing a low enthalpy reservoir at say 75°C and rejecting the water after use at 30°C could yield some 5 MW; a hot dry rock doublet system developing a reservoir at 200°C for electricity generation could yield about 2 MW of electrical power or perhaps 12 MW of electrical power from 5 doublets at a single site (ETSU, 1982). These outputs are significant on a local scale but not when considered in terms of national energy demand. In some localities and situations at the present time, low enthalpy resources could be developed but in many areas it remains a strategic resource, a resource for the future for, as already mentioned, the feasibility of using geothermal energy is closely related to the price of other fuels, which in the long-term are likely to increase. In any consideration of the use of geothermal energy it is worth bearing in mind that about 60% of the energy delivered in the UK is used to produce heat and nearly 40% is at temperatures of less than 120°C;

about 30% of the total is used for water and space heating. Clearly low enthalpy geothermal resources could have a role to play in meeting this demand.

At present uncertainty exists as to whether or not an individual reservoir can produce over a long period of time and whether reinjection wells in sandstones can be effective and economic. These must still be regarded as the major risks in initiating geothermal schemes particularly as developing a geothermal resource is capital intensive in its early stages. Despite all the uncertainties it must not be overlooked that the use of low enthalpy hot water is a commercial process currently being exploited in France and other countries and that the related sub-surface technical problems likely to be encountered have been overcome successfully by the hydrocarbon industry. Further experience is necessary, however, to indicate the feasibility of geothermal schemes within the framework of the use of energy in the UK and to give confidence in the resource assessments.

It may be concluded that geothermal energy resources in the form of hot water at temperatures of over 40°C and in some areas over 60°C, occur in the Permo-Triassic sandstones, and possibly in some Palaeozoic aquifers, in several deep sedimentary basins in the UK. The Permo-Triassic sandstones can yield the high flow rates required for successful development and economic schemes are feasible in favourable situations. The total resources are considerable and all the evidence indicates that they could provide local energy supplements given the right economic conditions and the development of an appropriate industrial base. The Hot Dry Rock resource is more widespread and if the engineering problems involved in developing the concept are solved this form of energy could be exploited for both electricity generation and direct heating applications, probably towards the end of the century.

References

Allen, D. J., Barker, J. A., and **Downing, R. A.** 1983. The production test and resource assessment of the Southampton (Western Esplanade) Geothermal Well. *Invest. Geotherm. Potent. UK Inst. Geol. Sci.*

Allen, D. J., and **Holloway, S.** 1984. The Wessex Basin. *Invest. Geotherm. Potent. UK Br. Geol. Surv.*

Allsop, J. M., 1977. In **Dawson, J.** and others. A mineral reconnaissance survey of the Doon-Glenkens area, southwest Scotland. *Miner. Reconnaissance Programme Rep. Inst. Geol. Sci.*, No. 18.

Allsop, J. M., Holliday, D. W., Jones, C. M., Kenolty, N., Kirby, G. A., Kubala, M., and **Sobey, R. A.** 1982. Palaeogeological maps of the floors beneath two major unconformities in the Oxford-Newbury-Reading area. *Rep. Inst. Geol. Sci.*, No. 82/1.

Anderson, E. M. 1940. The loss of heat by conduction from the earth's crust in Britain. *Proc. Roy. Soc. Edinburgh.*, Vol 60, 192–209.

Anderton, R., Bridges, P. H., Leeder, M. R and **Sellwood, B. W.** 1979. *A dynamic stratigraphy of the British Isles.* (London: Allen and Unwin).

Andrews, J. G., Richardson, S. W., and White, A. A. K. 1981. Flushing geothermal heat from moderately permeable sediments. *J. Geophys. Res.*, Vol 86, 9439–9450.

Andrews, J. N. 1983. Dissolved radioelements and inert gases in geothermal investigations. *Geothermics*, Vol 12. 67–82.

Andrews, J. N. 1985. Radiochemical and inert gas analyses. *Invest. Geotherm. Potent. UK Br. Geol. Surv.*

Andrews, J.N., and **Kay, R.L.F.** 1982. Natural production of tritium in permeable rocks. *Nature, London*, Vol. 298, 339–343.

Andrews, J. N., Burgess, W. G., Edmunds, W. M., Kay, R. L. F., and **Lee, D. J.** 1982. The thermal springs of Bath. *Nature, London*, Vol. 298, 361–363.

Andrews, J. N., and **Kay, R. L. F.** 1983. The uranium contents and $^{234}U/^{238}U$ activity ratios of dissolved uranium in groundwaters for some Triassic sandstones in England. *Isotope Geoscience*, Vol. 1, 101–117.

Andrews-Speed, C. P., Oxburgh, E. R., and **Cooper, B. A.** 1984. Temperatures and depth-dependent heat flow in Western North Sea. *Bull. Am. Assoc. Petrol Geol.*, Vol. 68, 1764–1781.

Arthurton, R. S., and **Wadge, A. J.** 1981. Geology of the country and around Penrith. *Mem. Geol. Surv. GB*, Sheet 24.

Audley-Charles, M. G. 1970. Triassic palaeogeography of the British Isles. *Q. J. Geol. Soc. London*, Vol. 126, 49–90.

Bamford, D., Nunn, K., Prodehl, C., and **Jacob, B.** 1978. LISPB-IV. Crustal structure of Northern Britain. *Geophys. J. R. Astron. Soc.*, Vol. 54, 43–60.

Barelli, A., and **Palama, A.** 1981. A new method for evaluating formation equilibrium temperatures in holes during drilling. *Geothermics*, Vol. 10, 95–102.

Barritt, S. D. 1983. The controls of radioelement distribution in the Etive and Cairngorm granites: implications for heat production. Unpublished PhD thesis, Open University.

Batchelor, A. S. 1982. An introduction to Hot Dry Rock Geothermal Energy. *J. Camborne Sch. Mines,* Vol. 82, 26–30.

Batchelor, A. S. 1983. Hot Dry Rock reservoir stimulation in the UK: an extended summary. In *Third International Seminar: European Geothermal Update*, 693–750. EUR 8853 EN. (Brussels: CEC.)

Bath, A. H., and **Darling, W. G.** 1981. Stable isotope analyses of water, carbonate and sulphate from the pump test of Marchwood No. 1 Borehole. *Stable Isot. Tech. Rep, Br. Geol. Surv.,* No. 9 (unpublished).

Bath, A. H., Edmunds, W. M., and **Andrews, J. N.** 1979. Palaeoclimatic trends deduced from the hydrogeochemistry of a Triassic sandstone aquifer, United Kingdom. In *Isotope Hydrology 1978,* 545–568. (Vienna: IAEA).

Bath, A. H., and **Edmunds, W. M.** 1981. Identification of connate water in interstitial solution of Chalk sediment. *Geochim. Cosmochim. Acta, Vol. 45,* 1449–1461.

Beard, D. C., and **Weyl, P. K.** 1973. Influence of texture on porosity and permeability of unconsolidated sand. *Bull. Am. Assoc. Pet. Geol.*, Vol. 57, 349–369.

Beck, A. E. 1965. Techniques of measuring heat flow on land. In *Terrestrial Heat Flow*, 24–57, Geophysical Monograph Series, No. 8. **W. H. K. Lee** (editor). American Geophysical Union.

Benfield, A. E. 1939. Terrestrial heat flow in Great Britain. *Proc. R. Soc., A*, Vol. 173, 430–450.

Bennett, J. R. P. 1976. The Lagan Valley hydrogeological study. *Open File Rep. Geol. Surv. Irel.*, No. 57.

Bennett, J. R. P. 1983. The sedimentary basins in Northern Ireland. *Invest. Geotherm. Potent. UK Inst Geol. Sci.*

Bernard, D., Bosch, B., and **Caulier, P.** 1980. Acquisition et rassemblement des données géothermiques dispoinibles et nouvelles dans la zone Franco-Belge de Saint-Ghiglain à Saint-Amand-les-Eaux (Norm). *Report 80 SGN 406 NPL/MGA/GTH.* (Orléans: BRGM).

Birch, F. 1948. The effects of Pleistocene climatic variations upon geothermal gradients. *Am. J. Sci.*, Vol. 246, 729–760.

Birch, F. 1950. Flow of heat in the Front Range Colorado. *Bull. Geol. Soc. Am.*, Vol. 61. 567–630.

Birch, F., and **Clark, H.** 1940. The thermal conductivity of rocks and its dependence upon temperature and composition. *Am. J. Sci.*, Vol. 238, 529–538 and 613–635.

Birch, F., Roy, R. F., and **Decker, E. R.** 1968. Heat flow and thermal history in New England and New York. In *Studies in Appalachian Geology,* 437–451. An-Zen, E. (editor). (New York: Wiley Interscience.)

Black, J. H., and **Barker, J. A.,** 1983. Hydrogeological reconnaissance study of the Worcester Basin. *ENPU Rep. Inst. Geol. Sci,* No. 81–3.

Black, J. H., Martin, B. A., Brightman, M. A., Robins, N. S., and **Barker, J. A.** 1981. The geology and hydrogeology of West Cumbria. *ENPU Rep. Inst. Geol. Sci.*, No. ILWGDS (81).

Bloomer, J. R. 1981. Thermal conductivities of mudrocks in the United Kingdom. *Q. J. Eng. Geol. London*, Vol. 14, 357–362.

Bloomer, J. R., Richardson, S. W., and **Oxburgh, E. R.** 1979. Heat flow in Britain: an assessment of the values and their reliability. In *Terrestrial heat flow in Europe*, 293–300. **Čermák, V.,** and **Rybach, L.** (editors). (Berlin: Springer-Verlag.)

Bloomer, J. R., Kuckes, A. F., Oxburgh, E. R., and **Richardson, S. W.** 1982. Heat flow studies in the Winterborne Kingston Borehole, Dorset. In *The Winterborne Kingston Borehole, Dorset, England*, 176–183. **Rhys, G. H., Lott, G. K.,** and **Calver, M. A.** (editors). *Rep. Inst. Geol. Sci.*, No. 81/3.

Bott, M. H. P. 1967. Geophysical investigations of the northern Pennine basement rocks. *Proc. Yorkshire Geol. Soc.*, Vol. 36, 139–168.

Bott, M. H. P. 1974. The geological interpretation of a gravity survey of the English Lake District and the Vale of Eden. *J. Geol. Soc. London*, Vol. 130, 309–331.

Bott, M. H. P. 1982. *The interior of the Earth: its structure, constitution and evolution* (2nd edition). (London: Edward Arnold.)

Bott, M. H. P., Johnson, G. A. L., Mansfield, J., and **Wheildon, J.** 1972. Terrestrial Heat Flow in North-east England. *Geophy. J. R. Astron. Soc.*, Vol 27, 277–288.

Bott, M. H. P., and **Masson Smith D.** 1960. A gravity survey of the Criffell Granodiorite and the New Red Sandstone deposits near Dumfries. *Proc. Yorkshire Geol. Soc.*, Vol. 32, 317–332.

Bowie, S. H. U., Simpson, P. R., and **Rice, C. M.** 1973. Application of fission-track and neutron activation methods to geochemical exploration. In *Geochemical Exploration*, 359–372. **Jones, M. J.** (editor). (London: Institution of Mining and Metallurgy.)

Brook, M., Pankhurst, R. H., and **Simpson, P. R.** 1982. Cairngorm Granite date. In *Igneous rocks of the British Isles.* **Sutherland, D. S.** (editor). (London: J. Wiley & Sons Ltd.)

Brown, P. E., Miller, J. A., and **Grasty, R. L.** 1968. Isotopic ages of late Caledonian intrusions in the British Isles. *Proc. Yorkshire Geol. Soc.*, Vol. 36, 251–276.

Brown, G. C., Plant, J., and **Lee, M. K.** 1979. Geochemical and geophysical evidence on the geothermal potential of Caledonian granites in Britain. *Nature, London,* Vol. 280, 129–131.

Brown, G. C., Webb, P. C., Lee, M. K., Wheildon, J., and **Cassidy, J.** 1982. Development of HDR reconnaissance in the United Kingdom. In *International Conference on Geothermal Energy*, Florence, Italy, 353–367. BHRA Fluid Engineering, Cranfield, England.

Browne, M. A. E., Hargreaves, R. L., and **Smith, I. F.** 1985. The Upper Palaeozoic basins of the Midland Valley of Scotland. *Invest. Geotherm. Potent. UK Br. Geol. Surv.*

Bullard, E. C. 1939. Heat flow in South Africa. *Proc. R. Soc., A,* Vol. 173, 474–502.

Bullard, E.C., 1947. The time necessary for a borehole to attain temperature equilibrium. *Monthly Notes R. Astron. Soc. Geophys. Suppl.*, No. 5, 127–130.

Bullard, E. C., and **Niblett, E. R.** 1951. Terrestrial heat flow in England, *Monthly Notes. Astron. Soc. Geophys. Suppl.*, No. 6, 222–238.

Bunker, C. M., Bush, C. A., Munroe, R. J., and **Sass, J. H.** 1975. Abundances of uranium, thorium and potassium for some Australian crystalline rocks. *Open File Rep. US Geol. Surv.*, No. 75–393.

Burgess, W. G. 1979. Hydrogeological studies in the Ballymacilroy borehole, Co. Antrim, to investigate the geothermal potential of the Permo-Triassic sandstones. *Rep. Hydrogeol. Dept. Inst. Geol. Sci.* No. WD/SW/79/17 (unpublished).

Burgess, W. G. 1982. Hydraulic characteristics of the Triassic Sherwood Sandstone and the Lower Jurassic Bridport Sands intervals, as derived from drill stem tests, geophysical logs and laboratory tests. In *The Winterborne Kingston Borehole, Dorset, England.* **Rhys, G. H., Lott, G. K.,** and **Calver, M. A.** (editors). *Rep. Inst. Geol. Sci.*, No. 81/3.

Burgess, W. G., Burley, A. J., Downing, R. A., Edmunds, W. M. and **Price, M.** 1981. The Marchwood Borehole — a preliminary assessment of the resource. *Invest. Geotherm. Potent. UK Inst. Geol. Sci.*

Burgess, W. G., Edmunds, W. M., Andrews, J. N., Kay, R. L. F., and **Lee, D. J.** 1980. The hydrogeology and hydrochemistry of the thermal waters in the the Bath-Bristol basin. *Invest. Geotherm. Potent. UK Inst. Geol. Sci.*

Burgess, W. G., Edmunds, W. M., Andrews, J. N., Kay, R. L. F., and **Lee, D. J.** 1982. The origin and circulation of groundwater in the Carnmenellis granite: the hydrogeochemical evidence. *Invest. Geotherm. Potent. UK Inst. Geol. Sci.*

Burley, A. J., Edmunds, W. M., and **Gale, I. N.** 1984. Catalogue of geothermal data for the land area of the United Kingdom. Second revision April 1984. *Invest. Geotherm. Potent. UK Br. Geol. Surv.*

Burley, S. D. 1984. Patterns of diagenesis in the Sherwood Sandstone Group (Triassic), United Kingdom. *Clay Minerals*, Vol. 19, 403–440.

Carruthers, R. M. 1978. An assessment of the Bouguer anomaly map of Northern Ireland with reference to the geothermal potential of sedimentary basins. *Rep. Appl. Geophys. Unit Inst. Geol. Sci.*, No. 58 (unpublished).

Cassidy, J. 1979. Gamma-ray spectrometric surveys of Caledonian granites: method and interpretation. Unpublished PhD Thesis, University of Liverpool.

Ĉermák, V. 1979. Heat flow map of Europe. In *Terrestrial heat Flow in Europe*, 3–40. **Ĉermák, V.,** and **Rybach, L.** (editors). (Berlin: Springer-Verlag.)

Challinor, J. 1978. The 'Red Rock Fault', Cheshire; a critical review. *Geol. J.*, Vol. 13, 1–10.

Chisholm, J. I., and **Dean, J. M.** 1974. The Upper Old Red Sandstones of Fife and Kinross: a fluviatile sequence with evidence of marine incursion. *Scott. J. Geol.*, Vol. 10, 1–29.

Clarke, L. 1981. Groundwater investigations in the Vale of Strathmore. *Water Services*, Vol. 85, 719–720.

Colter, V. S., and **Barr, K. W.** 1975. Recent developments in the geology of the Irish Sea and Cheshire Basins. In *Petroleum and the Continental Shelf of North West Europe.* 1. Geology. 61–73. **Woodland, A.W.** (Editor). (Barking: Applied Science Publishers.)

Colter, V. S., and **Ebburn, J.** 1978. The petrography and reservoir properties of some Triassic sandstones from the Northern Irish Sea Basin. *J. Geol. Soc. London*, Vol. 135, 57–62.

Colter, V. S., and **Havard D. J.** 1981. The Wytch Farm oilfield, Dorset. In *Petroleum Geology of the Continental Shelf of North West Europe.* 494–503. (London: Institute Petroleum).

Coplen, T. B., and **Hanshaw, B. B.** 1973. Ultrafiltration by a compacted clay membrane. 1. Oxygen and hydrogen isotope fractionation. *Geochim. Cosmochim. Acta*, Vol. 37, 2295–2310.

Cradock-Hartopp, M. A., and **Holliday, D. W.** 1984. The Fell Sandstone Group of Northumberland. *Invest. Geotherm. Potent. UK Br. Geol. Surv.*

Crook, J. H., and **Howell, F.T.**. 1970. The characteristics and structure of the Permo-Triassic sandstone aquifer of the Liverpool and Manchester region, NW England. *International Syposium on Groundwater,* Palermo, Sicily.

Cummings, A. D., and **Wyndham, C. L.** 1975. The geology and development of the Hewett Gas-field. In *Petroleum and the continental shelf of North West Europe.* 1. Geology. 313–325. **Woodland, A.W.** (Editor). (Barking: Applied Science Publishers).

Delmer, A., Dom, P., Leclercq, V., Hiroux, P., and **Le-grand, R.** 1980. Geothermics in Hainant (Belgium). In *Advances in European Geothermal Research,* 133–115. **Strub, A. S.** and **Ungemach, P.** (editors). (Dordrecht: D. Reider.)

Department of Energy. 1983. Digest of United Kingdom Energy Statistics 1983. (London: HMSO.)

Dewey, J. F. 1982. Plate tectonics and the evolution of the British Isles. *J. Geol. Soc. London.* Vol, 139, 371–412.

Di Pippo, R. 1984. Worldwide geothermal power development. *Bull. Geotherm. Res. Counc.*, Vol. 13, 4–16.

Downing, R. A. 1967. The geochemistry of groundwater in the Carboniferous Limestone in Derbyshire and the East Midlands. *Bull. Geol. Surv. GB.*, No. 27, 289–307.

Downing, R. A., and **Howitt, F.** 1969. Saline groundwaters in the Carboniferous rocks of the English East Midlands in relation to the geology. *Q. J. Eng. Geol.*, Vol. 1, 241–269.

Downing, R. A., Land, D. H., Allender, R., Lovelock, P. E. R., and **Bridge, L. R.** 1970. The Hydrogeology of the Trent River Basin. *Hydrogeol. Rep. Inst. Geol. Sci.*, No. 5.

Downing, R. A., Smith, D. B., Pearson, F. J., Monkhouse, R. A. and **Otlet, R. L.** 1977. The age of groundwater in the Lincolnshire Limestone and its relevance to the flow mechanism. *J. Hydrol.*, Vol. 33, 201–216.

Downing, R. A., Allen, D. J., Burgess, W. G., Smith, I. F. and **Edmunds, W. M.** 1982a. The Southampton (Western Esplanade) Geothermal Well — a preliminary assessment of the resource. *Invest. Geotherm. Potent. UK. Inst. Geol. Sci.*

Downing, R. A., Burgess, W. G., Smith, I. F., Allen, D. J., Price, M and **Edmunds, W. M.** 1982b. Geothermal aspects of the Larne No. 2 Borehole. *Invest Geotherm. Potent. UK. Inst. Geol. Sci.*

Downing, R. A., Allen, D. J., Barker, J. A., Burgess, W. G., Gray, D. A., Price, M. and **Smith, I. F.** 1984. Geothermal exploration at Southampton in the UK — a case study of a low enthalpy resource. *Energy Explor. Exploit.* Vol. 2, 327–342.

Drury, M. J. 1984. Perturbations to temperature gradients by water flow in crystalline rock formations. *Tectonophysics*, Vol. 102, 19–32.

Drury, M. J., Jessop, A. M., and **Lewis, T. J.** 1984. The detection of ground water flow by precise temperature measurements in boreholes. *Geothermics*, Vol. 13, 163–174.

Dunham, K. C., 1974a. Geothermal energy for the United Kingdom — geological aspects. Unpublished report, Institute of Geological Sciences.

Dunham, K. C. 1974b. Granite beneath the Pennines in north Yorkshire. *Proc. Yorkshire Geol. Soc.*, Vol. 40, 191–194.

Dunham, K. C., Bott, M. H. P., Johnson, G. A. L. and **Hodge, B. L.** 1961. Granite beneath the northern Pennines. *Nature, London, Vol. 190, 899.*

Dunham, K. C., Dunham, A. C., Hodge, B. L. and **Johnson, G. A. L.** 1965. Granite beneath Viséan sediments with mineralization at Rookhope, northern Pennines. *Q. J. Geol. Soc. London,* Vol. 121, 383–417.

Earp, J. R. 1955. The geology of the Bowland Forest Tunnel, Lancashire. *Bull. Geol. Surv. GB.* No. 7, 1–12.

Earlougher, R. C. 1977. *Advances in Well Test Analysis* SPE Monograph Series No. 5. (Dallas: Society of Petroleum Engineers of AIME).

Edmunds, W. M. 1971. Hydrogeochemistry of ground waters in the Derbyshire Dome with special reference to trace constituents. *Rep. Inst. Geol. Sci.* , No. 71/7.

Edmunds, W. M. 1973. Trace element variations across an oxidation-reduction barrier in a limestone aquifer. In *Proceedings Symposium Hydrogeochemistry and Biogeochemistry*, 500–526, Tokyo, 1970. (Washington: Clarke Co.)

Edmunds, W. M. 1975. Geochemistry of brines in the Coal Measures of north-east England. *Trans. Inst. Min. Metall.*, Vol. 84, 339–352.

Edmunds, W. M., and **Bath, A. H.** 1976. Centrifuge extraction and chemical analysis of interstitial waters. *Environ. Sci. Technol.*, Vol. 10, 467–472.

Edmunds, W. M., Burgess, W. G., Bath, A. H., Miles, D. L., and **Andrews, J. N.** 1980. Geochemical sampling and analysis of geothermal fluids in sedimentary basins. In *Advances in European Geothermal Research*, 410–421. (Dordrecht: D. Reidel.)

Edmunds, W. M., Andrews, J. N., Bath, A. H., Miles, D. l., and **Darling, W. G.** 1983. Geochemical evaluation during low enthalpy drilling and testing. In *Third International Seminar*: *European Geothermal Update*, 406–414, EUR 8853 EN (Brussels: CEC.)

Edmunds, W. M., Andrews, J. N., Burgess, W. G., Kay, R. L. F., and **Lee, D. J.** 1984. The evolution of saline and thermal groundwaters in the Carnmenellis granite. *Min. Mag.*, Vol. 48, 407–424.

Edmunds, W.M., Kay, R. L. F., and **McCartney, R. A.** 1985. Origin of saline groundwaters in the Carnmenellis granite; natural processes and reaction during Hot Dry Rock reservoir circulation. *Chem. Geol.* Vol. 49.

Edmunds, W. M., Lovelock, P. E. R., and **Gray, D. A.** 1973. Interstitial water chemistry and aquifer properties in the Upper and Middle Chalk of Berkshire, England. *J. Hydrol.*, Vol. 19, 21–31.

Edmunds, W. M., Taylor, B. J., and **Downing, R. A.** 1969. Mineral and thermal waters of the United Kingdom. In *Mineral and Thermal Waters of the World, A — Europe, Int. Geol. Congr., Prague 1968*, Vol. 18, 135–158.

Edwards, W. 1967. Geology of the country around Ollerton. *Mem. Geol. Surv. GB.*, Sheet 113.

Ellis, A. J., and **Mahon, W. A. J.** 1977. *Chemistry and geothermal systems.* (London: Academic Press).

Energy Technology Support Unit. 1982. Strategic review of the renewable energy technologies — an economic assessment. (London: Department of Energy.)

England, P. C., Oxburgh, E. R., and **Richardson, S. W.** 1980. Heat refraction and heat production in and around granite plutons in north-east England. *Geophys. J. R. Astron. Soc.*, Vol. 62, 439–455.

Evans, W. B., Wilson, A. A., Taylor, B. J., and **Price, D.** 1968. Geology of the country around Macclesfield, Congleton, Crewe and Middlewich. *Mem. Geol. Surv. GB*, Sheet 110.

Foster, S. S. D., Stirling, W. G. N., and **Paterson, I. B.** 1976. Groundwater storage in Fife and Kinross — its potential as a regional resource. *Rep. Inst. Geol. Sci.*, No. 76/9.

Fournier, R. O. 1973. Silica in thermal waters: laboratory and field investigations. In *Proceedings of Symposium Hydrogeochemistry and Biogeochemistry, Tokyo 1970.*, 122–139. (Washington: Clarke Co.)

Gale, I. N., Smith, I. F., and **Downing, R. A.** 1983. The post-Carboniferous rocks of the East Yorkshire and Lincolnshire Basin. *Invest. Geotherm. Potent. UK Inst. Geol. Sci.*

Gale, I. N., Evans, C. J., Evans, R. B., Smith, I. F., Houghton, M. T., and **Burgess, W. G.** 1984a. The Permo-Triassic aquifers of the Cheshire and West Lancashire Basins. *Invest. Geotherm. Potent. UK Br. Geol. Surv.*

Gale, I. N., Carruthers, R. M., and **Evans, R. B.** 1984b. The Carlisle Basin and adjacent areas. *Invest. Geotherm. Potent. UK Br. Geol. Surv.*

Gale, I. N., Holliday, D. W., Kirby, G. A., and **Arthur, M. J.** 1984c. The Carboniferous rocks of Lincolnshire, Nottinghamshire and southern Humberside *Invest. Geotherm. Potent. UK Br. Geol. Surv.*

Gale, I. N., Rollin, K. E., Downing, R. A., Allen, D. J., and **Burgess, W. G.** 1984d. An assessment of the geothermal resources of the United Kingdom. *Invest. Geotherm. Potent UK Br. Geol. Surv.*

Garnish, J. D. 1976. Geothermal energy: the case for research in the United Kingdom. Energy Paper No. 9. (London: HMSO.)

George, T. N., Johnson, G. A. L., Mitchell, M., Prentice, J. E., Ramsbottom, W. H. C., Sevastopulo, G. D., and **Wilson, R. B.** 1976. A correlation of Dinantian rocks in the British Isles. *Spec. Rep. Geol. Soc. London.* No. 7. 87pp.

Geothermics, 1982. Geothermal news and views: Los Alamos Hot Dry Rock Development Program. *Geothermics*, Vol. 11, 201–214.

Glennie, K. W., Mudd, G. C., and **Nagtegaal, P. J. C.** 1978. Depositional environment and diagenesis of Permian Rotliegendes sandstones in Leman Bank and Sole Pit area of the UK southern North Sea. *J. Geol. Soc. London*, Vol. 135, 25–34.

Golbet, P. 1980. Influence of stratified heterogeneities of permeability on the life span of a geothermal doublet. In *Advances in European geothermal research*, 720–734. **Strub, A. S.** and **Ungemach, P.** (editors). (Dordrecht: Reidel.)

Golabi, K., Scherer, C. R., Tsang, C. F., and **Mozumder, S.** 1981. Optimal energy extraction from a hot water geothermal reservoir. *Water Resour. Res.*, Vol. 17, 1–10.

Gray, D. A. 1955. The occurrence of a Corallian limestone in east Yorkshire, south of Market Weighton. *Proc. Yorkshire Geol. Soc.*, Vol. 30, 25–34.

Gray, D. A., Allender, R., and **Lovelock, P. E. R.** 1969. The groundwater hydrology of the Yorkshire Ouse River Basin (Hydrometric area 27). *Hydrogeol. Rep. Inst. Geol. Sci.*, No. 4.

Gringarten, A. C., and **Sauty, J. P.** 1975. A theoretical study of heat extraction from aquifers with uniform regional flow. *J. Geophys. Res.*, Vol. 80, 4956–4962.

Gringarten, A. C. 1979. Reservoir lifetime and heat recovery factor in geothermal aquifers used for urban heating. *Pure Appl. Geophys.*, Vol. 117, 297–308.

Halliday, A. N., Aftalion, M., Van Breemen, O., and **Jocelyn, J.** 1979. Petrogenetic significance of Rb-Sr and U-Pb systems in the c. 400 Ma old British Isles granitoids and their hosts. In *The Caledonides of the British Isles—Reviewed*, 653–661. **Harris, A. L.** and others (editors). (London: Geological Society of London.)

Harrison, R., Charmillon, R., Palama, A., and **Ungemach, P.** 1983. Economics and optimization of geothermal district heating in EC member states. (Brussels: CEC.)

Hawkesworth, C. J. 1974. Vertical distribution of heat production in the Eastern Alps. *Nature, London*, Vol. 249, 435–436.

Hawkins, P. J. 1978. Relationship between diagenesis, porosity reduction and oil emplacement in late Carboniferous sandstone reservoirs, Bothamsall Oilfield, E. Midlands. *J. Geol. Soc. London*, Vol. 135, 17–24.

Heier, K. S. 1978. The distribution and redistribution of heat producing elements in the continents. *Philos. Trans. R. Soc., A,* Vol. 288, 393–400.

Hemingway, J. E. 1974. Jurassic. In *The Geology and mineral resources of Yorkshire.* 161–223. **Raynor, D. H.,** and **Hemingway, J. E.** (editors). (Leeds: Yorkshire Geological Society.)

Hennessy, J. 1979. Uranium, thorium and Caledonian granite magmatism. University of Liverpool unpublished PhD thesis.

Henson, M. R. 1970. The Triassic rocks of south Devon. *Proc. Ussher Soc.*, Vol. 2, 172–177.

Hitchon, B., and **Friedman, I.** 1969. Geochemistry and origin of formation waters in the western Canada sedimentary basin. I. Stable isotopes of hydrogen and oxygen. *Geochem. Cosmochim. Acta*, Vol. 33, 1321–1349.

Holland, J. G., and **Lambert, R. St J.** 1970. Weardale Granite. *Trans. Nat. Hist. Soc. Northumberland*, Vol. 41, 103–123.

Holliday, D. W., and **Smith, I. F.** 1981. The geothermal potential of the Devonian and Carboniferous rocks of the United Kingdom — a preliminary appraisal. *Invest. Geotherm. Potent. UK Inst. Geol. Sci.*

House, M. R., Richardson, J. B., Chaloner, W. G., Allen, J. R. L., Holland C. H., and **Westoll, T. S.** 1977. A correlation of the Devonian rocks in the British Isles. *Spec. Rep. Geol. Soc. London*, No. 8. 110 pp.

Ineson, J. 1953. The hydrogeology of parts of Derbyshire and Nottinghamshire with particular reference to the Coal Measures. *Hydrogeol. Rep. Inst. Geol. Sci.*, No. WD/53/2 (unpublished).

Ineson, J. 1967. Ground-water conditions in the Coal Measures of the South Wales Coalfield. *Hydrogeol. Rep. Inst. Geol. Sci.*, No. 3.

International Energy Agency. 1980. *Man-made geothermal energy systems.* (Paris: International Energy Agency.)

Jaupart, C. 1983. Horizontal heat transfer due to radioactivity contrasts: causes and consequences of the linear heat flow relation. *Geophys. J. R. Astron. Soc.*, Vol. 75, 411–435.

Jeffreys, H.1938. Disturbances of the temperature gradient in the earth's crust by inequalities of height. *Monthly Notes R. Astron. Soc. Geophys. Suppl.*, No. 4, 309–312.

Jessop, A. M. 1983. The essential ingredients of a continental heat flow determination. *Zbl. Geol. Paläontol.* Teil 1, 70–79.

Johnson, G. A. L. 1982. Geographical change in Britain during the Carboniferous period. *Proc. Yorkshire Geol. Soc.*, Vol. 44, 181–203.

Kappelmeyer, O. 1979. Implications of heat flow studies for geothermal energy prospects. In *Terrestrial Heat Flow in Europe.* 126–135. Čermák, V., and Rybach, L.O. (editors). (Berlin: Springer-Verlag.)

Kappelmeyer, O., and **Haenel, R.** 1974. Geothermics with special reference to application. *Geoexplor. Monogr.,* Ser. 1, No. 4. (Berlin: Gebrüder Borntraeger.)

Kent, P. E. 1975. The tectonic development of Great Britain and the surrounding seas. In *Petroleum and the continental shelf of North West Europe. 1. Geology.* **Woodland, A. W.** (editor). (Barking: Applied Science Publishers.)

Kent, P. E. 1980. *British regional geology: Eastern England from the Tees to the Wash* (2nd edition). (London: HMSO for the Institute of Geological Sciences.)

Kessler, L. G. 1978. Diagenetic sequence in ancient sandstones deposited under desert climatic conditions. *J. Geol. Soc. London*, Vol. 135, 41–49.

Kharaka, Y. K., and **Berry, F. A. F.** 1974. Isotopic composition of oilfield brines from Kettleman North Dome, California, and their geologic implications. *Geochim. Cosmochim. Acta*, Vol. 37, 1899–1908.

Kinniburgh, D. G., and **Miles, D. L.** 1983. Extraction and chemical analysis of interstitial water from soils and rocks. *Environ. Sci. Technol.*, Vol. 17, 362–368.

Knox, R. W. O'B., Burgess, W.G., Wilson, K. S., and **Bath, A. H.** 1984. Diagenetic influences on reservoir properties of the Sherwood Sandstone (Triassic) in the Marchwood Geothermal Borehole, Southampton, England. *Clay Minerals*, Vol. 19, 441–456.

Lachenbruch, A. H. 1970. Crustal temperature and heat production: implications of the linear heat flow relation. *J. Geophys. Res.*, Vo. 75, 3291–3300.

Lachenbruch, A. H., and **Brewer, M. C.** 1959. Dissipation of the temperature effect in drilling a well in Arctic Alaska. *Bull. US Geol. Surv.*, No. 1083-C. 73–109.

Lachenbruch, A.H., and **Bunker, C.M.** 1971. Vertical gradients of heat production in the continental crust 2: some estimates from borehole data. *J. Geophys. Res.*, Vol. 76, 3852–3860.

Lavigne, J. 1978. Les ressources géothermiques francaises possibilités de mise en valeur. *Ann. Mines.* April 1978, 1–16.

Lee, M. K. 1978. Crystalline rock study: a preliminary assessment of the prospects for hot rock. *Invest. Geotherm. Potent. UK Inst. Geol. Sci.*

Lee, M. K. 1984a. The three-dimensional form of the Lake District granite batholith. *Invest. Geotherm. Potent. UK Br. Geol. Surv.*

Lee, M. K. 1984b. Analysis of geophysical logs from the Shap, Skiddaw, Cairngorm, Ballater, Mount Battock, and Bennachie heat flow boreholes. *Invest. Geotherm. Potent. UK Br. Geol. Surv.*

Lee, M. K., Brown, G. C., Wheildon, J., Webb, P. C., and **Rollin, K. E.** 1983. Hot Dry Rock exploration techniques in the British Caledonides. In *Third International Seminar: European Geothermal Update*, 775–784. EUR 8853 EN. (Brussels: CEC.)

Lee, M. K., Wheildon, J., Webb, P. C., Brown, G. C., Rollin, K. E., Crook, C. N., Smith, I. F., King, G., and **Thomas-Betts, A.** 1984. Hot Dry Rock prospects in Caledonian granites. *Invest. Geotherm. Potent. UK Br. Geol. Surv.*

Leeder, M. R. 1973. Sedimentology and palaeogeography of the Upper Old Red Sandstone in the Scottish Border Basin. *Scott. J. Geol.*, Vol. 9, 117–144.

Leeder, M. R. 1982. Upper Palaeozoic basins of the British Isles — Caledonide inheritance versus Hercynian plate margin processes. *J. Geol. Soc. London*, Vol. 139, 479–491.

Lott, G. K., and **Strong, G. E.** 1981. The Petrology and Petrography of the Sherwood Sandstone (? Middle Triassic) of the Winterborne Kingston Borehole, Dorset. In *The Winterborne Kingston Borehole, Dorset, England.* **Rhys, G. H., Lott, G. K.**, and **Calver, M. A.** (editors). Rep. Inst. Geol. Sci., No. 81/3.

Lovelock, P. E. R. 1977. Aquifer properties of Permo-Triassic sandstones in the United Kingdom. *Bull. Geol. Surv. GB.*, No. 56.

Lovelock, P. E. R., Price, M., and **Tate, T. K.** 1975. Groundwater conditions in the Penrith Sandstone at Cliburn, Westmorland. *J. Inst. Water. Eng. Sci.*, Vol. 29, 157–174.

Lucazeau, F., Vasseur, G., and **Bayer, R.** 1984. Interpretation of heat flow data in the French Massif Central. *Tectonophysics*, Vol. 103, 99–119.

Marie, J. P. P. 1975. Rotliegendes stratigraphy and diagenesis. In *Petroleum and the continental shelf of North West Europe. 1. Geology.* 205–210. **Woodland, A. W.** (editor). (Barking: Applied Science Publishers.)

Mather, J. D., Gray, D. A., and **Jenkins, D. G.** 1969. The use of tracers to investigate the relationship between mining subsidence and groundwater occurrence at Aberfan, South Wales. *J. Hydrol.*, Vol. 9, 136–154.

McCartney, R. A. 1984. A geochemical investigation of two hot dry rock geothermal reservoirs in Cornwall, UK. Unpublished PhD thesis, Camborne School of Mines, Cornwall.

McKenzie, D. 1978. Some remarks on the development of sedimentary basins. *Earth & Planet. Sci. Lett.*, Vol. 40, 25–32.

Mitchell, M. 1981. The age of the Dinantian (Lower Carboniferous) rocks proved beneath the Kent Coalfield. *Geol. Mag.*, Vol. 118, 703–711.

Monkhouse, R. A. 1982. The groundwater potential of the Monmouth area of the Wye catchment. Report WD/82/4 (unpublished). Inst. Geol. Sci.

Muffler, L. J. P. 1979. Assessment of geothermal resources of the United States — 1978. *US Geol. Surv. Circ.* 790.

Muffler, L. J. P., and **Cataldi, R.** 1978. Methods for regional assessment of geothermal resources. *Geothermics*, Vol. 7, 53–89.

Mullins, R., and **Hinsley, F. B.** 1957. Measurement of geothermic gradient in boreholes. *Trans. Inst. Min. Eng.* Vol. 117, 380–396.

Nichol, I., Thornton, I., Webb, J. S., Fletcher, W. K., Horsnail, R. S., Khaleelee. J., and **Taylor, D.** 1970. Regional geochemical reconnaissance of the Derbyshire area. *Rep. Inst. Geol. Sci.*, No. 70/2.

Oxburgh, E. R., Richardson, S. W., Turcotte, D. L., and **Hsui, A.** 1972. Equilibrium borehole temperatures from observation of thermal transients during drilling. *Earth & Planet. Sci. Lett.*, Vol. 14, 47–49.

Oxburgh, E. R. 1982. Oxford University's Heat Flow Group: heat flow results for the Department of Energy Geothermal programme. Unpublished document presented to the Energy Technology Support Unit.

Oxley, N. C. 1981. Limestone groundwater studies in South Wales. *Paper* given to the Institution of Water Engineers and Scientists. (South Western section) (unpublished).

Penn, I. E., Holliday, D. W., Kirby, G. A., Kubala, M., Sobey, R. A., Mitchell, W. I., Harrison, R. K., and **Beckinsale, R. D.** 1983. The Larne No. 2 Borehole: discovery of a new Permian volcanic centre. *Scott. J. Geol.*, Vol. 19, 333–346.

Plant, J., Brown, G. C., Simpson, P. R., and **Smith, R. T.** 1980. Signatures of metalliferous granites in the Scottish Caledonides. *Trans. Inst. Min. Metall.*, Vol. 89, B198–B210.

Plummer, L. N., Jones, B. F., and **Truesdell, A. H.** 1976. WATEQF — A Fortran IV version of WATEQ, a computer program for calculating chemical equilibria of natural waters. *USGS Water Resources Investigations* 76–13, 61pp.

Pollack, H. N., and **Chapman, D. S.** 1977. On the regional variation of heat flow, geotherms and lithospheric thickness. *Tectonophysics*, Vol. 38, 279–296.

Poole, E. G. 1977. Stratigraphy of the Steeple Aston Borehole, Oxfordshire. *Bull. Geol. Surv. GB*, No. 57, 1–85.

Poole, E. G., 1978. The stratigraphy of the Withycombe Farm Borehole near Banbury, Oxfordshire. *Bull. Geol. Surv. GB*, No. 68, 1–63.

Poole, E. G., and **Whiteman, A. J.** 1955. Variations in thickness of the Collyhurst Sandstone in the Manchester area. *Bull. Geol. Surv. GB*, No. 9, 33–41.

Price, M., and **Allen, D. J.** 1982. The production test and resource assessment of the Marchwood Geothermal Borehole. *Invest. Geotherm. Potent. UK Inst. Geol. Sci.*

Pugh, D. T. 1977. Geothermal gradients in British lake sediments. *Limnol. Ocean*, Vol. 22, 581–596.

Rae, G. 1978. Mine drainage from coalfields in England and Wales, a summary of its distribution and relationship to water resources. *Technical Note 24.* (Reading: Central Water Planning Unit.)

Ramingwong, T. 1974. Hydrogeology of the Keuper Sandstone in the Droitwich Syncline area, Worcestershire. Unpublished PhD thesis, University of Birmingham.

Ramsbottom, W. H. C., Calver, M. A., Eager, R. M. C., Hodson, F, Holliday, D. W., Stubblefield, C. J., and **Wilson, R. B.** 1978. A correlation of Silesian rocks in the British Isles. *Spec. Rep. Geol. Soc. London*, No. 10. 82pp.

Rayner, D. H. 1981. *The stratigraphy of the British Isles*, 2nd Edition. (Cambridge: Cambridge University Press.)

Richardson, S. W. and **Jones, M. Q. W.** 1981. Measurements of thermal conductivity of drill cuttings. In **Burgess, W. G.** and others, The Marchwood Geothermal Borehole — a preliminary assessment of the resource. *Invest. Geotherm. Potent. UK Br. Geol. Surv.*

Richardson, S. W. and **Oxburgh, E. R.** 1978. Heat flow, radiogenic heat production and crustal temperatures in England and Wales. *J. Geol. Soc. London.*, Vol. 135, 323–337.

Richardson, S. W. and **Oxburgh, E. R.** 1979. The heat flow in mainland UK. *Nature, London*, Vol. 282, 565–567.

Rodda, J. C., Downing, R. A., and **Law, F. M.** 1976. *Systematic Hydrology,* (London: Newnes-Butterworths).

Rollin, K. E. 1980. Caledonian magmatism in the Grampian area: preliminary interpretation of the Bouguer gravity anomaly map. *Rep. Appl. Geophys. Unit Inst. Geol. Sci.*, No. 110.

Rollin, K. E. 1983. An estimate of the Accessible Resource Base for the United Kingdom. In *Third International Seminar: European Geothermal Update*, 32–41. EUR 8853EN. (Brussels: CEC.)

Rollin, K. E. 1984. Gravity modelling of the Eastern Highlands granites in relation to heat flow studies. *Invest. Geotherm. Potent. UK Br. Geol. Surv.*

Roy, R. F., Blackwell, D. D., and **Birch, F.** 1968a. Heat generation of plutonic rocks and continental heat flow provinces. *Earth & Planet Sci. Lett.*, Vol. 5, 1–12.

Roy, R. F., Decker, E. R., Blackwell, D. D., and **Birch, F.** 1968b. Heat flow in the United States. *J. Geophys. Res.*, Vol. 73, 5207–5222.

Rundle, C. C. 1979. Ordovician intrusions in the English Lake District. *J. Geol. Soc. London*, Vol. 136, 29–38.

Rundle, C. C. 1981. The significance of isotopic dates from the English Lake District for Ordovician–Silurian timescale. *J. Geol. Soc. London*, Vol. 138, 569–572.

Rybach, L. 1976. Radioactive heat production; a physical property determined by the chemistry of rocks. In *The physics and chemistry of minerals and rocks.* **Strens, R. G. J.** (editor). (London: John Wiley & Sons.)

Rybach, L., and **Muffler, L. J. P.** (editors). 1981. *Geothermal systems: principles and case histories.* (London: John Wiley & Sons).

Sage, R. C., and **Lloyd, J. W.** 1978. Drift deposit influence on the Triassic Sandstone aquifer of NW Lancashire as inferred by hydrochemistry. *Q. J. Eng. Geol.*, Vol. 11, 209–218.

Sass, J. H., Lachenbruch, A. H., and **Munroe, R. J.** 1971. Thermal conductivity of rocks from measurements on fragments and its application to heat flow determinations. *J. Geophys. Res.*, Vol. 76, 3391–3401.

Sauty, J. P., Gringarten, A. C., Landel, P. A., Menjoz, A. 1980. Lifetime optimization of low enthalphy geothermal doublets. In *Advances in European geothermal research,* 706–719. **Strub, A** and **Ungemach, P.** (editors). (Dordrecht: Reidel.)

Sclater, J. G., Jaupart, C., and **Galson, D.** 1980. Heat flow through oceanic and continental crust and the heat loss of the earth. *Rev. Geophys. Space Phys.*, Vol. 18, 269–311.

Simpson, I. M., and **Broadhurst, F. M.** 1969. A boulder bed at Treak Cliff, north Derbyshire. *Proc. Yorkshire Geol. Soc.*, Vol. 37, 141–151.

Simpson, P. R., Brown. G. C., Plant, J., and **Ostle D.** 1979. Uranium mineralization and granite magmatism in the British Isles. *Philos. Trans. R. Soc. London., A,* Vol. 291, 385–412.

Skinner, A. C. 1977. Groundwater in the regional water supply strategy of the English Midlands. In *Optimal Development and Management of Groundwater,* A1-A12. Memoir of the 13th Congress of the International Association of Hydrogeologists, Birmingham, England.

Smith, D. B. 1974. Permian. In *The geology and mineral resources of Yorkshire. 115–144.* **Raynor, D. H.,** and **Hemingway, J. E.** (editors). (Leeds: Yorkshire Geological Society.)

Smith, D. B., Brunstrom, R. G. W., Manning, P. I., Simpson, S., and **Shotton, F. W.** 1974. A correlation of Permian rocks in the British Isles. *Spec. Rep. Geol. Soc. London*, No. 5.

Smith, I. F., Houghton, M. T., Burgess, W. G., and **Freshney, E. C.** 1979. An appraisal of the geothermal potential of the Southampton-Portsmouth area. *Rep. Appl. Geophys. Unit Inst. Geol. Sci.* (unpublished).

Smith, I. F., and **Burgess W. G.** 1984. The Permo-Triassic rocks of the Worcester Basin. *Invest. Geotherm. Potent. UK Br. Geol. Surv.*

Smith, K., Cripps, A. C., and **Evans, R. B.** 1984. The geothermal potential of Upper Palaeozoic rocks in the western Pennines-eastern Cheshire Basin of north-west England. *Invest. Geotherm. Potent. UK Br. Geol. Surv.*

Smith, M. C. 1975. The potential for the production of power from geothermal resources. Report LA-UR-73-926, Los Alamos Scientific Laboratory. Los Alamos, USA.

Smith, M. C., Nunz, G. J., and **Ponder, G. M.** 1983. Hot Dry Rock Geothermal Energy Development Programme. Annual Report for Fiscal Year 1982. Los Alamos National Laboratory, Los Alamos, USA.

Squirrell, H. C., and **Downing, R. A.,** 1969. Geology of the South Wales Coalfield, Part 1. The Country around Newport (Mon.). *Mem. Geol. Surv. GB*, Sheet 249.

Storey, B. C., and **Lintern, B. C.** 1981. The geochemistry of the rocks of the Strath Halladale–Altnabreac district. *ENPU Rep. Inst. Geol. Sci.*, No. 81–12.

Swanberg, C. A. 1972. Vertical distribution of heat generation in the Idaho batholith. *J. Geophys. Res.*, Vol. 77, 2508–2513.

Tammenagi, H. Y., and **Wheildon, J.** 1974. Terrestrial heat flow and heat generation in south-west England. *Geophys. J. R. Astron. Soc.*, Vol. 38, 83–94.

Tester, J. 1982. Energy conversion and economic issues for geothermal energy. In *Handbook of Geothermal Energy.* **Edwards, L. M., Chillingar, G. U., Rieke, H. H.,** and **Fertl, W. H.** (editors). (London: Gulf Publishing Company.)

Thomas, L. P., Evans, R. B., and **Downing, R. A.** 1983. The geothermal potential of the Devonian and Carboniferous rocks of South Wales. *Invest. Geotherm. Potent. UK. Inst. Geol. Sci.*

Thompson, D. B. 1970. The stratigraphy of the so-called Keuper Sandstone Formation (Scythian–? Anisisan) in the Permo-Triassic Cheshire Basin. *Q. J. Geol. Soc. London*, Vol. 126, 151–179.

Tindle, A. G. 1982. Petrogenesis of the Loch Doon granite intrusion, Southern Uplands of Scotland. Unpublished PhD thesis, Open University.

Tombs, J. M. C. 1977. A study of the space form of the Cornubian granite batholith and its application to detailed gravity surveys in Cornwall. *Miner. Reconn. Programme Rep. Inst. Geol. Sci.*, No. 11.

Tsang, C. F., Lippmann, M. J., and **Witherspoon, P. A.** 1977. Production and reinjection in geothermal reservoirs. *Trans. Geotherm. Resourc. Counc.*, Vol. 1, 301–303.

Tsang, C. F. Bodvarsson, G. S., Lippmann, M. J., and **Rivera, R. J.** 1980. Some aspects of the response of geothermal reservoirs to brine reinjection with application to the Cerro Prieto field. *Geothermics*, Vol. 9, 213–220.

Tweedie, J. R. 1979. Origin of uranium and other metal enrichments in the Helmsdale Granite, eastern Sutherland, Scotland. *Trans. Inst. Min. Metall., Sect. B, Appl. Earth Sci.* Vol. 88, B 145–153.

Veatch, R. W. 1983. Overview of current hydraulic fracturing design and treatment technology — Part 1. *J. Pet. Tech.*, Vol. 35, 677–687.

Von Herzen, R. P., and **Maxwell, A. E.** 1959. The measurement of thermal conductivity of deep sea sediments by the needle probe method. *J. Geophys. Res.*, Vol. 64, 1557–1563.

Wadge, A. J., Gale. N. H., Beckinsale, R. D., and **Rundle, C. C.** 1978. A Rb-Sr isochron for the Shap granite. *Proc. Yorkshire Geol. Soc.*, Vol. 42, 297–305.

Walkden, G. M. 1974. Palaeokarstic surfaces in Upper Viséan (Carboniferous) Limestone of the Derbyshire Block, England. *J. Sediment. Petrol.*, Vol. 44, 1232–1247.

Warrington, G. 1974. Trias. In *The The geology and mineral resources of Yorkshire*, 145–160. **Raynor, D. H.,** and **Hemingway, J. E.** (editors). (Leeds: Yorkshire Geological Society.)

Warrington, G., Audley-Charles, M. G., Elliott, R. E., Evans, W. B., Ivimey-Cook, H. C., Kent, P. E., Robinson, P. L., Shotton, F. W., and **Taylor, F. M.** 1980. A correlation of Triassic rocks in the British Isles. *Spec. Rep. Geol. Soc. London*, No. 13.

Waugh, B. 1970. Petrology, provenance and silica diagenesis of the Penrith Sandstone (Lower Permian) of north-west England. *J. Sediment. Petrol.*, Vol. 40, 1226–1240.

Webb, P. C., and **Brown, G. C.** 1984a. The Lake District Granites: their heat production and related geochemistry. *Invest. Geotherm. Potent. UK Br. Geol. Surv.*

Webb, P. C., and **Brown, G. C.** 1984b. The Eastern Highlands Granites: their heat production and related geochemistry. *Invest. Geotherm. Potent. UK Br. Geol. Surv.*

Wheildon, J., Francis M. F., Ellis, J. R. L., and **Thomas-Betts, A.** 1980. Exploration and interpretation of the SW England geothermal anomaly. In 2nd International Seminar on the results of the EC Geothermal Energy Research, 456–465. **Strub, A. S.,** and **Ungemach, P.** (editors). (Dordrecht, Holland: Reidel.)

Wheildon, J. Francis, M. F., Ellis, J. R. L., and **Thomas-Betts, A.** 1981. Investigation of the south-west England thermal anomaly zone. Report EUR 7276EN. (Brussels: CEC.)

Wheildon, J., King, G., Crook, C. N., and **Thomas-Betts, A.** 1984a. The Lake District granites: heat flow, heat production and model studies. *Invest. Geotherm. Potent. UK Br. Geol. Surv.*

Wheildon, J, King, G., Crook, C. N., and **Thomas-Betts, A.** 1984b. The Eastern Highlands granites: heat flow, heat production and model studies. *Invest. Geotherm. Potent. UK Br. Geol. Surv.*

Wheildon, J., Gebski, J. S., and **Thomas-Betts, A.** 1984c. Further investigations of the UK heat flow field 1981–1984. *Invest. Geotherm. Potent. UK Br. Geol. Surv.*

White, A. A. L. 1983. Sedimentary formations as sources of geothermal heat. *J. Volcanol. and Geotherm. Res.*, Vol. 15, 269–284.

White, P. H. N. 1949. Gravity data obtained in Great Britain by Anglo-American Company Ltd. *Q. J. Geol. Soc. London.*, Vol. 104, 339–364.

Williams, B. P. J., Downing, R. A., and **Lovelock, P. E. R.** 1972. Aquifer properties of the Bunter Sandstone in Nottinghamshire, England. *24th Int. Geol. Congr., Montreal 1972.* Section 11, 169–176.

Wills, L. J. 1973. A palaeogeological map of the Palaeozoic floor below the Permian and Mesozoic formations in England and Wales. *Mem. Geol. Soc. London,* No. 7. 23pp.

Wilson, A. A., and **Cornwell, J. D.** 1982. The Institute of Geological Sciences borehole at Beckermonds Scar, north Yorkshire. *Proc. Yorkshire Geol. Soc.,* Vol. 44, 59–88.

Young, B., and **Monkhouse, R. A.** 1980. The geology and hydrogeology of the Lower Greensand of the Sompting Borehole, West Sussex. *Proc. Geol. Assoc.,* Vol. 91, 307–313.

Appendix 1 Conversion factors for selected energy units and equivalent values of some principal sources of energy

Conversion factors for selected energy units

	Joule	kWh	MWa	Btu
Joule	1	2.778×10^{-7}	3.18×10^{-14}	9.48×10^{-4}
kWh	3.6×10^{6}	1	1.14×10^{-7}	3.41×10^{3}
MWa	3.15×10^{13}	8.75×10^{6}	1	2.99×10^{10}
Btu	1055	2.931×10^{-4}	3.35×10^{-11}	1

Approximate equivalent values of some energy sources

	COAL (tonne)	OIL (barrel)	(tonne)	NATURAL GAS (10^3 ft^3)	(m^3)	ENERGY* (joules)
1 tonne of coal	1	4.43	0.607	25.4	708	2.69×10^{10}
1 barrel of oil	0.226	1	0.137	5.65	160	6.08×10^{9}
1 tonne of oil	1.65	7.3	1	41.2	1.17×10^{3}	4.43×10^{10}
10^3 ft^3 natural gas	3.94×10^{-2}	0.18	2.4×10^{-2}	1	28.31	1.065×10^{9}
m^3 natural gas	1.41×10^{-3}	6.25×10^{-3}	8.57×10^{-4}	3.53×10^{-2}	1	3.8×10^{7}

* This represents the energy content of the fuel. Electrical energy produced would be approximately 30% of these values.

Other useful equivalents (Department of Energy, 1983)

Primary energy consumption in UK in 1982 (temperature corrected)	= 311.7 mtce †	≡ 8.4×10^{18} joules
Total electricity generated in UK in 1982	= 255.439 GWh	≡ 0.92×10^{18} joules
Total coal consumption in UK in 1982	= 111.0×10^{6} tonnes	≡ 3.0×10^{18} joules
Total oil products used for energy in UK in 1982	= 59.6×10^{6} tonnes	≡ 2.6×10^{18} joules
Reserves of a medium sized oil field	= 200×10^{6} barrels	≡ 1.2×10^{18} joules

† million tonnes of coal equivalent

Appendix 2 Observed heat flow values in the UK

National Grid 100km square	Borehole	National Grid Ref. E N	Latitude	Longitude	Depth interval† Top	Base	No.conductivity measurements	No.temperature measurements	Heat flow mW/m^2
HY	Warbeth	3235 10089	N58 57 39	W 3 19 49	19	247	45	50*	46
ID	Port More	I 3069 4435	N55 13 43	W 6 19 13	442	579	16	193	80
ID	Larne No. 2	I 3407 4022	N54 50 54	W 5 48 33	100	2000	119	200	59
IH	Killary Glebe	I 2869 3679	N54 33 21	W 6 41 10	0	1158	54	1	60
IJ	Ballymacilroy	I 3057 3976	N54 47 15	W 6 19 50	100	494	30	160	59
IJ	Annalong Valley	I 3343 3244	N54 9 20	W 5 45 2	0	66	20	20	87
IJ	Seefin Quarry	I 3361 3230	N54 8 30	W 5 55 3	0	149	48	47	84
NC	Altnabreac A	2999 9453	N58 23 5	W 3 42 43	0	0	U	U	43
ND	Altnabreac B	3023 9417	N58 21 12	W 3 40 11	0	0	U	U	53
ND	Achanarras	3152 9545	N58 28 15	W 3 27 14	7	92	16	5†	42
ND	Houstrie of Dunn	3203 9546	N58 28 22	W 3 22 0	17	87	31	5†	45
ND	Yarrows	3310 9445	N58 23 2	W 3 10 48	10	99	27	5†	52
NH	Cairngorm	2989 8063	N57 8 12	W 3 40 14	100	290	44	93	70
NH	Loch Ness 1	2396 8104	N57 9 24	W 4 39 8	0	0			73
NH	Loch Ness 2	2428 8145	N57 11 40	W 4 36 7	0	0			64
NH	Loch Ness 3	2463 8184	N57 13 50	W 4 32 47	0	0			62
NH	Loch Ness 4	2482 8208	N57 15 10	W 4 30 59	0	0			57
NH	Loch Ness 5	2500 8223	N57 16 1	W 4 29 15	0	0			82
NH	Loch Ness 6	2501 8229	N57 16 20	W 4 29 11	0	0			67
NH	Loch Ness 7	2518 8249	N57 17 24	W 4 27 33	0	0			55
NH	Loch Ness 8	2536 8276	N57 18 56	W 4 25 52	0	0			43
NH	Loch Ness 9	2560 8309	N57 20 46	W 4 23 36	0	0			43
NJ	Tilleydesk	3957 8364	N57 25 5	W 2 4 18	0	0	U	U	29
NJ	Bennachie	3669 8211	N57 16 46	W 2 32 57	100	290	45	93	76
NN	Ballachulish	2034 7564	N56 39 29	W 5 12 29	0	0	U	U	53
NO	Balfour	3323 7003	N56 11 26	W 3 5 27	543	1205	NS	U	36
NO	Montrose	3715 7603	N56 44 0	W 2 28 0	301	751	40	U	46
NO	Mount Battock	3543 7905	N 57 0 13	W 2 45 9	100	260	42	86	59
NO	Ballater	3400 7985	N57 4 26	W 2 59 23	100	290	47	95	71
NR	Meall Mhor	1834 6747	N55 55 0	W 5 28 0	21	130	81	5†	57
NS	Clachie Bridge	2645 6837	N56 1 36	W 4 10 30	30	300	U	U	55
NS	South Balgray	250 675	N55 56 41	W 4 24 8	0	137	NS	11	64
NS	Blythswood	2500 6682	N55 53 1	W 4 23 52	18	106	NS	6	52
NS	Kipperoch	2373 6774	N55 57 43	W 4 36 24	40	300	47	5†	54
NS	Barnhill	2427 6757	N55 56 55	W 4 31 10	320	355	36	5†	60
NS	Hurlet	2511 6612	N55 49 16	W 4 22 37	95	295	40	5†	60
NS	Maryhill	2572 6686	N55 53 22	W 4 17 1	100	303	82	99	63
NT	Marshall Meadows	3980 6569	N55 48 18	W 2 1 56	152	183	15	U	51
NT	Boreland	3304 6942	N56 8 8	W 3 7 12	0	1006	6	3	40
NT	Livingston	3018 6691	N55 54 18	W 3 34 15	59	641	81	5†	62
NX	Castle Douglas	2717 5550	N54 52 24	W 3 59 59	102	318	45	5†	61
NY	Rookhope	3938 5428	N54 46 48	W 2 5 50	427	792	21	19	92
NY	Silloth No. 2	3124 5544	N54 52 36	W 3 21 55	100	340	77	110	55
NY	Shap	3559 5087	N54 28 18	W 2 40 50	100	300	46	100	78
NY	Skiddaw	3314 5314	N54 40 22	W 3 3 50	100	281	45	88	101
NY	Becklees	3352 5716	N55 2 5	W 3 0 50	100	584	90	190	43

National Grid 100km square	Borehole	National Grid Ref. E N	Latitude	Longitude	Depth interval† Top	Base	No. conductivity measurements	No. temperature measurements	Heat flow mW/m^2
NY	Lake Windermere 2	3382 5006	N54 23 49	W 2 57 7	0	0			69
NY	Lake Windermere 3	3382 5010	N54 24 2	W 2 57 7	0	0			74
NZ	Woodland	4091 5277	N54 38 39	W 1 51 32	198	488	27	4	96
NZ	Kirkleatham	4588 5213	N54 34 59	W 1 5 25	71	935	54*	21	48
NZ	Tocketts	4631 5180	N54 33 11	W 1 1 27	143	906	54*	14	49
NZ	Boulby	4761 5184	N54 33 17	W 0 49 23	799	1087	27	2	47
NZ	South Hetton	4381 5452	N54 48 2	W 1 24 25	0	529	NS	7	61
SD	Raydale	3903 4847	N54 15 29	W 2 8 58	520	593	50	5†	65
SD	Kirkham	3432 4325	N53 47 8	W 2 51 44	20	405	NS	U	71
SD	Beckermonds Scar	3864 4802	N54 13 0	W 2 12 33	53	440	63	33	69
SD	Swinden No. 1	3860 4505	N53 57 1	W 2 12 48	95	685	57	5†	66
SD	Weeton Camp	3389 4359	N53 48 56	W 2 55 41	160	297	84	96	52
SD	Thornton Clevely	3331 4441	N53 53 19	W 3 1 5	0	290	81	94	52
SD	Clitheroe MHD2	3686 4463	N53 54 42	W 2 28 41	100	341	35	110	84
SD	Rosebridge Coll	3578 4059	N53 32 52	W 2 38 13	0	745	U	1	43
SD	Lake Windermere 1	3394 4979	N54 22 22	W 2 55 59	0	0			69
SE	North Duffield	4691 4352	N53 48 31	W 0 57 0	875	960	41	2	60
SE	Skipwith	4664 4371	N53 49 33	W 0 59 28	10	210	NS	5†	54
SE	Skipwith Bridge	4654 4407	N53 51 29	W 1 0 20	10	165	NS	5†	59
SE	Approach Farm	4628 4388	N53 50 29	W 1 2 44	10	160	NS	5†	54
SE	Farnham	4347 4600	N54 2 3	W 1 28 13	177	322	U	2	40
SE	Booth Ferry	4739 4258	N53 43 23	W 0 52 48	0	0	U	U	57
SE	Towthorpe	4618 4591	N54 1 26	W 1 3 24	22	947	NS	5†	56
SE	Marsden	4050 4119	N53 36 12	W 1 55 28	170	297	80	99	50
SH	Mochras	2553 3259	N52 48 40	W 4 8 48	78	440	38	5†	57
SH	Bryn Teg	2699 3321	N52 52 14	W 3 55 58	280	340	44	17	41
SH	Coed-y-Brenin	2747 3258	N52 48 53	W 3 51 33	200	450	23	5†	42
SH	Parys Mountain	2441 3906	N53 23 20	W 4 20 40	104	498	45	5†	59
SJ	Bradley Mill	3531 3767	N53 17 6	W 2 42 13	70	190	NS	5†	59
SJ	Clotton	3528 3636	N53 10 2	W 2 42 22	0	305	NS	5†	33
SJ	Organsdale	3551 3683	N53 12 34	W 2 40 21	70	470	NS	5†	25
SJ	Priors Heyes	3512 3664	N53 11 32	W 2 43 50	10	340	NS	5†	34
SJ	Holford	3667 3820	N53 20 1	W 2 30 0	61	168	6	11	31
SJ	Crewe	3683 3545	N53 5 11	W 2 28 24	100	296	78	96	57
SK	Papplewick	4547 3521	N53 3 47	W 1 11 2	240	695	26*	4	71
SK	Ranby Camp	4664 3808	N53 19 9	W 1 012	246	985	26*	10	83
SK	Ranby Hall	4649 3824	N53 20 2	W 1 1 32	154	975	26*	8	77
SK	Scaftworth	4676 3917	N53 25 2	W 0 58 57	225	1146	26*	9	75
SK	Eyam	4210 3760	N53 16 50	W 1 41 8	82	612	NS	5†	17
SK	Woodlands Farm	4769 3323	N52 52 56	W 0 51 26	0	351	NS	5†	51
SK	Leicester Forest	4525 3028	N52 37 12	W 1 13 28	35	170	14	5†	53
SK	Eady's Farm	4796 3371	N52 55 30	W 0 48 57	0	260	NS	5†	54
SK	Goosedale	4564 3494	N53 2 19	W 1 9 32	191	534	26*	4	64
SK	Misson	4695 3958	N53 27 15	W 0 57 12	787	1192	26*	6	85
SK	Eakring 5	4677 3611	N53 8 34	W 0 59 14	305	599	54*	3	114
SK	Eakring 6	4670 3614	N53 8 43	W 0 59 51	305	662	54*	8	115
SK	Eakring 64	4683 3592	N53 7 32	W 0 58 45	428	611	54*	5	82
SK	Eakring 141	4671 3629	N53 9 30	W 0 59 47	305	606	54*	3	120
SK	Caunton 11	4735 3603	N53 8 4	W 0 54 3	244	650	54*	8	70
SK	Kelham Hills	4759 3576	N53 6 36	W 0 51 55	305	667	54*	4	62
SK	Long Bennington	4838 3416	N52 57 54	W 0 45 8	35	230	NS	5†	88
SK	Corringham	4899 3936	N53 25 53	W 0 38 48	40	385	NS	5††	63
SK	Welby Church	4723 3208	N52 46 47	W 0 55 42	40	410	31	5†	47
SK	Twycross	4339 3056	N52 38 49	W 1 29 57	45	293	48	5†	41
SK	Grove No. 3	4763 3813	N53 19 22	W 0 51 16	0	2933	51	3	54
SM	Treffgarne No. 2	1931 2238	N51 52 26	W 5 0 22	100	180	26	59	39
SM	Treffgarne No. 3	1943 2246	N51 52 53	W 4 59 21	100	193	37	64	43
SN	Glanfred	2630 2881	N52 28 24	W 4 0 59	281	396	39	5†	59
SN	Betws	2654 2069	N51 44 39	W 3 56 59	100	550	167	180	34
SO	Malvern Gasworks	3788 2492	N52 8 25	W 2 18 35	35	245	NS	5†	34
SO	Worcester	3862 2576	N52 12 58	W 2 12 7	100	298	68	99	41

National Grid 100km square	Borehole	National Grid Ref. E N	Latitude	Longitude	Depth interval† Top	Depth interval† Base	No.conductivity measurements	No.temperature measurements	Heat flow mW/m²
SP	Steeple Aston	4469 2259	N51 55 43	W 1 19 5	229	440	21	22	46
SP	Withycombe Farm	4432 2402	N52 3 28	W 1 22 12	850	1060	56	5†	60
SP	Thorpe-by-Water	4886 2965	N52 33 30	W 0 41 36	280	360	47	5†	56
SP	Croft Quarry	4513 2964	N52 33 46	W 1 14 35	222	324	30	5†	37
SP	Home Farm	4432 2731	N52 21 14	W 1 21 56	28	251	13	5†	36
SS	South Molton	2723 1323	N51 4 31	W 3 49 23	9	73	14	6†	55
SS	Honeymead 2	2779 1393	N51 8 22	W 3 44 44	10	286	15	46	54
ST	Currypool Farm	3227 1387	N51 8 30	W 3 6 18	9	182	24	58	61
ST	Cannington Park	3248 1401	N51 9 17	W 3 4 31	100	760	159	234	45
ST	West Lavington	3990 1563	N51 18 19	W 2 0 52	80	152	NS	5†	42
ST	St Fagans	3117 1781	N51 29 40	W 3 16 20	102	150	13	5†	50
ST	Chard	3343 1065	N50 51 13	W 2 56 0	100	289	83	95	51
SU	Bunkers Hill	4304 1150	N50 55 58	W 1 34 2	20	186	NS	5†	60
SU	Fair Cross	4697 1632	N51 21 48	W 0 59 55	75	310	NS	5†	59
SU	Barton Stacey	4437 1428	N51 10 56	W 1 22 29	84	270	NS	5†	42
SU	Clumphill	4066 1064	N50 51 23	W 1 54 22	110	400	NS	U	67
SU	Shrewton	4031 1420	N51 10 35	W 1 57 18	20	1060	U	5†	51
SU	Vernham Dean	4343 1565	N51 18 22	W 1 30 28	60	115	NS	U	25
SU	Marchwood	4399 1112	N50 53 53	W 1 25 56	0	1667	243	1	61
SU	Humbly Grove No. 1	4712 1448	N51 11 51	W 0 58 51	0	1609	22	7	51
SU	Harwell No. 3	4468 1864	N51 34 27	W 1 19 29	60	360	22	98	44
SU	Ramnor Inclosure	4311 1048	N50 50 29	W 1 33 30	45	340	U	U	61
SU	Chalgrove	4654 1963	N51 39 40	W 1 33 16	100	324	68	107	48
SU	Southampton No. 1	4416 1120	N50 54 20	W 1 24 30	100	1818	88	182	71
SU	Godley Bridge No.	4952 1366	N51 7 14	W 0 38 21	0	2584	31	3	54
SW	Wheal Jane E	1761 0425	N50 14 22	W 5 8 25	20	143	19	6†	136
SW	Wheal Jane P	1784 0438	N50 15 7	W 5 6 32	20	268	49	164	126
SW	Wheal Jane O	1782 0436	N50 15 0	W 5 6 42	20	300	15	6†	113
SW	Longdowns	1737 0346	N50 10 2	W 5 10 14	30	182	50	51	112
SW	Medlyn Farm	1708 0340	N50 9 40	W 5 12 34	0	100	32	8	114
SW	Grillis Farm	1680 0385	N50 12 1	W 5 15 5	0	100	33	20	113
SW	Trerghan Farm	1735 0303	N50 7 44	W 5 10 10	0	100	32	18	113
SW	Trevease Farm	1719 0318	N50 8 30	W 5 11 34	0	100	33	20	112
SW	Predannack	1690 0163	N50 0 6	W 5 13 25	0	304	61	100	61
SW	Troon	1657 0368	N50 11 2	W 5 16 56	0	122	40	36	123
SW	Rosemanowas A	1735 0346	N50 10 1	W 5 10 18	0	303	52	99	106
SW	Rosemanowas D	1735 0346	N50 10 2	W 5 10 18	0	292	52	97	106
SW	Polgear Beacon	1693 0366	N50 11 2	W 5 13 56	0	100	23	22	122
SW	Newmill	1461 0343	N50 9 14	W 5 33 18	0	100	32	23	124
SW	Bunker's Hill	1402 0273	N50 5 18	W 5 37 57	0	100	31	23	124
SW	Newlyn East	1815 0539	N50 20 37	W 5 4 18	0	103	34	34	105
SW	Belowda Beacon	1979 0625	N50 25 38	W 4 50 45	0	141	31	20	85
SW	Kennack Sands	1732 0165	N50 0 16	W 5 9 53	0	152	22	50	73
SW	Merrose Farm	1656 0435	N50 14 39	W 5 17 17	0	100	23	23	79
SW	Kestle Wartha	1753 0258	N50 5 20	W 5 8 28	0	150	41	47	96
SW	Gaverigan	1932 0592	N50 23 42	W 4 54 38	0	325	30	105	98
SW	Geevor Mine	1377 0348	N50 9 15	W 5 40 20	124	402	31	7	129
SW	South Crofty	1666 0413	N50 13 30	W 5 16 20	440	650	57	7	129
SX	Wilsey Down	2179 0891	N50 40 20	W 4 34 40	30	726	42	200	67
SX	Hemerdon	2573 0585	N50 24 30	W 4 0 29	0	128	12	42	108
SX	Bray Down	2191 0818	N50 36 25	W 4 33 26	0	100	31	18	113
SX	Blackhill	2184 0782	N50 34 28	W 4 33 56	0	100	34	20	119
SX	Pinnockshill	2189 0745	N50 32 29	W 4 33 21	0	100	33	13	121
SX	Browngelly	2192 0725	N50 31 24	W 4 33 1	0	100	32	21	108
SX	Gt Hammet Farm	2189 0699	N50 29 59	W 4 33 16	0	100	34	20	119
SX	Tregarden Farm	2055 0595	N50 24 7	W 4 44 12	0	100	32	20	126
SX	Colcerrow Farm	2068 0576	N50 23 10	W 4 43 5	0	100	32	20	127
SX	Winter Tor	2612 0916	N50 42 23	W 3 57 58	0	100	34	29	107
SX	Blackingstone	2785 0859	N50 39 35	W 3 43 9	0	100	34	31	105
SX	Soussons Wood	2673 0797	N50 36 5	W 3 52 29	0	100	34	27	132
SX	Laughter Tor	2656 0755	N50 33 47	W 3 53 51	0	100	34	31	114

National Grid 100km square	Borehole	National Grid Ref. E N	Latitude	Longitude	Depth interval† Top	Depth interval† Base	No. conductivity measurements	No. temperature measurements	Heat flow mW/m^2
SX	Foggin Tor	2566 0733	N50 32 29	W 4 1 24	0	100	34	31	111
SX	Lanivet	2022 0641	N50 26 34	W 4 47 12	0	86	0	29	93
SX	Meldon	2568 0922	N50 42 40	W 4 1 44	0	61	25	17	114
SX	Bovey Tracey	2827 0793	N50 36 3	W 3 39 27	0	95	33	35	95
SX	Callywith Farm	2089 0678	N50 28 42	W 4 41 39	0	150	47	43	101
SY	Winterborne Kingston	3847 0979	N50 46 47	W 2 13 1	324	1803	600	5†	70
SY	Seabarn Farm	3626 0805	N50 37 22	W 2 31 42	18	415	53	80	56
SY	Withycombe Raleigh	3033 0841	N50 38 53	W 3 22 5	100	263	46	87	50
SY	Venn Ottery	3066 0911	N50 42 41	W 3 19 23	100	308	58	101	56
TF	Burton Lodge	5114 3438	N52 58 47	W 0 20 26	8	735	NS	U	58
TF	Donington-on-Bain	5240 3819	N53 19 9	W 0 8 18	30	195	10	5†	75
TF	Nettleton Bottom	5125 3982	N53 28 6	W 0 18 18	0	520	400	U	67
TF	Welton No. 1	5036 3768	N53 16 40	W 0 26 45	0	2562	66	3	65
TF	Tydd St Mary	5431 3175	N52 44 9	E 0 7 11	100	295	79	96	57
TG	Trunch	6293 3345	N52 51 31	E 1 24 23	530	650	47	5†	63
TL	Huntingdon	5237 2714	N52 19 35	W 0 11 5	152	244	18	U	38
TL	Stowlangtoft	5947 2688	N52 16 57	E 0 51 17	100	277	86	91	35
TL	Cambridge	5432 2595	N52 12 52	E 0 5 44	130	236	16	29	54
TQ	Fetcham Mill	5158 1565	N51 17 43	W 0 20 19	152	268	14	U	53
TQ	Hankham Colliery	562 105	N50 49 17	E 0 18 1	0	235	8	2	30

† Indicates the top and the base of the depth interval used for calulating heat flow

* Indicates thermal conductivities were measured in nearby boreholes

† Indicates the interval in metres at which the temperature was logged

NS = not sampled

U = Unknown

A more comprehensive list of the data for each site is given in the Catalogue of geothermal data for the land area of the United Kingdom, by A. J. Burley, W. M. Edmunds and I. N. Gale, published in 1984.

Appendix 3 Glossary of Technical Terms

Accessible Resource Base—the thermal energy stored in rocks and the fluids they contain at a temperature of more than the mean surface temperature and at a depth of less than 7 kilometres (which is taken as the practical limit of current economic drilling technology).

Aquifer—a rock formation that is sufficiently permeable to yield a significant quantity of water to a well. Sometimes referred to as a groundwater reservoir or just a reservoir, which is a term more commonly used in the oil industry for a formation containing hydrocarbons.

Doublet—a well system comprising two wells at a single site, one for production and the other for reinjection of the geothermal brine after heat has been extracted.

Drawdown of water level in an aquifer or reduction of reservoir pressure—the difference between the rest water level (or static reservoir pressure) and the water level (or reservoir pressure) caused by pumping from a well.

Drill stem test—a test carried out during the drilling of a well whereby formation fluids (i.e. groundwater or in some cases hydrocarbons) are allowed to flow into the drill pipe and to the surface for a relatively short period (usually up to 24 hours); it provides measurements of the reservoir properties and of the properties of the fluid the reservoir contains.

Enthalpy—a thermodynamic function of a system equal to the sum of its heat energy and the product of its pressure and volume. It is the total heat content of a system. A high enthalpy system refers to a high grade heat system (i.e. high temperature) and a low enthalpy system is a low grade heat system. The unit is J/g.

Geothermal gradient—rate at which the temperature changes with depth; in °C/km. It is directly proportional to the heat flow and inversely proportional to thermal conductivity.

Geothermal Resources—that part of the Accessible Resource Base that could possibly be extracted economically at some specified time in the future.

Heat exchanger—equipment for transferring heat from a primary geothermal fluid to a secondary fluid circulating in, for example, a heating network.

Heat flow—the heat flux through the earth that is derived mainly from trace amounts of radioactive elements in rocks but also partly from a slight cooling of the earth; the practical unit is mW/m^2.

Heat production (by radioactivity)—during radioactive decay mass is converted into radiation energy which is in turn converted into heat. Significant contributions are derived only from uranium-238, uranium-235, thorium-232 and potassium-40; the practical unit is $\mu W/m^3$.

Heat pump—a machine that extracts heat from a lower temperature source and upgrades it to a higher temperature. This is done by circulating a refrigerant, such as a halocarbon, which is evaporated by contact with a lower temperature water source. The resultant gas is compressed, raising the pressure and hence the temperature.

Hot Dry Rock—impermeable rock that occurs at depths where temperatures are high enough to be of economic value in the future, if the Hot Dry Rock concept is successfully developed.

Hot Dry Rock Accessible Resource Base—heat stored in rocks at temperatures of more than 100°C and at a depth of less than 7 kilometres (which is taken as the practical limit of current economic drilling technology).

Hydraulic conductivity—the rate of flow of water through a unit cross-sectional area of a porous medium under unit hydraulic gradient (assuming an isotropic medium and a homogeneous fluid). It is related to intrinsic permeability by: $K = k\varrho g/\mu$ where k is intrinsic permeability, ϱ is the density of the liquid, μ is the dynamic viscosity and g is the acceleration due to gravity. The unit is m/d.

Hydraulic fracturing—a process during which water and chemicals are forced under high pressure into a rock formation in order to create a fracture or a series of interconnected fractures in the rock, thereby increasing the permeability.

Identified Resources—that part of the Geothermal Resources that is more likely to be available for economic exploitation.

Permeability or intrinsic permeability—a measure of the relative ease with which a porous medium can transmit a fluid under a potential gradient. It is the property of the medium only and is independent of the fluid. It has the dimensions of L^2. Units include μm^2 and the darcy. (See also *hydraulic conductivity*).

Porosity—a measure of the void space in a rock. It is expressed as the ratio of the volume of the interstices to the total volume expressed as a fraction or a percentage. *Effective porosity* is the interconnected pore space available for the transmission of fluids expressed as the ratio of the interconnected interstices to the total volume.

Specific capacity of a well—rate of discharge of water from a well divided by the drawdown of the water level in the well; units include m^3/d m and l/s m.

Specific heat or specific heat capacity—the heat required to raise the temperature of a unit mass of a substance through one degree kelvin; J/kg K.

Specific yield—the ratio of the volume of water, which saturated rock will yield by gravity, to the volume of the rock, expressed as a fraction or percentage.

Storage coefficient—the volume of water an aquifer releases from or takes into storage per unit surface area of the aquifer per unit change in head.

Thermal conductivity—the rate of flow of heat through unit cross-sectional area of a substance under unit temperature gradient assuming no heat is lost or gained in transit; W/m K.

Transmissivity—the product of the hydraulic conductivity of an aquifer and its saturated thickness. It relates to the ability of an aquifer to transmit water through its entire thickness. It has the dimensions of L^2/T and the unit is m^2/d. The product of the intrinsic permeability and the saturated thickness of the aquifer is the *intrinsic transmissivity*; the unit is the darcy-metre (D m).

Appendix 4 Reports issued by the British Geological Survey on the Geothermal Potential of the United Kingdom

The British Geological Survey publishes the results of its geothermal energy programme in a series entitled 'Investigation of the Geothermal Potential of the UK'. Titles in the series are given below. They may be obtained from the British Geological Survey, Sales Desk, Keyworth, Nottingham NG12 5GG. Photocopies can be made available at cost for those titles out of print.

General reports:

Catalogue of geothermal data for the land area of the United Kingdom (1978). By A. J. Burley and W. M. Edmunds. £10.00*

A preliminary assessment (1980). £7.10.

Catalogue of geothermal data for the land area of the United Kingdom — First revision August 1981 (1982). By I. N. Gale and A. J. Burley. Out of print.

The geothermal energy programme in the United Kingdom of the British Geological Survey (1984). £2.80.

An assessment of the geothermal resources of the UK (1984). By I. N. Gale, K. E. Rollin, R. A. Downing, D. J. Allen and W. G. Burgess. £7.00.

Atlas of the geothermal resources of the United Kingdom (1984). Compiled by I. N. Gale. £20.00.

Catalogue of geothermal data for the land area of the United Kingdom — Second revision April 1984 (1984). By A. J. Burley, W. M. Edmunds and I. N. Gale. £7.50.

Summary of the geothermal prospects for the United Kingdom (1985).

*Published by the Department of Energy.

Low enthalpy reports

The hydrogeology and hydrochemistry of the thermal water in the Bath-Bristol Basin (1980). By W. G. Burgess, W. M. Edmunds, J. N. Andrews, R. L. F. Kay and D. J. Lee. £5.80.

A preliminary appraisal of the geothermal potential of the Devonian and Carboniferous rocks of the United Kingdom (1981). By D. W. Holliday and I. F. Smith. Out of print.

The Marchwood Geothermal Borehole — a preliminary assessment of the resource (1981). By W. G. Burgess, A. J. Burley, R. A. Downing, W. M. Edmunds and M. Price. Out of print.

The production test and resource assessment of the Marchwood Geothermal Borehole (1982). By M. Price and D. J. Allen. £5.50.

Geothermal aspects of the Larne No. 2 Borehole (1982). By R. A. Downing, W. G. Burgess, I. F. Smith, D. J. Allen, M. Price and W. M. Edmunds. Out of print.

The Southampton (Western Esplanade) Geothermal Well — a preliminary assessment of the resource (1982). By R. A. Downing, D. J. Allen, W. G. Burgess, I. F. Smith and W. M. Edmunds. Out of print.

The sedimentary basins in Northern Ireland (1983). By J. R. P. Bennett. £3.50.

The post-Carboniferous rocks of the East Yorkshire and Lincolnshire Basin (1983). By I. N. Gale, I. F. Smith and R. A. Downing. £4.70.

The geothermal potential of the Devonian and Carboniferous rocks of South Wales (1983). By L. P. Thomas, R. B. Evans and R. A. Downing. £6.00.

The production test and resource assessment of the Southampton (Western Esplanade) Geothermal Well (1983). By D. J. Allen, J. A. Barker and R. A. Downing. £5.40.

Heat flow measurements in north-west England (1984). By W. G. Burgess, J. Wheildon, J. S. Gebski, A. Sartori, A. A. Wilson and D. V. Frost. £5.10.

The Permo-Triassic rocks of the Worcester Basin (1984). By I. F. Smith and W. G. Burgess. £3.75.

The Carlisle Basin and adjacent areas (1984). By I. N. Gale, R. M. Carruthers and R. B. Evans. £3.60.

The Wessex Basin (1984). By D. J. Allen and S. Holloway. £6.00.

The Fell Sandstone Group of Northumberland (1984). By M. A. Cradock-Hartopp and D. W. Holliday. £2.90.

The Permo-Triassic aquifers of the Cheshire and West Lancashire basins (1984). By I. N. Gale, C. J. Evans, R. B. Evans, I. F. Smith, M. T. Houghton and W. G. Burgess. £3.90.

The Carboniferous rocks of Lincolnshire, Nottinghamshire and southern Humberside (1984). By I. N. Gale, D. W. Holliday, G. A. Kirby and M. J. Arthur (confidential report — not published).

The Upper Palaeozoic rocks in the West Pennines-East Cheshire Basin (1984). By K. Smith, A. C. Cripps and R. B. Evans. £4.18.

Interpretation of a deep seismic reflection profile in the Glasgow area (1984). By I. E. Penn, I. F. Smith and S. Holloway. £2.70.

The Upper Palaeozoic basins of the Midland Valley of Scotland (1985). By M. A. E. Browne, R. L. Hargreaves and I. F. Smith. £3.30.

Radiochemical and inert gas analyses (1985). By J. N. Andrews. £2.55.

Further investigations of the U K heat flow field, 1981–1984 (1985). By J. Wheildon, J. S. Gebski and A. Thomas-Betts.

Hot Dry Rock reports

Crystalline rock study: a preliminary assessment of the prospects for hot dry rock (1978). By M. K. Lee. Out of print.

The origin and circulation of groundwater in the Carnmenellis Granite: the hydrogeochemical evidence (1982). By W. G. Burgess, W. M. Edmunds, J. N. Andrews, R. L. F. Kay and D. J. Lee. £7.50.

A review of data relating to hot dry rock and selection of targets for detailed study (1982). By K. E. Rollin. £7.00.

The three dimensional form of the Lake District granite batholith (1984). By M. K. Lee. £2.90.

Analysis of geophysical logs from the Shap, Skiddaw, Cairngorm, Ballater, Mount Battock and Bennachie heat flow boreholes (1984). By M. K. Lee. £2.90.

Gravity modelling of the Eastern Highlands granites in relation to heat flow studies (1984). By K. E. Rollin. £2.90.

The Lake District granites: heat production and related geochemistry (1984). By P. C. Webb and G. C. Brown. £4.57.

The Eastern Highland granites: heat production and related geochemistry (1984). By P. C. Webb and G. C. Brown. £3.92.

The Lake District granites: heat flow, heat production and model studies (1984). By J. Wheildon, G. King, C. N. Crook and A. Thomas-Betts.

The Eastern Highlands granites: heat flow, heat production and model studies (1984). By J. Wheildon, G. King, C. N. Crook and A. Thomas-Betts.

Hot Dry Rock prospects in Caledonian granites: evaluation of results from the BGS-IC-OU research programme, 1981–84 (1984). By M. K. Lee, J. Wheildon, P. C. Webb, G. C. Brown, K. E. Rollin, C. N. Crook, I. F. Smith, G. King and A. Thomas-Betts. £4.00.

Index

Abbotsbury-Ridgeway fault 54
Abriachan Granite 36
Abstraction temperature 127, 128
Accessible Resource Base 6, 132, 133, 135–138, 153; map *136*
Aeolian sandstones 43, 50, 51, 54, 63, 67, 69, 70, 73, 77, 82
Aeromagnetic maps 6; surveys 77, 81
Africa 2
Air-lift tests 149
Albite 123
Alpine orogeny 2, 10, 54, 119, 120
Alston Block 30, 31
Alternative energy sources 5
Ampthill Clay 46
Anhydrite 14, 54, 79, 100
Anisian 43
Annalong Valley 81
Annan 79
Antrim, County 80, 81, 147
Appleby 79
Arden Sandstone 43
Arenaceous Coal Group 89, 96
Armagh, County 80
Arsenic 101
Australian Shield 10
Avon 116
Aylesbeare Group 54
Ayrshire 98, 107

Bakewell 101
Ballachulish Granite 36
Ballater Granite 6, 25, 32–34, 134
Ballymacilroy 81–83, 118, 146
Ballymena 147
Barite 117, 123
Barrow-in-Furness 69
Basal Grit 94, 98
Basal Permian Sands and Breccia 44, 47, 50, 140–142
  isopachytes *51*; temperature *52*
Basalt 80–82, 86
Basement 40, 41
Basin development 42–44, 87
Bath 4, 103, 104, 111, 116, 120, 122, 157
Bath-Bristol area 88, 101, 103, 104, 109, 111–114, 157;
  map *103*
  water flow path *117*
Becklees *79*
Belfast 80, 82
Belgium 88, 100, 110
Bennachie Granite 6, 25, 32–34, 135
Berkshire 96
Bibliography 162–169
Big Vein 95
Binary-fluid systems 153
Blackpool 69
Blandford Forum 61
Boreholes
  Annalong Valley 81
  Ballymacilroy 82, 83, 118, 146
  Becklees 79
  Carnmenellis 26
  Cooles Farm 65
  Cranbourne 54, 57
  Crewe 72
  Devizes 57
  Eldersfield 65
  Eyam 101
  Formby 69, 70
  Glenburn 107
  Holford 72
  Kempsey 64–68, 118
  Knutsford 69, 70
  Langford Lodge 80, 82
  Larne 82, 83, 118, 146, 156
  Magilligan 80
  Malvern Link 65
  Marchwood 54
  Netherton 64, 65
  Newmill 83, 146
  Port More 146
  Prees 69, 70
  Raydale 31
  Rookhope 6,30–32, 40
  Seefin Quarry 81
  Shap 27, 28, 40
  Silloth 77, 79; log *78*
  Skiddaw 27, 28, 40
  Sompting 62
  Steeple Aston 106
  Stowell Park 65
  Thornton Cleveleys 14, 16–18, 72, 73
  Trunch 118, 119
  Twyning 65
  Weeton Camp 72
  Winterborne Kingston 54, 57, 58, 118
  Woodland 31, 32
  Worcester 65
Bottom-hole temperatures 4, 5, 12, 16, 18, 22, 60, 61, 65, 87
Bouguer gravity anomalies 6, 32, 65, 78, 79
  maps *23, 63, 78*
Bournemouth 52, 58–60, 136, 143, 155
Bowland Forest 18, 98
Brampton 77
Breakthrough time 128, 130, 131
Breccias 42, 43, 54, 56, 64, 77, 79
Bridgnorth Sandstone 63, 64, 67, 68, 144, 145, 156
Bridlington 47, 141
Bridport 143, 155
Bridport Sands Formation 44, 118, 119
Brigantian 98
Brine disposal 5, 122, 123, 141, 143, 146, 147, 149, 151, 160
Bristol 4, 101, 102, 104, 116, 117, 157
Bristol-Somerset Coalfield 95, 96
BGS geothermal energy programme 5,16, 24; Reports 177, 178
Brockrams 79, 80
Bromsgrove Sandstone Formation 64, 65, 67, 144, 145, 156
Bulk permeability 89
Bunter Pebble Beds 45

Bunter Series 54
Burnley 98, 102
Butterknowle Fault 31
Buxton 4, 101, 117

Cairngorm Granite 6, 25, 32–34, 134, 135
Caithness 109
Caldew 27, 29
Caledonian granites 6, 8, 14, 16, 18, 20, 22, 24, 32–41, 87
  Orogeny 3, 37, 78, 85, 86, 105
Caliche 57, 100
Camborne School of Mines 2, 6, 21, 26
Canada 3, 119
Carbon-14 116
Carboniferous 4, 6, 16, 20, 30–32, 42, 50, 52, 68, 69, 71, 77, 78, 80, 84–110, 113, 117–120, 134, 154, 156–158
  groundwater analysis *91*; map *84*
Carboniferous Limestone 4, 32, 86–90, 98–104, 109, 110, 113–115, 117, 157
Carboniferous-Permian boundary 64
Cardiff 113
Carlisle 77, 156
Carlisle Basin 6, 42, 63, 77–80, 156; map *76*
Carnmenellis Granite 20, 21, 26, 27, 30, 40, 120, 122
  groundwater circulation *121*
Carsphairn Granite 36
Catalogue of Geothermal Data 16
Celestite 118
Cementation 3, 49, 50, 51, 57, 58, 60, 63, 64, 67, 73, 76, 79, 80, 89, 91, 94, 106, 142, 158, 159
Central America 2
Central Channel Basin 54, 61
Chalcedony 122, 123
Chalk 44, 47, 81, 113, 118, 119
Cheltenham 62, 145
Chemical parameters 112
Cheshire 102, 109, 158
Cheshire Basin 4, 6, 42, 63, 68–77, 91, 132, 145, 156, 158
  depth and thickness 71
  gravity *70*
  map *69*
  stratigraphy *69*
  structure contours and temperatures *70–72*
Chester 69
Chester Pebble Beds Formation 70, 73, 74, 76
Cheviot Hills 104, 106
Cirencester 62, 144, 145, 156, 158
Clackmannan 109
Clastic rocks, Dinantian 104
Cleethorpes 5, 49, 51, 52, 142, 148, 158, 159
Cleveland Basin 46, 47, 50
Cleveland Uplift 48
Climate 14, 86
Clyde 107, 108
Coal 134, 156
Coal Measures 87, 88–96, 104, 116, 117, 134
  isopachytes, structure contours, temperatures *90*
Coefficient of Performance 160
Coleraine 147
Collyhurst Sandstone 63, 69, 70, 74, 76, 145, 146
Concealed granites 36, 37
Conduction 1, 2, 26, 40
Conglomerates 51, 54, 56, 64, 67, 86, 106, 107
Convection 1, 8, 9, 12, 15, 18, 20, 26, 39
Conversion factors 170
Cooles Farm 65
Corallian 44, 46
Core analysis 5, 6, 33, 67, 73, 83, 89, 91, 96, 97, 100, 108, 142, 149, 159
Cornbrash 44
Cornubian batholith 14, 22, 24–27, *26*, 37–41, 137, 153
  groundwater chemistry *121*
Cornwall 1, 2, 6, 18, 21, 87, 105, 109, 111, 120, 137, 152
Corrosion 5, 122, 123
Cotswold Hills 44, 62, 68
Cranborne 54, 57
Cranborne Fault 54, 57, 58, 60
Cranborne-Fordingbridge High 54, 56–58
Craven Basin 102
Craven Fault 96
Crediton 54
Cretaceous 44, 46, 52, 54, 61, 62, 80, 100, 118–120, 140, 147, 152, 156
Crewe 69, 71–73, 145, 146, 156
Crewkerne 20
Criffel Granite 36, 78
Cross-bedding 43, 63
Crustal heat production *10*
Crustal model *134*
Culm Basin 100, 102
Culm facies 86, 87
Cumbria 91
Cwm Berem Sandstone 94
Cynheidre Colliery 94, 95

Dalradian 32, 34, 41, 80
Dartmoor 12
  granite 11
Data acquisition 16
Delamere Member 73
Deltaic deposits 44, 46, 86, 96
Denmark 122
Derbyshire 4, 86, 89, 99–102, 109, 157
Derbyshire Dome 96, 102, 113, 117
Devizes 57
Devon 52, 58, 86, 87, 105, 109
Devonian 3, 6, 20, 27, 41, 57, 84–110, 158
  map *84*
  stratigraphy *85*
Dinantian 86, 87, 90, 96, 99–105, 158
Diorite 34
Dissolved gases 103, 111, 114, 117
Dogger 139
Dolomite 43, 45, 79, 100, 102, 117, 120, 123
Dorchester 58
Dorset 6, 54, 118, 143, 158
Dorset Basin 54, 57, 58, 60, 61
Doublet System 2, 120, 127, 139, 142, 144, 150, 154–156, 158, 160
  diagram *127*
Down, County 80
Dowsing Fault Line 47
Drawdown 57, 62, 104, 124–127, 129–131, 141, 143, 151, 156, 157, 159, 160
Drill-stem tests 5, 6, 51, 58, 66–68, 83, 89, 91, 97, 98, 101, 111, 113, 118, 149, 159
Drilling technology 148
Droitwich 62, 67
Dumfries 77–79, 146
Dundee 108
Durham 48, 91, 113, 117
Durham Coalfield 50
Dyfed 102

Eakring 4, 97, 101
East Anglia 3, 86, 105, 113
East Midlands 4, 6, 44, 87, 89, 96, 100–102, 109, 117, 119, 152, 156, 156
  Millstone Grit
    structure contours, isopachytes, groundwater head *97*
  oilfield 97
East Yorkshire and Lincolnshire Basin 4, 6, 42, 44–52, 67, 118, 136, 137, 140–142, 153, 155, 158
  map *45*
  potential geothermal fields *141*
  stratigraphy 46
  unit power *141*
Eastern Highlands 6, 20, 24, 25, 32–37, 39, 153

Economics 148–151
Eden Shales 79, 80
Edinburgh 108
Eldersfield 65
Electricity generation 2, 4, 21, 153, 160
Energy costs 4
Engineering 148–151
English Channel 52
Ennerdale Granite 27, 30
Eskdale Granite 27, 29, 30
Estimated heat flow 18
Etive batholith 32, 36
Evaporites 43, 45, 70, 80, 100, 119
Evesham 62, 156
Exeter 54
Explosive fracturing 97
Eyam 16, 101

Falkirk 109
Farewell Rock 98
Fearn Granite 36
Fell Sandstone 88, 104, 109, 157
  facies, thickness, structure contours, temperatures *105*
Fennoscandia 3
Fenton Hill 1, 21, 22
Fife 107–109
Finite element models 15, 29
Fish farming 2
Fissure permeability 44, 58
Flamborough Head 52, 141
Fleet Granite 36
Fleetwood 73
Fluid inclusions 122
Fluid potential 15
Forest of Dean 95, 102
Forest of Wyre 63
Forfar 108
Formation waters 67, 68, 83, 113, 117–119
Formby 69, 70
Fracture permeability 26, 89
Fracture systems 2, 21
Fracturing 98, 126, 127, 129–131
France 2, 54, 88, 100, 122, 139, 161
Fordsham Member 73
Fuller's Earth 44
Fylde 71–73

Gainsborough 90, 97
Galena 123
Gas-lift tests 83, 113, 142, 149
Geochemistry 111–123
  applications 111
Geological structure, UK *3*, 5
Geological Survey of Northern Ireland 81, 82
Geophysical logs 5, 6, 14, 33, 57, 58, 66–68, 70, 71, 73, 74, 77, 83, 89, 90, 108, 142, 149
Geothermal fields 2
Geothermal fluids 113
Geothermal gradients 3–5, 8, 10, 11, 111, 114, 133, 150
  Bath-Bristol area 114
  Cheshire Basin 72, 156
  Derbyshire Dome 117
  Eakring 101
  East Yorkshire and Lincolnshire 50
  Eastern Highlands 34
  Northern Ireland 81, 109, 146
  Orcadian Basin 109
  Rookhope 31
  Shap 30
  South Wales 102, 117
  Southern Uplands 36
  UK 22, 41, 87, 88, 152
  Wessex Basin 62
  West Lancashire 156
  Worcester Basin 156
Geothermal Map 7, 8
Geothermal potential 152–161
  East Yorkshire and Lincolnshire Basin 47–52
  Fell Sandstone 104
  Orcadian Basin 109
  Upper Palaeozoic 109, 110
  Variscan Fold Belt 109
Geothermal provinces 10
Geothermal Resource Base 132
Geothermal Resources 132, 138–140
  Carlisle Basin 146
  Cheshire Basin 145
  Cretaceous 156
  East Yorkshire and Lincolnshire 140, 141
  Lower Greensand 147
  Northern Ireland 147
  Permo-Triassic sandstones 154, 158
  Wessex Basin 142, *143*
  West Lancashire Basin 145
  Worcester Basin 144, 145
Geothermal waters 111–123
Geothermometers 111
Germany 43, 122
Glaciation 14, 15, 80
Glasgow 6, 109
Glen Gairn Granite 34
Glenburn 107
Glossary 176
Gloucestershire 158
Gondwanaland 3
Gower 102
Grainsgill *29*
Granite 2–4, 6, 7, 9–11, 14, 15, 18, 20, 22–26, 78, 81, 87, 106, 109, 120–122, 133, 134, 137, 152, 153
  heat production *24*
Granodiorite 34, 78
Gravity anomalies 20, 24, 25, 32, 36, 65, 78, 79, 82
  maps *23, 63, 78*
Gravity surveys 5, 11, 27, 30, 39, 57, 71, 77, 81
Great Glen 135
Great Oolite 44, 46
Great Orton 77, 78
Great Whin Sill 32
Greenland 3
Grimsby 89, 141, 142, 154
Groundwater 1, 2, 15, 16, 20, 67, 68, 80, 83, 96, 102, 111, 113, 116, 117, 118, 119, 153, 155–157
  analysis 68, *91*
Gypsum 43, 79, 123

Haematite 28, 103
Halite 14, 43, 68, 80, 83, 118–120, 123
Hampshire 152
Hampshire-Dieppe High 54, 56–58
Hardegsen Disconformity 43, 70
Heat exchangers 2, 21
Heat flow 1–6, 8–21, 133
  Alston Block 31
  anomalies 22
  boreholes 14, 16, 27, 30, 32, 40
  calculation 14
  central England 40
  Cheshire and West Lancashire basins 72, 73, 156
  conductive 9, 16, 26
  contours 7
  convective 9, 16
  Cornubian batholith 26, 153
  Devonian and Carboniferous 87, 88
  Eakring 101
  East Yorkshire and Lincolnshire Basin 50
  Eastern Highlands 32, 34, 134
  field 16
  Lake District batholith 28, 30, 31

Loch Ness 135
mantle 39
map 18, *19*, 39
measurement techniques 11–14
Northern Ireland 81, 109
Nottinghamshire and Lincolnshire 140
Orcadian Basin 109
provinces 3, 37, 39
relationship to heat production 37–40
Southern Uplands 36
surface 22, 39
UK 152, 171–174
Upper Palaeozoic 157
Variscan Fold Belt 109
Wessex Basin 60, 61
Worcester Basin 65
Heat production 6, 9–11, 15, 20, 133
basement 41
Caledonian granites 35, 36
Cornubian batholith 26, 153
crustal *10*
Eastern Highlands 32–36, 134
granites *24*
Highlands of Scotland 36
Lake District batholith 28, 29
Northern Ireland 81
relationship to heat flow 37–40
Southern Uplands 36
surface 22, 29
Weardale Granite 31
Heat pumps 2, 62, 139, 142, 144–147, 150, 156, 158, 160
Heat store, East Midlands *157*
Helium-4 103, 114, 116, 117, 120, 122
Helmsdale Granite 36
Helsby Sandstone Formation 70, 73, 76, 145
Hercynian granites 14, 18
Hercynian Orogeny 3, 4, 6, 22, 38, 42, 78, 113, 156
Hewett Field 49
High enthalpy systems 1, 2, 4, 111, 122
Highland Boundary Fault 3, 108
Highlands of Scotland 36, 86, 87, 135
Hill of Fare Granite 34
Holderness 47, 52, 141
Holford, 72
Horticulture 2
Hot Dry Rock 1–4, 6, 8, 20–41, 85, 120, 122, 132, 137, 152, 153, 138, 160, 161
Accessible Resource Base 132, 136–138, 153, 160; map *139*
Hotwells 104, 116, 117
Howardian-Flamborough Fault Belt 47
Humber 46, 50, 51, 96, 141, 155
Humberside 5, 50, 141, 148
Hungary 2
Hydraulic barriers 125, 127, 131
Hydraulic characteristics, Permo-Triassic sandstones 58
Hydraulic conductivity 130, 131, 175
Hydraulic fracturing 2, 126, 129, 157
Hydrocarbon boreholes 8, 16, 101, 102, 111, 117, 158, 159
Hydrothermal alteration 29
Hydrothermal systems 1

Iapetus Ocean 3, 85, 86
Suture 3, 39
Idaho batholith 10
Identified Resources 132, 139, 140
Carlisle Basin 146
Cheshire Basin 145, 146
East Yorkshire and Lincolnshire Basin 140, 141, 155
Lower Greensand 147
Northern Ireland 147, 156
Permo-Triassic sandstones 154, 155, 158, 160
Wessex Basin 142, 155
West Lancashire Basin 145
Worcester Basin 144, 156
Imperial College 16, 24
Inferior Oolite 44, 46, 118
Injection wells 5, 21, 149, 150, 158
Inkberrow-Haselor Hill Fault Belt 64, 65
Insulating sediments 4, 44, 144, 154
Intergranular permeability 44, 57, 58, 67, 73, 74, 79, 82, 83, 88, 94
dolomitised rocks 100
Millstone Grit 96
Intergranular porosity
Millstone Grit 96
Intergranular transmissivity 79
Interstitial waters 112, 113, 118, 119
Irish Sea 4, 68, 69, 73
Isle of Wight 54, 56, 58, 60–62, 147
Italy 1, 2

Japan 2
Jurassic 2, 6, 14, 22, 43–47, 55, 62, 68, 71, 80, 88, 118, 119, 122, 144, 146

Karst 86–88, 100–102, 109, 114
Keele Beds 91
Kellaways Beds 46
Kempsey 64–68, 118
Kent 88
Kent Coalfield 96, 102
structure contours *95*
Keuper Series 54
Keuper Waterstones 47
Kidderminster Formation 64, 67, 68
Killary Glebe 83
Kimmeridge Bay 61
Kimmeridge Clay 44
Kingston-upon-Hull 141
Kinnerton Sandstone 69, 73, 74, 76, 145
Kinnesswood Formation 107
Kirkham 72
Kirkham Abbey Formation 45
Kirklinton Sandstone 77–80
Knox Pulpit Formation 107, 109, 157
Knutsford 69, 70

Lake District 6, 18, 24, 25, 27–34, 36–41, 69, 77, 86, 104, 134, 136, 153
three-dimensional model *27*
Lancashire 89, 96
Lancashire Coalfield *69*, 91
Langford Lodge 80, 82
Lardarello 1
Larne 5, 80, 82, 83, 118, 146–148, 156, 158
Leaching 11, 28, 29
Lewisian 39
Lias 77, 78, 81, 104, 116, 118, 119, 120
Limestone Coal Group 98
Limestone Shales 102
Lincoln 90, 102
Lincolnshire 4, 5, 20, 41, 46, 47, 49–52, 96, 140–142, 153–155
Liverpool 68, 69, 77
Llanelli 93
Loch Doon Granite 36
Loch Ness 135
Lochmaben 77–79, 146
Lochnagar Granite 34
London 4
London Platform 43–45, 52, 54, 56
Londonderry, County 80
Los Alamos 1, 2
Los Alamos National Laboratory 21
Lough Foyle 80
Lough Neagh 80–82, 146, 147, 157
Lough Neagh Clays 81

Low enthalpy systems 1, 2, 4, 6–8, 42, 88, 109, 111, 122, 132, 138, 140, 144, 153, 158
modelling 124–131
summary 159–161
Lower Greensand 4, 44, 61, 62, 118, 147
depth and isopachytes *61*
Lower Lias 77
Lower Mottled Sandstone 45
Ludlovian 86

Magilligan 80
Magnesian Limestone 45, 82, 83
Major element analyses 111, 114, 115
Malvern Hills 62
Malvern Link 65
Manchester 68, 77, 87, 91
Manchester Marl 69, 71–74, 145
Manganese 101
Mansfield 48, 98
Mantle heat flow 39
Marchwood 5, 41, 54, 56–58, 60, 62, 106, 113, 118, 120, 122, 123, 148–150
Marine transgressions, Permian 43, 45
Market Weighton 37, 46
Marls, Permian 43, 45, 51, 63, 80, 82
Massif Central 11
Matlock 4, 101, 117
Matrix permeability 94, 158
Maximum recovery factor 132, 139
Mendip Hills 101, 102, 104, 105, 109, 114, 157
Mercia Mudstone Group 14, 16, 43, 45, 54, 63–68, 70–73, 76, 79, 80, 119, 144, 146
Mere Fault 54, 56, 58, 60, 61
Mersey 77
Merthyr Tydfil 98, 102
Mesozoic Basins 42–83, 132, 133, 154, 156
Mesozoic Valley of Scotland 3, 6, 80, 82, 88, 91, 96, 98, 105, 107–109, 153, 157, 158
Coal Measures
isopachytes, structure contours, temperatures *93*
Old Red Sandstone
aquifer properties 108
structure contours, isopachytes, temperatures *107*
Midlands 38, 64, 86, 87
isopachytes, structure contours, temperatures *92, 99*
Midlands Microcraton 3
Millstone Grit 32, 87, 89, 90, 96–99, 101, 102, 104, 117
Mining activity 26, 89, 90, 94, 101, 111, 120
Minor and trace elements 111
Miocene 120
Moho 20
Moinian 32, 34, 41
Monadhliath Granite 32, 34
Moray Firth 109
Morecambe Bay 73, 76, 77
Moreton 65
Mount Battock Granite 6, 25, 32, 34, 134
Mourne Mountains 81
Mull Tertiary Centre 36
Multiple-well schemes 131

Namurian 86, 87, 90, 96–100
Netherton 64, 65
New Mexico 1, 2, 21
New Zealand 2
Newark 97,98
Newent Coalfield 63
Newmill 83, 146
Norfolk 49, 118, 119
North American Craton 3
North Atlantic Shield 3
North Channel 80, 82
North Sea 3, 4, 16, 47, 51, 141
Basin 117, 118
North Staffordshire 89, 91, 102
Northern Ireland 4–6, 80–84, 89, 109, 118, 136, 148, 153, 156, 158
Basins 42, 80–83, 118, 146,
map *80*
Northern Irish Sea Basin 69, 70, 73, 74, 76
Northumberland 88, 91, 96, 104, 106, 109, 113, 117, 158
Northumberland Trough 104
Northwich 145, 146
Nottingham 47, 90
Nottinghamshire 20, 48, 89, 94, 117, 140
Nottinghamshire-Lincolnshire-South Humberside region 88

Oil exploration 5, 6
Old Red Sandstone 3, 6, 86, 88, 104–109, 115, 116, 157
One-dimensional flow 15, 26, 37
Open University 24, 31
Operational parameters 124
Orcadian Basin 3, 109
Ordovician 52
Orkney 3, 106, 109
Ormskirk 71
Oxford Clay 44
Oxford University 16
Oxfordshire 89, 95, 96, 106
Oxfordshire Coalfield 117
aquifer properties 95
Oxygen enrichment 119

Pacific 2
Palaeogeography
Berkshire and Oxfordshire *95*
Devonian and Carboniferous 85–87
Palaeotemperatures 116
Pangea 4
Pannonian Basin 2
Paris Basin 2, 122, 150, 160
Passage Group 98
Pembrokeshire 88
Penarth Group 45, 54, 62, 71, 80
Pennant measures 88, 89, 93, 94, 96
Pennines 3, 16, 20, 31, 38, 43, 69, 77, 86, 87, 89, 91, 96, 104, 109, 117, 134, 136, 137
Penrith 79
Penrith Sandstone 63, 77–80,, 146
Permeability 3–5, 8, 131, 149, 150, 158–160
Basal Permian Sands
East Yorkshire and Lincolnshire 51
Bridgnorth Sandstone 67
Bromsgrove Sandstone 67
Carboniferous
Devon and Cornwall 109
Carboniferous Limestone 113
East Midlands 101
South Lancashire 102
Chester Pebble Beds 73
Coal Measures 89–91
South Wales 95
Collyhurst and Kinnerton sandstones 74, 76
Cretaceous
Wessex Basin 62
Dinantian clastics, Scotland 105
Fell Sandstone 104
Helsby Sandstone 73
Millstone Grit
East Midlands 97, 98, 99
North-west England 98
South Wales 98
Old Red Sandstone
Devon and Cornwall 109
Oxfordshire 106
Scotland 107, 108
South Wales 106
Passage Group, Scotland 98

Pennant Measures 94
Penrith Sandstone 79
Permo-Triassic sandstones 82
Cheshire Basin 156
probability distributions *75*
Sherwood Sandstone
Dorset 58
East Yorkshire and Lincolnshire 155
Hewett Field 49
Lincolnshire 154
Northern Ireland 83, 146, 156
Nottinghamshire 48
Wessex Basin 60, 142
Worcester Basin 144
Tarporley Siltstone 73
Upper Palaeozoic 157
Wildmoor Sandstone 67
Wilmslow Sandstone 73
Permian 4, 41–45, 52, 54, 56, 62, 63, 77–83, 85, 87, 100, 102, 109, 118, 141, 144–146, 154–156, 159
structure contours *50*
Permo-Triassic 2, 4, 6–8, 14, 42, 54, 62, 66, 68, 69, 71–73, 76–82, 91, 109, 128, 140, 145–147, 152, 153, 156, 158–161
geothermal resources 140–147
Philippines 2
Physico-chemical parameters 111
Piezometric head 104, 114
Piezometric level 143
Piezometric surface 57, 58, 67, 76, 83
Pleistocene 14, 15, 71, 114, 116, 119
Plymouth 109
Poisson's equation 9
Pore fluids 7, 112, 113, 118
Porosity 3, 4, 14, 159
Basal Permian Sands
Lincolnshire 50, 51
Yorkshire 50, 51
Bridgnorth Sandstone 67
Carboniferous Limestone
East Midlands 101
South Lancashire 102
Chester Pebble Beds 74
Coal Measures 89–91
South Wales 95
Collyhurst and Kinnerton sandstones 74
Dinantian clastics, Scotland 105
dolomitised rocks 100
Fell Sandstone 104
Helsby Sandstone 73
Millstone Grit
East Midlands 97, 98
South Wales 98
Old Red Sandstone
Oxfordshire 106
Scotland 107, 108
South Wales 106
Pennant Measures 94
Penrith Sandstone 77, 79
Permo-Triassic sandstones 82
Cheshire Basin 156
St Bees Sandstone 78
Sherwood Sandstone
Cleethorpes 49
Dorset Basin 58, 60
East Yorkshire and Lincolnshire 47, 48
Northern Ireland 83
Wessex Basin 57, 142
Tarporley Siltstone 73
Wilmslow Sandstone 73
Porosity-permeability plots *74*
Port More 80, 81, 146
Portland Beds 44
Portsdown 118
Portsdown-Middleton Fault 54, 56, 58, 61
Portsmouth 52, 54, 61, 62, 147, 156
Potassium 1, 4, 9, 22, 117
Potential geothermal fields 132, *143, 159*, 158
Power generation 2
Precambrian 1, 3
granitic basement 21
Prees 67–71, 73
Production tests 58, 113, 118, 123, 142, 159
Pumping 124, 125, 141, 143, 150
Pumping tests 57, 62, 82, 83, 98, 104, 108, 149–151
Purbeck 54, 58, 60

Quartz 122, 123
Quartz Conglomerate 106
Quaternary 47, 81

Radioactivity 1, 4, 6
Radioisotope ratios 111
Radiothermal potential, Lake District 28
Radiothermal properties, summaries 28, 33, 35, 36
Rankine Cycle 2
Raydale Borehole 31
Recharge 1, 68, 89, 90, 97, 101, 111, 114, 117, 124, 125
Red Chalk 46
Red Marls 106
Red Rock Fault 71
Re-injection 2, 5, 124, 141, 143, 146, 149–151, 160
Reject temperature 2, 142, 144, 146, 147, 154–156, 158, 160
Repeat Formation Tests 91, 98
Reports 177, 178
Reserve 132
Resource 132
Resource Base 132
Rhaetic 43, 55, 62
Ribble 76
Risk factor 159
Rock mechanics 4, 21
Rookhope Borehole 6, 30–32, 40
Rosemanowes 21

St Bees Sandstone 73, 77–80
St Bees Shales 79, 80
St George's Land 89–89, 93, 96, 100, 102
St Valery-Bembridge line 54
Salinity
Carboniferous Limestone
East Midlands 101, 102, 117
South Lancashire 102
Coal Measures 90
Durham and Northumberland 117
South Wales 95
Lower Greensand 62
Millstone Grit
East Midlands 98
Permo-Triassic sandstones 118
Carlisle Basin 80
Sherwood Sandstone
Northern Ireland 83
Wessex Basin 57, 58, *59*, 60, 143
South Crofty Mine 120
Salisbury 62, 147
Saturation Indices 122, 123
Scaling 122, 123
Scarborough 50, 141
Scilly Isles 25
Scottish Borders 106
Scythian 43
Sea-level changes 86
Seafield Colliery 91
Seefin Quarry 81
Seismic surveys 5, 6, 39, 55, 61, 71, 77–79, 81, 156
Severn 68, 103
Shap Granite 6, 25, 27–32, 39, 40
Sheffield 98

Sherborne 143
Sherwood Sandstone Group 5, 43–45, 47–50, 52–61, 65, 70, 71, 77, 79–83, 113, 118, 119, 120, 122, 140–147, 155, 156, 158, 159
  East Yorkshire and Lincolnshire Basin
    isopachytes *47*
    structure contours *48*
    temperature *49*
  Northern Ireland
    temperatures *81*
  Wessex Basin 120
    facies *56*
    isopachytes *55*
    salinity *59*
    structure contours *55*
    temperature *60*
    transmissivity *59*
Shetland 3, 106, 109
Shropshire 74
Silesian 74
Silloth 77–79, 146, 156
Silurian 3, 86, 105
Single-well schemes 125, 141
Skene complex 34
Skiddaw Granite 6, 25, 27–32, 39, 40, 78
Skiddaw Slates 31
Solway-Carlisle Basin 69
Solway Firth 3, 77, 79
Solway Plain 77, 78
Somerset 58, 86, 102
Sompting 62
South Crofty Mine 120, 122
South Lancashire 102
South Midlands 105
South Staffordshire Coalfield 63
South Wales 3, 4, 86, 89, 93, 95, 96, 98, 102, 105, 109, 157
  aquifer properties 95
  Coal Measures
    isopachytes, structure contours, temperatures *94*
  Millstone Grit
    isopachytes, structure contours, temperatures *99*
  Old Red Sandstone 106
  thermal waters 116
South Wales Coalfield 87, 94, 106, 116
South-West England 4, 16, 18, 22, 24, 25, 26, 40
Southampton 5, 20, 56–61, 106, 113, 118–120, 122, 129, 148–151, 155, 158–160
Southern North Sea Basin 44
Southern Uplands 36, 40, 77, 86, 88, 104, 108
Southport 69, 71
Space heating 2, 21, 137, 148, 150, 151, 160
Spadeadam 104
Speeton Clay 46
Spilsby Sandstone 44, 46
Stable isotope ratios 111, 113, 114, 116, 118, 119
Stanwix Shales 79, 146
Steeple Aston 106
Stonehaven 108
Stoney Middleton 101
Stow-in-the-Wold 145
Stowell Park 65
Strath Halladale Granite 39
Stratheden Group 107
Strathmore 108
Stress-field 3
Strontian Granite 36
Strontianite 123
Strontium 118
Subduction 87
Subsidence 54, 55, 71, 78, 94
Sub-surface temperatures 9–11, 133, 134, 137
  Carlisle Basin 79
  Cheshire and West Lancashire basins 72, 73
  Cornubian batholith 26
  Dartmoor 12
  Devonian and Carboniferous 87, 88
  Eastern Highlands 34, 40
  Knox Pulpit Formation 109
  Lake District 29, 30, 31, 40
  Lincolnshire 154
  maps *135, 138, 152*
  Northern Ireland 81
  Southern Uplands 36
  Wessex Basin 60, 62
  West Wales 12
  Worcester Basin 65
Sulphate reduction 117
Surface heat flow 22, 39
Surface heat production 22, 29
Sussex 118
Swansea 93, 94, 102
Sylvite 118

Taff Valley 93, 103, 113, 117
Taffs Well 102, 113, 116
Tarporley Siltstone 70, 73, 145
Tectonic provinces 39
Tectonics 1, 2, 42, 54, 55, 64, 85, 87
Tees 52
Temperature-depth profiles 133, 134, 153
  British granites *40*
  Carnmenellis *27*
  Eastern Highlands *34*
  Northern England 30
Temperature gradient measurements 12
Tertiary 4, 6, 36, 44, 52, 80, 81, 97, 100, 103, 109, 118, 119, 133
Thermal conductivity 1, 3, 4, 8, 9, *10,* 11–16, 18, 22, 128, 133, 153
  basement 41
  Cheshire and West Lancashire basins 72, 156
  Coal Measures 87
  Devonian and Carboniferous 88
  East Yorkshire and Lincolnshire Basin 50
  granites 14, 26, 28–32, 34
  Northern Ireland 81
  Orcadian Basin 109
  Permo-Triassic mudstones *15*
  UK 13
  Variscan Fold Belt 109
  Wessex Basin 60, 61
  Westphalian 133
  Worcester Basin 65
Thermal data, summaries 25, 31, 38, 66
Thermal resistance 31, 32
Thermal springs 4, 88, 101–104, 111, 114, 117
Thermal waters, analysis 116
Thermal yields 142
Thermometers 12
Thorium 1, 4, 9, 22, 29, 120
Thornton Cleveleys 14, 16–18, 72, 73
Three-dimensional models
  Cornubian batholith *26*
  Lake District batholith *27*
Thurstaston Member 73
Tin mining 11, 120
Tiverton 54
Torquay 109
Transmissivity 4, 79, 83, 85, 125, 127, 138, 149, 150, 151, 158, 160
  Basal Permian Sands 52, 142
  Bridgnorth Sandstone 67
  Carboniferous Limestone
    East Midlands 101
  Coal Measures 89, 90, 94, 95
  Collyhurst Sandstone 146
  Cretaceous 62
  Dinantian clastics 104

Fell Sandstone 104, 109
Lower Greensand 147
Mesozoic Basins 132
Millstone Grit
East Midlands 97, 98
Old Red Sandstone
Scotland 107, 108
South Wales 106
Permo-Triassic
Carlisle Basin 79, 80
Cheshire Basin 146, 156
Sherwood Sandstone
Cleethorpes 49
Dorset Basin 58
East Yorkshire and Lincolnshire 141, 142, 155
Northern Ireland 82, 83, 146, 156
Nottinghamshire 48
Southampton 150
Wessex Basin 57, *59*, 60, 143, 155
Worcester Basin 144, 145, 156
Upper Palaeozoic 157
Wildmoor Sandstone 67
Triassic 4, 42, 43, 45, 52, 54, 56–58, 62–64, 69, 76, 77, 80–83, 88, 104, 116, 118–120, 122, 140–142, 144–146
Tritium 101, 115, 117, 120
Trunch 118, 119
Tuffs 80
Turbidites 86, 87, 96
Twelve-Feet Sandstone 94
Twyning 65
Tyneside 104, 109, 157
Tyrone, County 80

Ulster White Limestone 81
Ultrafiltration 119, 120
Unit Power 140
East Yorkshire and Lincolnshire *141*
Wessex Basin *143*
Worcester Basin 144
Upper Coal Measures 41
Upper Limestone Group 98
Uranium 1, 4, 9, 22, 28, 29, 36, 109, 120
USA 2
USSR 2

Vale of Eden 77, 78, 79, 146
Vale of Pewsey 54, 56, 57
Vale Fault 56
Variscan Front 3, 87, 88, 96, 102
Variscan Orogeny 3, 42, 52, 54, 85, 87–89, 104, 105, 109
Volcanism 1, 2, 4, 43, 85–87, 89, 106, 107

Wash, The 37, 50, 52, 96, 155
Weald 52, 54, 61, 118
Weald Basin 54, 56, 57
Weardale Granite 6, 18, 22, 24, 25, 30–32, 36, 39–41, 153
Weathering 29
Weeton Camp 72
Well engineering 4
Welsh Borderlands 3, 68, 86, 105, 106
Wensleydale Granite 22, 31, 32, 39
Wessex Basin 4, 6, 20, 41, 42, 44, 52–62, 67, 111, 118–120, 122, 123, 137, 142, 143, 148, 153, 155, 158
stratigraphy 53
structure *54*
West Lancashire Basin 3, 6, 16, 42, 63, 68–77, 145, 156
West Pennines 88
West Wales 11, 12
Western Approaches 4, 54
Western Esplanade Well 118, 122, 129, 151
Western Highlands 40
Westphalian 87–96, 96, 134
Weymouth 60
White Lias 55
Wildmoor Sandstone Formation 64, 67, 68, 144, 145, 156
Wilmslow Sandstone Formation 70, 73, 76, 145
Winchcombe 65
Winterborne Kingston 54, 57, 58, 60, 118, 123
Woodland Borehole 31, 32
Worcester 62, 65, 144, 145
Worcester Basin 4, 6, 42, 62–68
hydrogeological units 66
isopachytes *65*
map *62* section *144*
stratigraphy *63*
structure contours and temperatures *64*
Worksop 98, 102
Worthing 54, 62, 147, 156
Wrexham Coalfield 69
Wytch Farm 54

Yeovil 155
York 48
Yorkshire 44, 46–48, 50, 89, 98, 141

Zechstein Sea 45

Printed for Her Majesty's Stationery Office by Commercial Colour Press.
Dd.737387 3/86 C15